Biometeorology for Adaptation
to Climate Variability and Change

BIOMETEOROLOGY

Volume 1

Chief Editor

Glenn R. McGregor, University of Auckland, New Zealand

Editorial Board

Richard de Dear, *Macquire University, Sydney, Australia*
Kristie L. Ebi, *ESS, LLC, Alexandria, VA, USA*
Daniel Scott, *University of Waterloo, Ontario, Canada*
Scott Sheridan, *Kent State University, Ohio, USA*
Mark D. Schwartz, *University of Wisconsin-Milwaukee, WI, USA*

For other titles published in this series, go to
www.springer.com/series/8101

Kristie L. Ebi • Ian Burton • Glenn R. McGregor
Editors

Biometeorology for Adaptation to Climate Variability and Change

Springer

Editors
Kristie L. Ebi
ESS, LCC Virginia
USA

Glenn R. McGregor
Department Geography, Geology
and Environment Science
University of Auckland
New Zealand

Ian Burton
Independent Scholar and Consultant
Ontario, Canada

ISBN 978-1-4020-8920-6   e-ISBN 978-1-4020-8921-3

Library of Congress Control Number: 2008940150

© Springer Science + Business Media B.V. 2009
No part of this work may be reproduced, stored in a retrieval system, or transmitted in any form or by any means, electronic, mechanical, photocopying, microfilming, recording or otherwise, without written permission from the Publisher, with the exception of any material supplied specifically for the purpose of being entered and executed on a computer system, for exclusive use by the purchaser of the work.

Printed on acid-free paper

springer.com

Foreword

Biometeorology continues to grow as a discipline. It is increasingly recognised for its importance in providing science of relevance to society and well being of the environment. This book is the first in a new book series on Biometeorology. The purpose of the new series is to communicate the interdisciplinary philosophy and science of biometeorology to as wide an audience as possible, introduce scientists and policy makers to the societal relevance of and recent developments in its subfields and demonstrate how a biometeorological approach can provide insights to the understanding and possible solution of cross-cutting environmental issues. One such cross-cutting environmental issue is climate change.

While the literature on the science of climate change, climate change mitigation and the impacts of climate change is voluminous, that on adaptation to climate change is meagre in comparison. The purpose of this book is to partly redress this imbalance by providing insights from a biometeorological perspective. The book acknowledges that society has a long history of adapting to the impacts associated with climatic variability and change but makes the point that climate change poses a real threat to already strained coping systems. Therefore there is a need to realign human use systems with changing climate conditions.

The topics covered in this volume, under the umbrella of adaptation to climate change, reflect the core areas of biometeorology, for example, the human thermal environment, early warning systems, plant and animal biometeorology, tourism, water resources and psychological aspects of weather and climate as well as theoretical aspects of adaptation. The book concludes with an assessment of the status and prospects for biometeorology The key messages that emerge from the 12 chapters are that:

- Driving forces of impacts, including climate, often interact in complex and surprising ways. Models are needed to study the behaviour of complex systems.
- Global environmental changes are increasing the complexity of challenges to which human, animal, and plant systems have to adjust
- Current levels of adaptation are uneven across vulnerable regions and sectors, with many poorly prepared to deal with projected changes in climate and climate variability

- Understanding of the interactions of climatic factors with human, animal, and plant systems can be used to increase adaptive capacity through for example early warning systems
- Effective adaptation to climate variability and change will come from adjustments in social and economic systems
- Significant opportunities exist for biometeorology to contribute to policy development so that societies can live effectively with climate variability and change

These messages bear significant implications for those involved in developing climate change adaptation policies and provide directions for research in the area of climate change adaptation science.

Chief Editor
Professor Glenn McGregor
University of Auckland
New Zealand

Contents

1 **Biometeorology for Adaptation to Climate Variability and Change**... 1
Ian Burton, Kristie L. Ebi, and Glenn McGregor

Section I Research Frontiers

2 **Adaptation and Thermal Environment** 9
Gerd Jendritzky and Richard de Dear

3 **Heat/Health Warning Systems: Development, Implementation, and Intervention Activities**...................................... 33
Laurence S. Kalkstein, Scott C. Sheridan, and Adam J. Kalkstein

4 **Malaria Early Warning Systems** 49
Kristie L. Ebi

5 **Pollen, Allergies and Adaptation** 75
Mikhail Sofiev, Jean Bousquet, Tapio Linkosalo, Hanna Ranta, Auli Rantio-Lehtimaki, Pilvi Siljamo, Erkka Valovirta, and Athanasios Damialis

6 **Plant Biometeorology and Adaptation** 107
Simone Orlandini, Marco Bindi, and Mark Howden

7 **Response of Domestic Animals to Climate Challenges**..................... 131
John Gaughan, Nicola Lacetera, Silvia E. Valtorta, Hesham Hussein Khalifa, LeRoy Hahn, and Terry Mader

8 **Adaptation in the Tourism and Recreation Sector** 171
Daniel Scott, Chris de Freitas, and Andreas Matzarakis

9 **Adaptation and Water Resources**.. 195
Chris R. de Freitas

10 **Psychological Perspectives on Adaptation to Weather and Climate** 211
Alan E. Stewart

Section II Perspectives

11 **Human Adaptation within a Paradigm of Climatic Determinism and Change** 235
Andris Auliciems

12 **The Status and Prospects for Biometeorology** 269
Kristie L. Ebi, Glenn McGregor, and Ian Burton

Index 279

Chapter 1
Biometeorology for Adaptation to Climate Variability and Change

Ian Burton, Kristie L. Ebi, and Glenn McGregor

1.1 Introduction

Biometeorology was first officially recognised as an emerging interdisciplinary science in August 1956, when Dr. Tromp and colleagues organized the first International Symposium of Biometeorology in the UNESCO Building in Paris. The meeting led to the establishment of the international society. The first issue of the societies' journal, The International Journal of Biometeorology, was published 5 years later in 1961. This volume was initiated to mark the 50th anniversary of the establishment of the International Society of Bioclimatology and Biometeorology.

In 1956, there were few interdisciplinary scientific societies and the idea of creating new disciplines at the interstices of well-established fields was not universally welcomed by established scholars and their scientific societies in meteorology and biology. One consequence of being viewed as somehow less scientific was that the emerging interdisciplinary groups tended to stress the importance of scientific rigour, sometimes at the expense of innovation and originality. This was reflected in the rapid crystallisation of biometeorology into specialised sub-fields,

---

I. Burton
Professor Emeritus, University of Toronto, Toronto, Canada

K.L. Ebi
ESS, LLC, Alexandria, VA

G. McGregor
Director, School of Geography, Geology and Environmental Science,
University of Auckland, Auckland New Zealand

particularly human, plant, and animal biometeorology in the early decades. These groups attracted the interest of some applied meteorologists and climatologists, with their partners and collaborators drawn from traditional disciplines, i.e. health sciences for human biometeorology, and botanists, zoologists, agronomists, and related sciences for plant and animal biometeorology. These three sub-disciplines tended to dominate the Congresses of the International Society of Biometeorology held every 3 years, and the papers published in the journal. The Society flourished and its scientific rigour and respectability was established, although the interdisciplinary ambitions were not entirely fulfilled in the way imagined.

In more recent decades, the Society has branched out and become more innovative. At the same time, it has tended to fragment even more. To some extent this represents the growing maturity and richness of the field of biometeorology. At its best, this is exemplified in the work of the several Commissions of the Society that organise collaborative research and scientific exchanges in the period between the Congresses. In addition there have been groups that have spun-off and created their own organizations, such as Urban Climatology.

The recent arrival on the international scientific and political agenda of the concept of adaptation to climate variability and change offers an opportunity for the international biometeorology community to demonstrate its relevance to one of the major global issues of the time. It also provides an opportunity for the diverse groups and specialisations within biometeorology to converge around a crosscutting and integrating theme.

1.2 Adaptation to Climate and Climate Change

Adaptation is a component of vulnerability, such that impacts on human activities and biophysical systems are, or can be, moderated by adaptation. As a technical term the word "adaptation" has a long and complex history in both the social and the natural sciences. In the natural sciences, adaptation is most closely associated with Darwinian evolutionary ideas and the process of natural selection. Those biological characteristics that have survival value in a hostile and changing environment get selected. This form of "adaptation" is strictly biological and usually occurs slowly over long time spans. Biological adaptation applies to humans as well as other species. It is clear, however, that some species also adapt in other ways. One obvious way is through migration. As environmental conditions decrease the desirability of a location, plants and animals can migrate to adjacent locations where conditions may be more favourable. The more mobile species, such as birds and some mammals, utilise this mode of adaptation more effectively than trees and plants that can only move from generation to generation by the spatially limited dispersal of seeds.

When applied to human society, the term adaptation covers a much wider range of possible actions. Human have not only adapted in the evolutionary biology sense of the word, but also in their culture, societal infrastructures, and technology.

Furthermore, they interfere with natural adaptation processes and adapt plants and animals by domestication and selective breeding.

Biological and social adaptations have long been the subjects of scientific enquiry. A feature of the processes under investigation has always been that change occurs slowly over generations and centuries, with biological adaptation requiring longer time periods than socio-cultural adaptation. Now, however, the pace of change is accelerating as species and systems respond to climate change. Anthropogenic interference with the chemical composition of the atmosphere is bringing about major shifts in climatic patterns and is leading to significant rise in sea level. The adaptations that are needed now and will be increasingly essential in the future is of a different order of magnitude and of a different character than any previously experienced.

Over historic and prehistoric periods, human beings have on balance adapted well to climate. This success is evidenced by the fact that the human species has spread widely over the planet and that more or less successful societies and livelihoods have been created in every climatic zone from the tropics to the arctic; from equatorial rainforests to arid regions; and from mountains to coasts, plains, river valleys, and deltas.

Much of this successful adaptation took place by a process of trial and error, and sometimes errors have led to the temporary or even permanent collapse of societies in particular localities. In modern times, understanding adaptation to environment, and specifically to climate, has become a matter for scientific and professional expertise. Architects and engineers design buildings and infrastructure to withstand extreme weather conditions such as heat, cold, and windstorms. Agronomists design and select cultivars suitable to the climate of particular farming regions. Medical scientists and public health authorities safeguard populations against diseases prevalent in particular climates. Meteorologists deliver forecasts and warnings of potentially adverse conditions. And so forth. These scientific, professional, and managerial tasks are not generally described as "adaptation". However, since the United Nations Framework Convention on Climate Change was negotiated, signed (1992), and ratified (1995), all these and similar activities are lumped together under the label of "adaptation".

This came about because the negotiators of the Convention found it convenient to divide responses to anthropogenic climate change into two types. The first and foremost response was called "mitigation" by which the negotiators meant any and all measures that could be taken to stabilise and eventually reduce the concentrations of greenhouse gasses in the atmosphere. This includes the development and deployment of energy sources other than fossil fuels, effective approaches to increase energy efficiency, and technologies to reduce greenhouse gas emissions, such as carbon capture and storage. The framers of the Convention understood that many different form of action could be taken to reduce vulnerability to climate, climate variability, and climate change. Adaptation was the convenient shorthand to refer to these potential actions.

The various special fields within biometeorology have long been concerned with the adverse impacts of climate and weather on human activities and the natural (biosphere) system, and how best to cope with them. Applied biometeorologists have tended to pursue their research in relative isolation from each other. There is now a need to better understand the commonalities as well as the differences. The driving force is that these specialists within biometeorology are faced with a common problem – how best to cope (adapt) with climate change. Adaptation is a cross cutting theme and range of research and practice in which the various branches of biometeorology have something to learn from each other and something to contribute to the wider agenda of managing the impacts of climate change.

The purpose of this collection of papers on biometeorology and adaptation should now be clear:

1. To communicate some of the basic ideas and concepts of the sub-fields of biometeorology as they relate to climate change
2. To explore ideas, concepts, and practice that may be developed in common
3. To begin to converge on a new vision for biometeorology that will help to communicate its understanding and expertise, as well as enhance its utility

1.3 Organization of This Volume

With these ideas in mind, the essays have been organised into two sections. After this introductory essay, the first section focuses on research frontiers in biometeorology in the fields of human, animal, and plant biometeorology.

Chapter 2 by Gerd Jendritzky and Richard de Dear begin the exploration with a discussion of the thermal environment, taking a physiological perspective and exploring the problems of developing a Universal Thermal Climate Index.

Chapter 3 by Larry Kalkstein and Scott Sheridan take the heat stress problem one step further, describing their work in developing and deploying heatwave early warning systems in major world cities.

Chapter 4 by Kristie L. Ebi explores the general question of the requirements for developing early warning systems for vectorborne diseases, then surveying the current status of early warning systems for one of the most important vectorborne diseases worldwide, malaria.

Chapter 5 by Mikhail Sofiev and colleagues takes up the issue of pollen, allergies, and adaptation. The basic mechanisms of allergies are presented and the paper describes the risks in terms of pollen ecology, seasonality, and forecasting. The paper concludes with a section on the implications of climate change and deployment of better prevention and treatment (adaptation) techniques.

Chapter 6 by Simone Orlandini and colleagues turn to plant biometeorology, discussing the main impacts and stresses of climate change, and identifying available adaptation options and conditions for their effective deployment. The chapter concludes with a discussion of future directions in plant (agricultural) biometeorology and research gaps.

Chapter 7 by John Gaughan and colleagues confront the problems of adaptation in domestic (farm) animals. The paper focuses upon the specific climate stresses of heat, heatwaves, and nutrition, and alternative ways in which these can be managed at the farm level. The paper concludes with an assessment of increasing climate stress and suggests some needed policy responses.

Chapter 8 by Daniel Scott and colleagues turn to the questions of the climate change risks and possible adaptation responses in the tourism-recreation sector. This sector is thought to have high adaptive capacity overall, with considerable variation from one part of the industry to another. The paper concludes with observations about the relative lack of attention to climate change, and the need for more consideration of adaptation in the sector.

Chapter 9 by Chris de Freitas turns to the problems of water resources and the need for adaptation in a sector that is being increasingly stressed by climate change, population growth, and infrastructure development.

Chapter 10 by Alan Stewart examines the psychological dimensions of adaptation to weather and climate. Psychological approaches to adaptation are discussed, followed by the presentation of a model that lends itself well to organizing the different psychological variables that can affect adaptation. The model components are discussed along with the cognitive and motivational biases that can affect the adaptive course that people might pursue.

The second and final section of the book provides perspectives on the previous chapters.

Chapter 11 by Andris Auliciems brings us back to some long debated ideas about adaptation and climatic influences on humans and human societies. He points us to a reconsideration of climatic determinism and presents a theoretical construction for thinking about the impacts of climate and human response. Adaptations adopted in the past have been more or less successful, but we are now faced with a new situation in which we know that change will occur, but how, when, and at what rate remains highly uncertain. In these circumstances is it not the well adapted that can expect to thrive best, but the most adaptable.

In the final chapter, Chapter 12, the editors take up the challenge of adaptive capacity. Given what we know and the long-standing perspectives of biometeorology and its current research directions, what more can be done to promote adaptation and enhance adaptive capacity? Human society now faces a considerable challenge in learning to live successfully with a changing climate. The essays in this book point the way to some of the answers and needed directions.

Section I
Research Frontiers

Chapter 2
Adaptation and Thermal Environment

Gerd Jendritzky and Richard de Dear

Abstract Due to the need for human beings to adapt their heat budget to the thermal environment in order to optimise comfort, performance and health the adaptation issue is a question of vital importance. Balancing the human heat budget, i.e. equilibration of the organism to variable environmental (atmospheric) and metabolic heat loads is controlled by a very efficient (for healthy people) autonomous thermoregulatory system that is additionally supported by behavioural adaptation which are driven by conscious sensations of thermal discomfort. These capabilities enable the (healthy) human being to live and to work in virtually any climate zone on earth, albeit with varying degrees of discomfort. Based on mortality studies a large number of publications show the evidence of adverse health impacts by thermal stresses, in particular during heat waves.

Based on thermo physiology and heat exchange theory an overview is given on different assessment approaches up to the development of the "Universal Thermal Climate Index" within ISB Commission 6 and the European COST Action 730. Selected applications from the weather/climate and human health field such as Heat Health Warning Systems HHWS and precautionary planning in urban areas illustrate the significance of thermal assessments with respect to short-term and long-term adaptation. A huge potential to save energy – and by this to avoid CO_2 emissions – without loosing acceptable thermal conditions indoors, also in a future warmer climate, results from a adaptive model which has been derived from thermal comfort investigations across the world.

G. Jendritzky
Meteorological Institute, University of Freiburg, Germany

R. de Dear
Division of Environmental & Life Sciences, Macquarie University, Sydney, Australia

2.1 Introduction

The close relationship of humans to the thermal component of the atmospheric environment is evident and belongs to everybody's daily experience. Thus, issues related to thermal comfort, discomfort, and health impacts are the reason that the assessment and forecast of the thermal environment in an effective and practical way is one of the fundamental topics of human biometeorology. In this context the term "thermal environment" encompasses both the atmospheric heat exchanges with the body (stress) and the body's physiological response (strain).

Balancing the human heat budget, i.e. equilibration of the organism to variable environmental (atmospheric) and metabolic heat loads is controlled by a very efficient (for healthy people) autonomous thermoregulatory system (see Section 2.2.1) that is additionally supported by behavioural adaptation (e.g. eating and drinking, activity and resting, clothing, exposure, housing, migration) which are driven by conscious sensations of thermal discomfort. These capabilities enable the (healthy) human being to live and to work in virtually any climate zone on earth, albeit with varying degrees of discomfort.

De Dear and Brager (2002) found that people who are able to adapt themselves (e.g. by clothing and manipulation of operable windows) to the prevailing weather outdoors will prefer different indoor temperatures, depending on their outdoor thermal exposure over the preceding couple of weeks. This linkage between outdoor weather exposures and indoor comfort preferences has been quantified in an adaptive model of thermal comfort (de Dear and Brager 1998, 2002) which, as noted by the IPCC Working Group III (Levine et al. 2007) carries enormous potential to reduce reliance on energy-intensive air conditioning in the built environment.

Furthermore, observed differences in temperature thresholds for thermal stress when, for example, mortality will increase significantly, indicate that societies at large can become acclimatized to their local climate ("acclimatization" being a special term for adaptation to climate), at least to a certain degree. However, the typical seasonal trends in time series of health data shows clearly that, at the population level, acclimatization is incomplete. In this context the oft-published statement that in a future warmer world the reduction in winter mortality will more than compensate the increase in summer mortality should be questioned due to the fundamentally different cause–effect relationships at play.

When looking at traditional, vernacular buildings (Roaf et al. 2005) different cultures always relied on local experience in climate related building design as a measure to adapt to the thermal environment. With the widespread introduction of air-conditioning in the twentieth century this connection with climate seemed less relevant or important to architects who were generally happy to hand over issues of thermal performance of buildings to heating, ventilating and air-conditioning (HVAC) engineers. On the other hand probably due to the climate change discussion in the last decade or so, and to the health impacts of extreme events such as the Chicago heat wave 1995 and the extreme summer 2003 in Europe, there is an increasing awareness of the significance of the thermal environment for sustainability, health, well-being, including for the need of adaptation. The main application areas are listed in the context of the UTCI (Universal Thermal Climate Index) development (see Section 2.3). Selected examples for the use of thermal assessment procedures to facilitate short- and long-term adaptation are given in Section 2.4. In particular the development of Heat Health Warning Systems. HHWS with locally adjusted intervention measures are now very popular despite the quite hesitant beginning to the WMO/WHO/UNEP HHWS showcase project initiative in 1997. Examples are the WHO coordinated European projects cCASHh, PHEWE, EuroHEAT and the development in USA (EPA 2006) and Canada.

In spite of the promising evolution of thermal environmental science and its application, there are still some weaknesses to be dealt with, including:

- Most of the assessment procedures used operationally do not meet the minimum requirements listed in Sections 2.2.1 and 2.2.2, i.e. the physiological and physical basics. Simple indices can only be of limited value and often lead to misrepresentations of the thermal environment. Therefore the development of the UTCI as an international standard intends to rectify this situation.
- When looking for dose–response relationships between thermal conditions and health outcomes, a large degree of "fuzziness" must be accepted because the data, usually from first order weather stations such as rural airports, are applied as a crude approximation of the exposure of people living indoors with unknown metabolic rate, in unknown buildings, in unknown floors under the influence of an unknown heterogeneous urban heat island. There is a strong research need to get a better idea on the *actual* thermal exposure which is a function of the heat budget and not only of the environmental stressors.
- Because the skill of the numerical weather forecast models differs between the meteorological variables there are some difficulties in the medium-range predictability of complex procedures, in particular with water vapour and clouds (which, in turn, impact latent heat transfer from the body, and mean radiant temperature *Tmrt*, which drives the net radiation exchange at the body surface).
- In spite of the vast amount of research literature in urban climatology there is still negligible uptake in urban planning and architecture, where the justifications of that urban climatological research are usually pitched.
- Very few approaches are available to quantitatively model the effects of acclimatization (e.g. the adaptive thermal comfort model or HeRATE (Health related adaptation to the thermal environment)). More quantitative adaptive research in thermal environments is needed to gain deeper insight at the population level in order to improve thermal environmental applications.
- The multidisciplinary interfaces between the participants (researchers, stakeholders, planners etc.) in the issues "thermal environment and adaptation" remains a challenge, but one that we need to deal with since the problems and applications never fall neatly into any single discipline.

2.2 Thermoregulation, Human Heat Budget, and Thermal Assessment Procedures

2.2.1 Thermoregulation

For the human being it is crucial to keep the body's core temperature at a constant level (37°C) in order to ensure functioning of the inner organs and of the brain. In contrast the temperature of the shell, i.e. skin and extremities, can vary strongly depending on the volume of blood it contains, which in turn, depends on metabolic and environmental heat loads. Heat is produced by metabolism as a result of activity, sometimes increased by shivering or slightly reduced by mechanical work where applicable, e.g. when climbing. The surplus heat must be released to the environment. The body can exchange heat by convection (sensible heat flux), conduction (contact with solids), evaporation (latent heat flux), radiation (long- and short-wave), and respiration (latent and sensible).

From the analytical point of view, the human thermoregulatory system can be separated into two interacting sub-systems: (1) the controlling active system which includes the thermoregulatory responses of shivering thermo genesis, sweat moisture excretion, and peripheral blood flow (cutaneous vasomotion) of unacclimatized subjects, and (2) the controlled passive system dealing with the physical human body and the heat transfer phenomena occurring in it and at its surface (Fig. 2.1). That accounts for local heat losses from body parts by free and forced convection, long-wave radiation exchange with surrounding surfaces, solar irradiation, and evaporation of moisture from the skin and heat and mass transfer through non-uniform clothing. Under comfort conditions the active system shows the lowest activity level indicating no strain. Increasing discomfort is associated with increasing strain and according impacts on the cardiovascular and respiratory system. The tolerance to thermal extremes depends on personal characteristics (Havenith 2001, 2005): age, fitness, gender, acclimatization, morphology, and fat thickness being among the most significant. Of these, age and fitness are the most important predictors and both are closely correlated. High age and/or low fitness level means low cardiovascular reserve which causes low thermal tolerance. The strain for the organism due to thermal stress can be quantified e.g. by an Physiological Strain Index PSI that is based on heart rate and T_{core} (Hyperthermia) (Moran et al. 1998) and on T_{skin} and T_{core} (Hypothermia) (Moran et al. 1999).

2.2.2 The Heat Budget

The heat exchange between the human body and the thermal environment (Fig. 2.2) can be described in the form of the energy balance equation which is nothing but the first theorem of thermodynamics applied to the body's heat sources

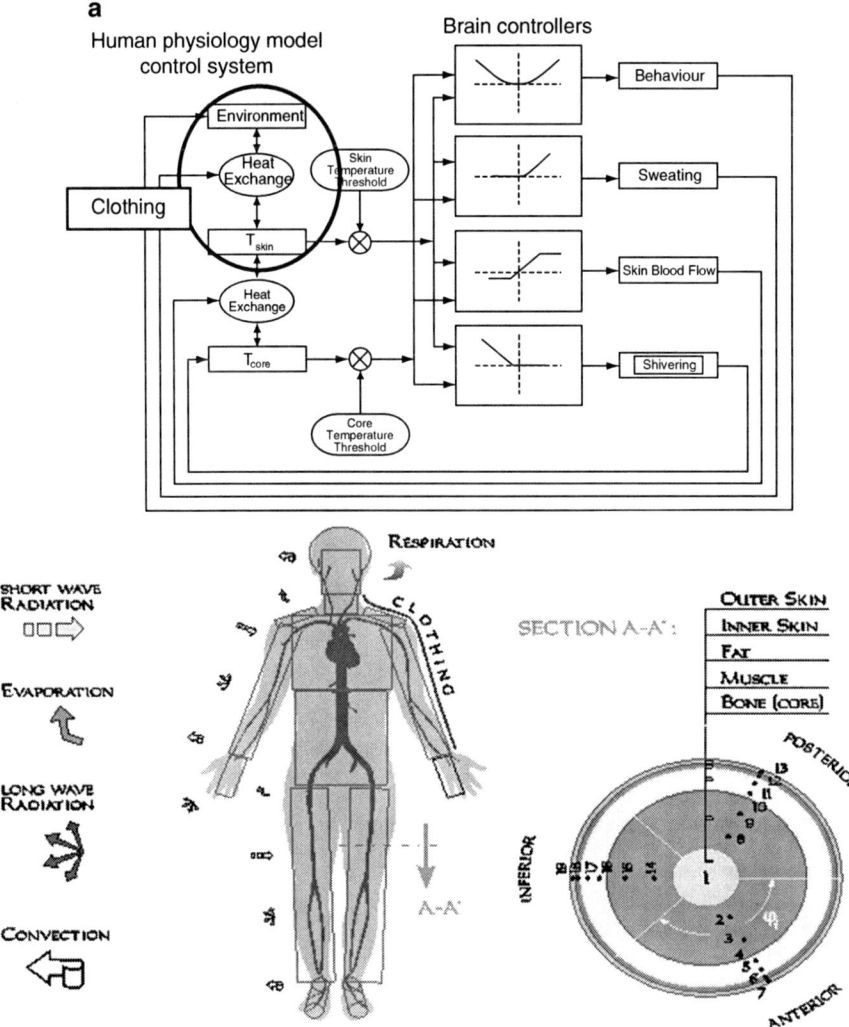

Fig. 2.1(a, b) Schematic representation of human physiological and behavioural thermoregulation (After Fiala et al. 2001; Havenith 2001)

(metabolism and environmental), and the various avenues of heat loss to environment (Büttner 1938):

$$M - W - \left[Q_H(T_a, v) + Q^*(T_{mrt}, v)\right] - \left[Q_L(e, v) + Q_{SW}(e, v)\right] - Q_{Re}(T_a, e) \pm S = 0 \quad (2.1)$$

- M Metabolic rate (activity)
- W Mechanical power
- S Storage (change in heat content of the body)

Avenues of Heat Exchange

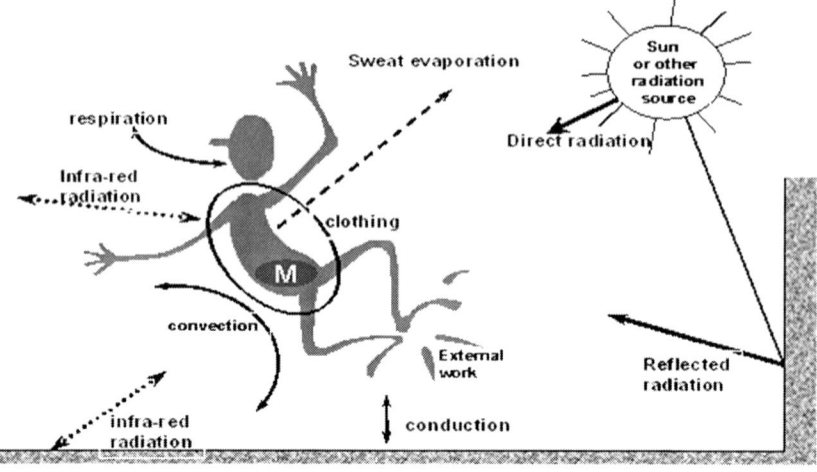

Fig. 2.2 The human heat budget (Havenith 2001)

Peripheral (skin) heat exchanges:
Q_H Turbulent flux of sensible heat
Q^* Radiation budget
Q_L Turbulent flux of latent heat (diffusion water vapour)
Q_{SW} Turbulent flux of latent heat (sweat evaporation)

Respiratory heat exchanges:
Q_{Re} Respiratory heat flux (sensible and latent)

Thermal environmental Parameters:
T_a Air temperature
T_{mrt} Mean temperature
v Air speed relative to the body
e Partial vapour pressure

The meteorological input variables include air temperature Ta, water vapour pressure e, wind velocity v, mean radiant temperature $Tmrt$ including short- and long-wave radiation fluxes, in addition to metabolic rate and clothing insulation. In Eq. (2.1) the appropriate meteorological variables are attached to the relevant fluxes. However, the internal (physiological) variables (Fig. 2.1), such as the temperature of the core and the skin, sweat rate, and skin wettedness interacting with the environmental heat exchange conditions are not explicitly mentioned here.

2.2.3 Thermal Assessment Procedures

In recognition that the human thermal environment cannot be represented adequately with just a single parameter, air temperature, over the last 150 years or so more than

100 simple thermal indices have been developed, most of them two-parameter indices. For warm conditions such indices usually consist of combinations of T_a and different measures for humidity, while for cold conditions the combination typically consists of T_a combined in some way with v. Simple indices are easy to calculate and therefore, easy to forecast. In addition they are readily communicated to the general public and stakeholders such as health service providers (Koppe et al. 2004). However, due to their simple formulation of the human heat balance as represented in Eq. (2.1) (i.e. neglecting significant fluxes or variables), these indices can never fulfil the essential requirement that for each index value there must always be a corresponding and unique thermo physiological state (strain), regardless of the combination of the meteorological input values. Thus their use is limited, results are often not comparable and additional features such as safety thresholds etc. have to be defined arbitrarily. Comprehensive reviews on simple indices can be found e.g. in Fanger (1970), Landsberg (1972), Driscoll (1992), and Parsons (2003).

Another approach based on synoptic climatology starts by identifying the various broad-scale weather types characterising a given locality. Several studies have identified that specific weather types (air masses) adversely affect mortality. Kalkstein et al. (1996) successfully extended this approach to heat health warning systems (HHWSs). The synoptic procedure classifies days that are considered to be meteorologically similar by statistically aggregating days in terms of a selection of meteorological variables such as air temperature, dew point, cloud cover, air pressure, wind speed and direction. The classification must be specifically derived for each particular locality where the synoptic approach is to be applied (see also Chapter 3).

Comprehensively characterising the thermal environment in thermo physiologically significant terms requires application of a complete heat budget model that takes all mechanisms of heat exchange into account, as described in Eq. (2.1). Such models (Fig. 2.3) possess the essential attributes that enable them to be universally utilised in virtually all biometeorological applications, across all climates zones, regions, and seasons.

This is certainly true for MEMI (Höppe 1984, 1999), and the Outdoor Apparent Temperature (Steadman 1984, 1994). However, it is not the case for the simple Indoor AT, which forms the basis of the US Heat Index, often used in outdoor applications by neglecting the prefix "Indoor". Other comprehensive heat balance indices include the Standard Effective Temperature (SET*) index (Gagge et al. 1986), and OUT_SET* (Pickup and De Dear 2000; De Dear and Pickup 2000), which translates Gagge's indoor version of the index to an outdoor setting by simplifying the complex outdoor radiative environment down to a mean radiant temperarture. Blazejczyk (1994) presented the man-environment heat exchange model MENEX, while the extensive work by Horikoshi et al. (1995, 1997) resulted in a Thermal Environmental Index.

Fanger's (1970) PMV (Predicted Mean Vote) equation can also be considered among the advanced heat budget models if Gagge's et al. (1986) improvement in the description of latent heat fluxes by the introduction of PMV* is applied. This approach is generally the basis for the operational thermal assessment procedure Klima-Michel-model (Jendritzky et al. 1979, 1990) of the German national weather

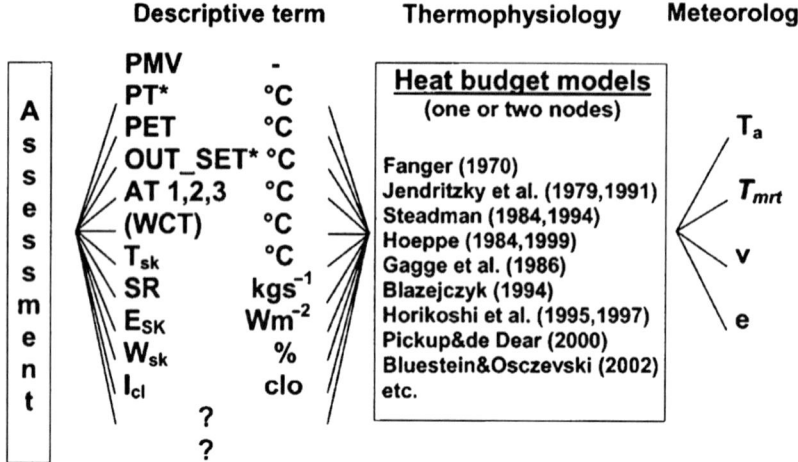

Fig. 2.3 Thermal physiological assessment of the thermal environment. PMV Predicted Mean Vote, PT* Perceived Temperature, PET Physiological Equivalent Temperature, OUT_SET* Outdoor Standard Effective Temperature, AT Apparent Temperature, WCT Wind Chill Temperature, *Tsk* mean skin temperature, *SR* sweat rate, *Esk* evaporative heat loss, *Wsk* wetness of the skin, *Icl* insulation value of clothing *clo*, *Ta* air temperature, *Tmrt* mean radiant temperature, *v* wind velocity, *e* water vapour pressure

service DWD (Deutscher Wetterdienst) with the output parameter "perceived temperature, PT" (Staiger et al. 1997) that considers a certain degree of adaptation by various clothing. This procedure is running operationally taking quantitatively the acclimatisation approach HeRATE (Koppe and Jendritzky 2005) into account. HeRATE is a conceptual model of short-term acclimatisation that modifies absolute PT thresholds by superimposition of the (relative) experience of the population in terms of PT of the previous weeks (Fig. 2.4). This procedure has the advantage that the index can be used without modification in different climate regions and during different times of the year without the need to artificially define seasons and to calibrate it to a particular city. Nevertheless, to date the German weather service (DWD) is the only national weather service to run a complete heat budget model (Klima-Michel-model) on a routine basis specifically for its applications in human biometeorology.

2.3 The Near Future: The Universal Thermal Climate Index UTCI

Although each of the published heat budget models is, in principle, appropriate for use in any kind of assessment of the thermal environment, none of the models is accepted as a fundamental standard, neither by researchers nor by end-users.

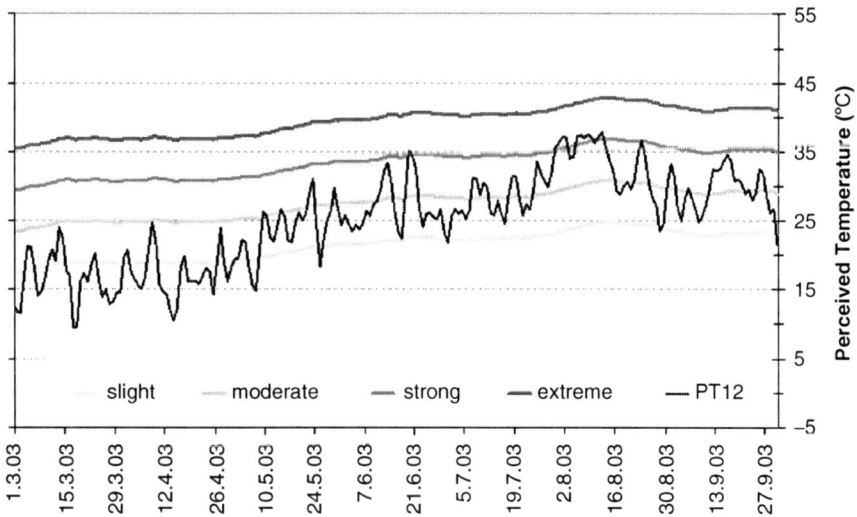

Fig. 2.4 Acclimatisation related thresholds for example Lisbon 2003 based on the HeRATE approach (Koppe and Jendritzky 2005)

On the other hand, it is surprising that after 40 years experience with heat budget modelling and easy access both to computational power and meteorological data, the oversimplified and thus unreliable indices are still widely used.

Some years ago the International Society on Biometeorology ISB recognised the issue presented above and established a Commission "On the development of a Universal Thermal Climate Index UTCI" (Jendritzky et al. 2002) (*www.utci.org*). Since 2005 these efforts have been reinforced by the COST Action 730 (*Co*operation in *S*cience and *T*echnical Development) of the European Science Foundation ESF that provides the basis that at least the European researchers plus experts from abroad can join together on a regular basis in order to achieve significant progress in deriving such an index (COST UTCI 2004). Aim is an international standard based on scientific progress in human response related thermo physiological modelling of the last 4 decades (Fiala et al. 2001, 2003) including the acclimatisation issue.

This work is performed under the umbrella of WMO's Commission on Climatology CCl, and will finally be made available in a WMO "Guideline on the Thermal Environment", probably by 2009, so that everybody dealing with biometeorological assessments, in particular NMS (National Meteorological and Hydrological Services), but also universities, public health agencies, epidemiologists, environmental agencies, city authorities, planners etc. can then easily apply the state-of-the-art procedure for their specific purposes. The guideline will provide numerous examples for applications and solutions for handling meteorological input data.

The Universal Thermal Climate Index UTCI (working title) must meet the following requirements:

1. Thermo physiologically significant across the entire range of heat exchange
2. Applicable for whole-body calculations but also for local skin cooling (frost bite)
3. Valid in all climates, seasons, and scales
4. Useful for key applications in human biometeorology

The following fields of applications are considered as particularly significant for users:

1. *Public weather service PWS.* The issue is how to inform and advice the public on thermal conditions at a short time scale (weather forecast) for outdoor activities, appropriate behavior, and climate-therapy.
2. *Public health system PHS.* In order to mitigate adverse health effects by extreme weather events (here heat waves and cold spells) it is necessary to implement appropriate disaster preparedness plans. This requires warnings about extreme thermal stress so that interventions can be released in order to save lives and reduce health impacts.
3. *Precautionary planning.* UTCI assessments provide the basis for a wide range of applications in public and individual precautionary planning such as urban and regional planning, and in the tourism industry. This is true for all applications where climate is related to human beings. The increasing reliability of monthly or seasonal forecasts will be considered to help develop appropriate operational UTCI products.
4. *Climate impact research in the health sector.* The increasing awareness of climate change and therewith related health impacts requires epidemiological studies based on cause–effect related approaches. UTCI will be the appropriate impact assessment tool. So also do scenario based calculations and down-scaling methods in the climate change and human health field need appropriate UTCI based procedures.

Mathematical modeling of the human thermal system goes back 70 years. In the past four decades more detailed, multi-node models of human thermoregulation have been developed, e.g. Stolwijk (1971), Konz et al. (1977), Wissler (1985), Fiala et al. (1999, 2001), Huizenga et al. (2001) and Tanabe et al. (2002). These models simulate phenomena of the human heat transfer inside the body and at its surface taking into account the anatomical, thermal and physiological properties of the human body (see Fig. 2.1). Environmental heat losses from body parts are modeled considering the inhomogeneous distribution of temperature and thermoregulatory responses over the body surface. Besides overall thermo physiological variables, multi-segmental models are thus capable of predicting 'local' characteristics such as skin temperatures of individual body parts. Validation studies have shown that recent multi-node models reproduce the human dynamic thermal behaviour over a wide range of thermal circumstances (Fiala et al. 2001, 2003; Havenith 2001; Huizenga et al. 2001). Many of these models have been valuable research tools contributing to a deeper understanding of the principles of human thermoregulation (Fiala et al. 2001). However, there is still a need for better understanding of adaptive responses and their physiological implications.

The passive system of the Fiala model (Fiala et al. 1999, 2001) is a multi-segmental, multi-layered representation of the human body with spatial subdivisions. Each tissue node is assigned appropriate thermo physical and thermo physiological properties. The overall data replicates an average person with respect to body weight, body fat content, and Dubois-area. The physiological data aggregates to a basal whole body heat output and basal cardiac output, which are appropriate for a reclining adult in a thermo-neutral environment of 30°C. In these conditions, where no thermoregulation occurs, the model predicts a basal skin wettedness of 6%; a mean skin temperature of 34.4°C; and body core temperatures of 37.0°C in the head core (hypothalamus) and 36.9°C in the abdomen core (rectum) (Fiala et al. 1999). Verification and validation work using independent experiments from air exposures to cold stress, cold, moderate, warm and hot stress conditions, and a wide range of exercise intensities revealed good agreement with measured data for regulatory responses, mean and local skin temperatures, and internal temperatures for the whole spectrum of boundary conditions considered (Richards and Havenith 2007).

The experts of the COST Action 730 WG on Thermo Physiological Modeling have agreed to base the UTCI model on the Fiala approach which will be substantially advanced by including as yet unused data from other research groups. The UTCI model must meet all the above listed requirements in application. From practical considerations the advanced Fiala multi-segmental model cannot be applied explicitly on a routine basis. Thus the future UTCI computations will make use of a statistical approach derived from simulations with the Fiala model that covers all conceivable combinations of air temperature, wind, humidity, and mean radiant temperature plus clothing.

In an operational procedure the non-meteorological variables metabolic rate MET and thermal resistance of clothing are of great importance. The UTCI Commission has defined a representative activity to be that of a person walking with a speed of 4 km/h. This provides a metabolic rate of 2.3 MET (135 W/m^2). Clothing isolation Icl will be considered as an intrinsic clo-value in the range of Icl = 0.4–1.7 clo (1 clo = 0.155 km^2/W) determined by air temperature. This should cover the kinds of clothing worn by people who are adapted to their local climate. The need to address specific characteristics of clothing, such as significant ventilation between body surface and inner surface of clothing is still subject of discussion.

2.4 Use for Adaptation (Selected Examples)

There are numerous epidemiological studies published which impressively show worldwide the health impact of extreme thermal conditions such as heat waves. Figure 2.5 shows as an example a detail of the daily mortality rate time series from south-west Germany that includes the hot summer 2003 (Schär and Jendritzky 2004). During this summer about 55,000 extra deaths attributable to heat occurred in Europe, and from these about 35,000 alone in August (Brücker 2005; Kosatsky 2005). Neither the NMSs nor the public health systems were sufficiently prepared.

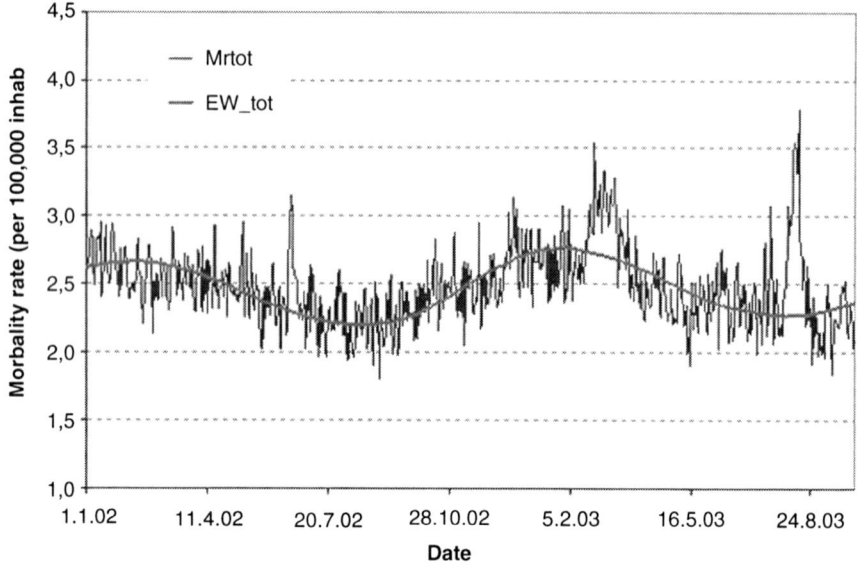

Fig. 2.5 Daily total mortality rates (MR) in SW – Germany. Smoothed line, i.e. expected value (EW_tot) based on Gauss-Filter. Evident: MR peak in June 2002 (short heat wave), episode in spring 2003 (related to an influenza epidemic), peaks in July and the August heat wave effect (Schär and Jendritzky 2004)

When considering the impact of the hot summer 2003 numerous questions arise, such as:

1. There is no general accepted definition of a heat wave. A conceptually adequate definition must be based on the physiological response, see Eq. (2.1).
2. There is no consensus on the definition of the mortality baseline. In Fig. 2.5 a time series filtering approach is used rather than just calculating monthly mean values.
3. There is no idea about the actual heat exposure of the population in different floors of different buildings in urban areas when applying meteorological data from usually rural measuring sites. It can be assumed that the urban heat island effect (UHI) has intensified the regional heat load. But how many?

Based on a climate change simulation with HIRAM (Beniston 2004) the distribution of the maximum temperatures of the summer 2003 (Fig. 2.6) indicates that – if the prediction were correct – this extreme summer is expected to be a fairly normal regular occurrence by the end of this century in Central Europe! This is basically compatible with the change of the annual mean of Perceived Temperature PT in a future climate (2041–2050) in central Europe compared to the control run (1971–1980), which here is taken as the "actual" climate, based on the MPI time-slice experiment with ECHAM4 in T106 resolution, assuming the "business-as-usual" scenario IS92a (Fig. 2.7a, b). The need for adaptation is evident. Short-term (I) and long-term (II) adaptation measures are crucial.

2 Adaptation and Thermal Environment

The heat wave 2003 in Europe: A unique feature?

IPCC WGI, 2001:

"Higher maximum temperatures and more hot days over nearly all land areas are very likely"

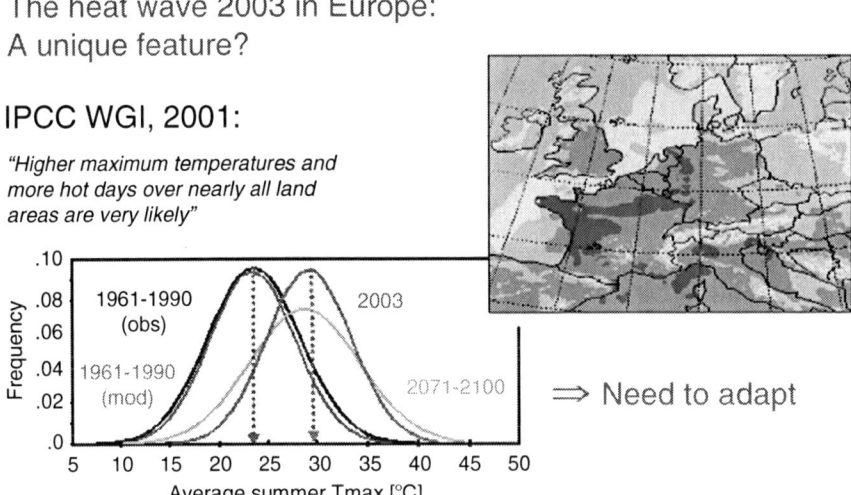

⟹ Need to adapt

Fig. 2.6 The heat wave 2003 in Europe compared to the current climatological distribution, and that predicted towards the end of the current century. What was an extreme event against today's climatology will become much more common in the future (Beniston 2004)

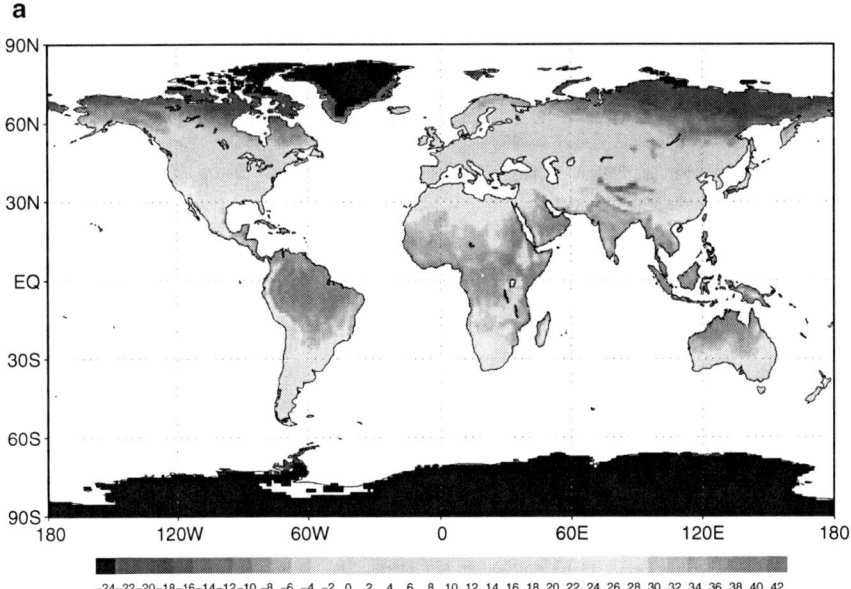

Fig. 2.7 (a) Annual mean of Perceived Temperature PT (°C) based on the control run (1971–1980)

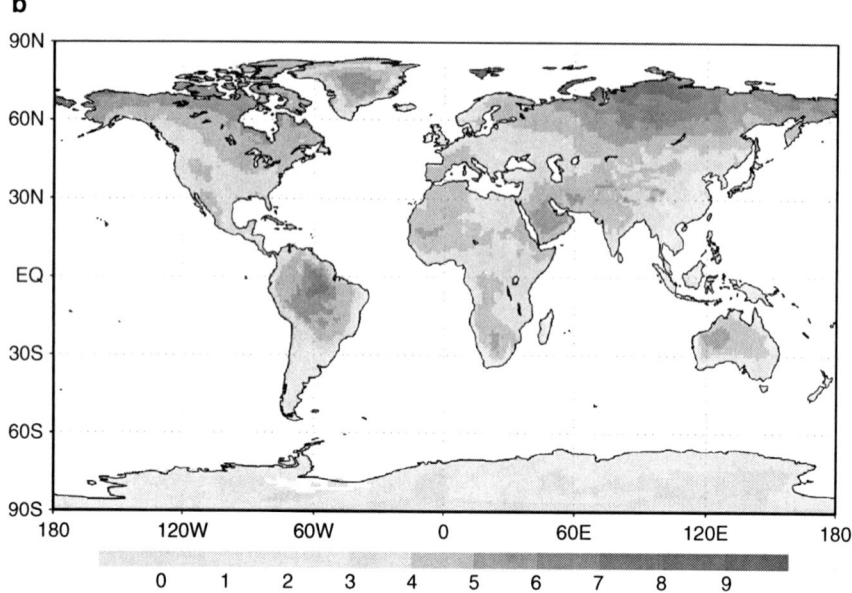

Fig. 2.7 (**b**) change of the annual mean of Perceived Temperature PT (K) in a future climate (2041–2050). ECHAM4/T106 (DKRZ, Hamburg)

2.4.1 Short-Term Adaptation

Lives would have been saved if adequate heat-health warning services (HHWS) had been activated in Europe in 2003, as promoted by the WMO/WHO/UNEP showcase projects in Rome and Shanghai. Such systems are based on biometeorological forecasts (Fig. 2.8) expecting exceeding of an agreed threshold (heat load forecast). The following interventions (based on a locally adjusted emergency response plan) are the responsibility of public health services PHS. HHWSs must be prepared in advance with complete descriptions of all processes and clear definition of the interface between NMHS and PHS (Koppe et al. 2004; WMO 2004; Kovats and Jendritzky 2006; EPA 2006; WMO 2007) (see also Chapter 3).

The whole HHWS procedure can be divided into four more or less independent modules:

1. The Public Health Issue.

The most important module is a locally adjusted disaster preparedness (emergency response) plan based on a specific mitigation strategy. This plan becomes active whenever a heat load event is expected. The scopes concerned, intervention measures, and responsible agencies, decision-makers, stakeholders, and other people etc. must be defined. The experience with existing approaches to implement HHWSs shows clearly that the development of an appropriate intervention strategy that takes into consideration local needs, such as political and urban infrastructure, is the most difficult step.

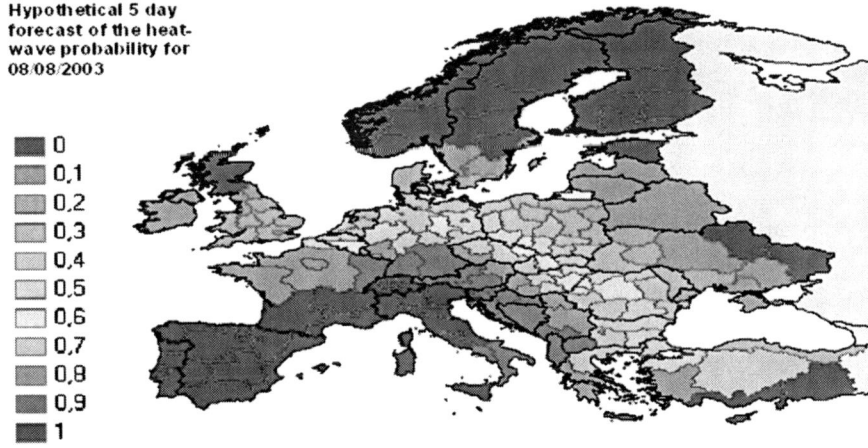

Fig. 2.8 Example hypothetical 5 day forecast of the heat-wave probability 08/08/2003 for Europe based on the ECMWF ensemble prediction system

2. What is Heat Load?

Hampering heat exchange from the human body to the atmosphere produces strain for the organism. People with limited adaptive capacity, i.e. people who are not fit, can die from manifold causes but the failure of thermoregulation is always implicated. So there is a need for a health related definition of thermal environmental stress that is thermo physiologically significant.

3. Heat Load Forecasts

The forecast for extreme heat load must be based on routine procedures of National Meteorological Services (NMSs) considering the situation of the next few days. Figure 2.8 shows a hypothetical 5 day forecast of heat load for Europe already demand-oriented to administrative borders. The forecast was based on the ensemble prediction system EPS of the European Centre for Medium Range Weather Forecast (ECMWF) which is capable of issuing probabilistic forecasts of up to 15 days leadtime. The public health authorities are responsible to define the kind of emergency information they want considering heat load intensity and time schedule.

4. Epidemiology

Correlation studies between the biometeorological indices and population health data (mortality/morbidity) are reasonable for calibrations, i.e. to define specific thresholds, but it should be noted that this "fine tuning" is only valid for the area under investigation. Frequently the paucity of health data, lack of expertise and resources are insurmountable obstacles. Scientifically it would be satisfactory to have reliable epidemiological results; however, from a practical point of view there is no urgent need (just "nice to have"). Whenever a heat wave occurs running a functioning HHWS is more important than the attempt to be perfect in any detail.

Heat Health Warning Systems (HHWS) can be realized in the short-term. The numerous successful systems established by Kalkstein and collaborators, some as

WMO/WHO/UNEP Showcase Projects, and the outcomes of the WHO coordinated European projects cCASHh, PHEWE, and EuroHEAT are the best exemplars of the value of this approach. For a more comprehensive consideration of HHWSs see Chapter 3.

2.4.2 Long-Term Adaptation

2.4.2.1 Adaptation to Urban Climates

The climates of cities present some of the most impressive examples of anthropogenic climate modification resulting from intentional or accidental changes in land-use. When looking for the thermal environment issue in the urban climate and human health field, the urban heat island (UHI) is the essential subject of anticipatory (or proactive) adaptation/mitigation measures both in short term and long term time-scales. Unfortunately till now the term UHI is based just on air temperature (actually on the difference between inner-city and rural temperatures) and not on the complex controls of the human body's heat balance.

There is no doubt that the urban heat island is relevant for human health. It causes adverse health effects from exposure to extreme thermal conditions. The urban heat island has an added effect on heat wave intensity, which may exacerbate the impact of weather on heat-related mortality. As the urban heat island is the result of urban density, form and materials, it is correspondingly also sensitive to future urban planning. But in spite of the impressive depth in knowledge about urban climate there is unfortunately still a need to bridge the gap between science and application at the relevant time scales. In the short term time scale HHWS intervention strategies are useful tools for mitigating adverse effects due to heat waves. In long term time scales there is a need to create urban development standards, to make existing knowledge accessible and intelligible, and to develop practical tools for urban planning. That support urban planners to reach their fundamental aim: creating and safeguarding of healthy environmental conditions for residence and work. The global warming problem increases the urgency of this prescription.

From a biometeorological point of view the atmospheric fields determining the thermal environment are significant in the urban canopy layer (Fig. 2.9), i.e. the settlement structure (including its interactions) as the result of planned or free development (Oke 1987; Ali-Toudert and Mayer 2005). While the air temperature shows almost no difference between the shaded and the sunny side of the street, the fully-developed heat-balance index, PT, clearly differentiates the two situations on thermo physiologically significant criteria. When wind is calm direct sun exposure affects the heat budget of the human being as an increase of air temperature of about 12 K.

For urban planning purposes modelling seems to be the appropriate method. Urban climate models with high resolutions that cover urban districts as well as towns and cities as a whole including its varying urban structure are computationally

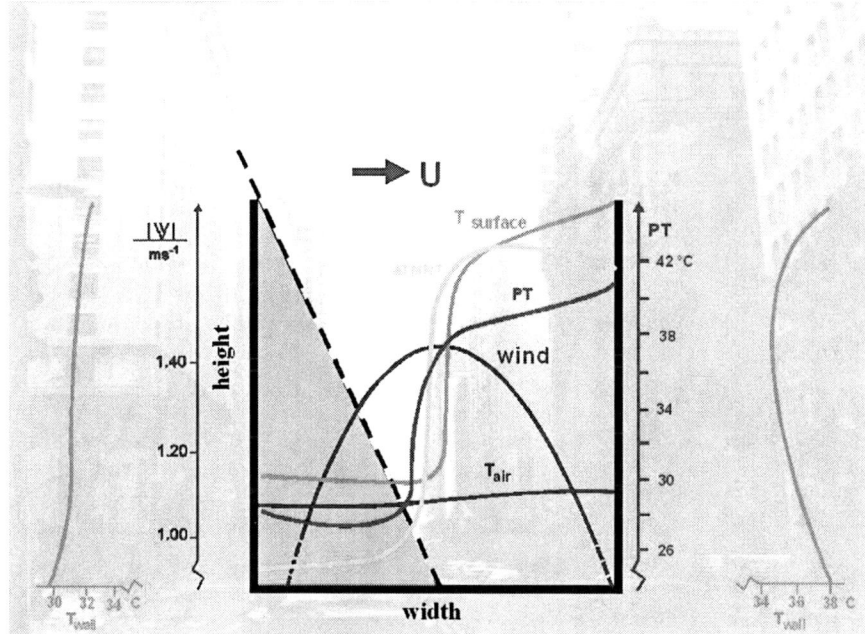

Fig. 2.9 Meteorological and biometeorological conditions in a cross section of a street (PT Perceived Temperature, $\Delta Tmrt$ means difference of mean radiant temperature from air temperature $Tair$)

demanding and need users with professional skill. For practical applications in urban planning the *U*rban *Bi*o-*c*limate *M*odel.

UBIKLIM (Graetz et al. 1992; Friedrich et al. 2001) was developed as an expert system that utilizes available knowledge in urban climate science in an objective procedure. Using GIS-techniques UBIKLIM simulates the thermal environment in the urban boundary layer at a given location in an urban area that depends on the kind of land use, i.e. the settlement structure (these are the planning variables to be transformed into boundary layer parameters). Interactions between neighbouring structures, topography (local scale), and meso- and macro-scale climate are taken into account. The example in Fig. 2.10 shows an urban area with a differentiated pattern of probabilities of the occurrence of heat load (in terms of Perceived Temperature, not of air temperature) resulting from different land uses (settlement structures).

2.4.2.2 Adaptation to Indoor Climates

Since we spend at least 90% of our daily lives inside built environments of one sort or another, indoor climates are probably more significant drivers of our state of thermal adaptation than the outdoor climate or local microclimate. The same basic physics of heat balance and also physiological responses to the thermal environment are as relevant to indoor settings as they are to the outdoor environment. The main

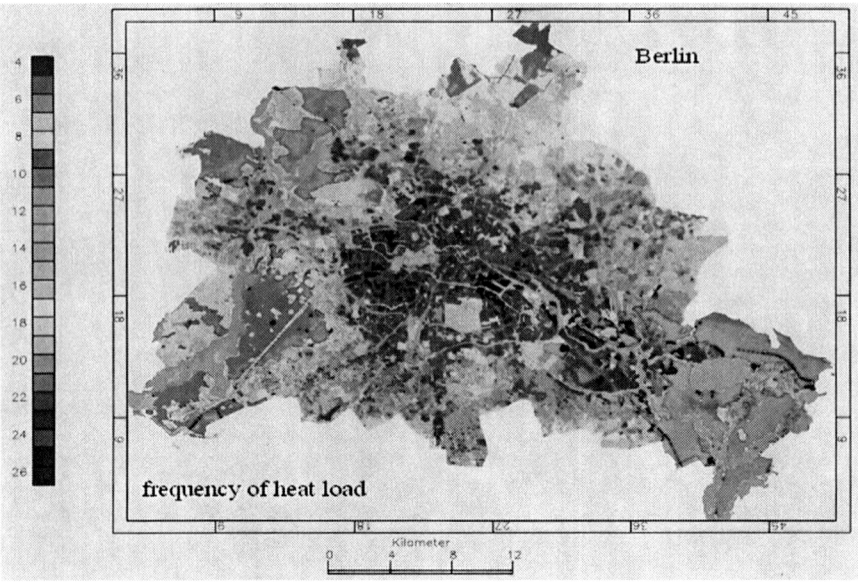

Fig. 2.10 The frequency of heat load conditions in days per year in Berlin, Germany based on UBIKLIM-simulations in 10 m grids

distinction between indoor and outdoor settings is one of extremes. Indoor climates are, in the vast majority of cases, best described as moderate thermal environments, whereas outdoor microclimates span a much wider range, in the spatial sense, but also temporally across all scales, from diurnal, through synoptic, and up to seasonal.

The fact that indoor climates account for a substantial component of energy end-use, and therefore greenhouse gas emissions, recently led the Intergovernmental Panel on Climate Change to identify the buildings sector as affording the highest likelihood of deep reductions in greenhouse gas reductions of all sectors looked at in the IPCC Fourth Assessment Report (Levine et al. 2007). That optimism, however, was based on gradual improvements in the energy efficiency of building envelopes and Heating, Ventilation and Air-Conditioning plant (HVAC). Although the potential of human thermal adaptation to indoor climate was recognised as highly relevant to energy savings, the IPCC focused its attention on market transformation that didn't rely on adjustments to life styles or comfort levels. Nevertheless, it is becoming clear that simply shifting building thermostat settings just a few degrees away from static targets like 23°C, without expensive retrofits of efficiency measures to plant or building envelope can effect a profound saving of energy and greenhouse gas emissions. For example, Ward and White (2007) measured a 14% reduction in HVAC energy consumption on identical summer days, just by shifting the set-point in a conventionally air-conditioned office building in Melbourne just one degree higher from the building's previous 22°C target. With HVAC energy typically accounting for up to 40% or 50% of total building energy end-use, the need to shift the comfort expectations of building occupants away from static HVAC set-points is becoming compelling, because it is an efficiency measure that is readily and immediately applicable across much of the existing building stock, not just new construction and refurbishments.

Shifting indoor comfort expectations is going to rely on human thermal adaptation. The so-called adaptive model of thermal comfort is premised on the widely reported relationship between the indoor temperature at which building occupants express thermal comfort, and the mean indoor temperature to which they have been exposed over periods ranging from a week to a month (Humphreys 1981; Auliciems 1981, 1986; Nicol and Humphreys 2002; De Dear and Brager 1998). If indoor temperatures are held constant, detached from the diurnal, synoptic and seasonal drifts outdoors, then indoor comfort temperatures will also remain fixed as well. However, in un-air conditioned or free-running buildings, especially with user-operable windows, comfort temperatures have been noted to be highly correlated with the outdoor climatic environment.

The graphs in Fig. 2.11, excerpted from an adaptive comfort research project commissioned by the American Society of Heating, Refrigerating and Air-Conditioning Engineers (ASHRAE) (De Dear and Brager 1998, 2002), compares indoor comfort temperatures based on the "static" PMV heat balance model with those from an adaptive model that was statistically fitted to actual observations of comfort in hundreds of office buildings located in various climate zones around the world. The static model's comfort temperature for each building was derived by inputting the building's mean v, rh, clo, met into the PMV model and then iterating for different operative temperatures until PMV = 0 i.e. *"neutral"*. The x-axis in both panels of Fig. 2.11 is the monthly mean outdoor temperature prevailing at the time of each building's comfort survey, with the left-hand panel showing results from buildings with centrally-controlled HVAC systems, and the right-hand panel showing results from naturally ventilated buildings. The modest adaptation (barely 2°C)

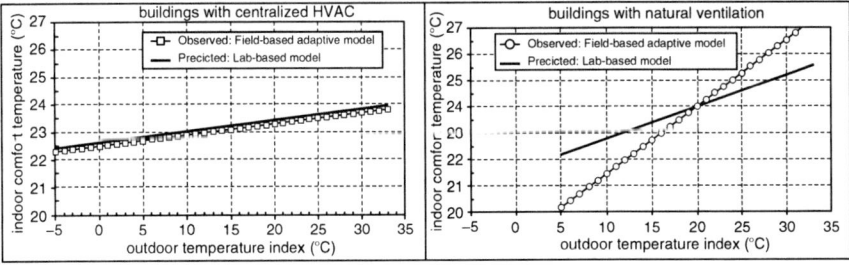

Fig. 2.11 The adaptive model of indoor thermal comfort, compared to comfort temperature predictions by the static heat-balance PMV model. While the PMV model's predictions compare well with the field observations in buildings with centralized HVAC, the classic heat-balance parameters underpinning PMV are inadequate at explaining the greater variance in climatically-correlated indoor comfort temperatures observed in free-running buildings (naturally ventilated) (After De Dear and Brager 1998)

to outdoor climate shown by occupants of centrally air-conditioned buildings (left panel in Fig. 2.11) is driven largely by adjustments to clothing insulation, and is well predicted by the PMV heat balance comfort model. However, occupants of naturally ventilated or free-running buildings (right-hand panel in Fig. 2.11) adapted to a much wider range of comfort temperatures than could be predicted by heat-balance parameters alone. The divergence between observed and PMV-predicted comfort in the right-hand panel was ascribed to shifting comfort expectations (De Dear and Brager 1998, 2002). The indoor temperature regimes prevailing in free-running buildings are themselves more closely correlated with outdoor weather and climate than in the central-HVAC buildings, therefore the indoor temperatures which free-running building occupants come to expect are more closely correlated with outdoor temperatures too.

This psychological dimension of comfort, expectation, is not one of the classic human heat-balance parameters (Eq. (2.1)), but it probably holds at least as much promise for carbon reductions in the buildings sector as do energy efficiency improvements in building envelope and HVAC plant – as long as building occupants are provided with adequate adaptive opportunity, especially by means of operable windows (Brager et al. 2004).

Having made its way into the 2004 revision of ASHRAE's comfort standard 55 "*Thermal environmental conditions for human occupancy*" (ASHRAE 2004), de Dear and Brager's adaptive comfort model (1998) is already being taken up in the design of new buildings. A recent example is the new Federal Building in San Francisco (McConahey et al. 2002), which features a natural ventilation façade, the first of its kind in an office building on the US west coast since the advent of air conditioning in the first half of the twentieth century. However, the question of how long it will take for occupants to adapt to variable indoor temperatures after they have been acclimatised to static HVAC indoor climates, remains yet to be answered.

2.5 Conclusions

The basic state of knowledge in the field of weather/ climate and human health allows for the delivery of a number of advisory services in order to enhance the capability of societies and individuals to properly adapt to climate and climate change. As regards risk factors, biometeorology has to inform and advise the public and decision makers in politics and administration with the aim of recognizing and averting health risks at an early stage, in the framework of preventive planning, for example by making recommendations for ambient environmental standards, by evaluation of location decisions, and by consultation on adaptive behaviour. Thus services for improving health and well-being of the population can be provided as a result of the work of the National Meteorological Services (NMSs).

Climate services can contribute to identify the most appropriate approaches, measures, technologies, and policies to improve the adaptive capacity to climate and climate change. Examples are given from the fields HHWSs and precautionary planning in urban areas. The significance of these issues also in the context of the climate change problem is obvious. Evidently, the services required for the good health, safety and well-being of national communities can be significantly improved if NMSs are ready to tap into the existing body of knowledge, practices, research and technology to design and deliver appropriate biometeorological information and advisories to the public in order to support people in proper adaptation.

References

Ali-Toudert F and Mayer H (2006) Thermal comfort in an east-west oriented street canyon in Freiburg (Germany) under hot summer conditions. *Theor. Appl. Climatol.* DOI 10.1007/s00704-005-0194-4

ASHRAE (2004) ASHRAE Standard 55-2004: Thermal Environmental Conditions for Human Occupancy. ASHRAE, Atlanta, GA

Auliciems A (1981) Towards a psycho-physiological model of thermal perception. *Int. J. Biometeorol.* 25: 109–122

Auliciems A (1986) Air conditioning in Australia III: Thermobile controls. *Archit. Sci. Rev.* 33: 43–48

Beniston M (2004) The 2003 heat wave in Europe: A shape of things to come? An analysis based on Swiss climatological data and model simulations. *Geophys. Res. Lett.* 31, L02202: 1–4

Blazejczyk K (1994) New climatological- and -physiological model of the human heat balance outdoor (MENEX) and its applications in bioclimatological studies in different scales. *Zeszyty IgiPZ PAN* 28: 27–58

Brager GS, Paliaga G and de Dear RJ (2004) Operable windows, personal control and occupant comfort. *ASHRAE Trans.* 110, 2: 510–526

Brücker G (2005) Vulnerable populations: lessons learnt from the summer 2003 heat waves in Europe. *Euro Surveill* 10, 7: www.eurosurveilance.org

Büttner K (1938) Physikalische Bioklimatologie. Probleme und Methoden. Akad. Verl. Ges., Leipzig, 155 pp.

COST UTCI (2004) Towards a Universal Thermal Climate Index UTCI for Assessing the Thermal Environment of the Human Being. MoU of COST Action 730. www.utci.org. pp. 17

Driscoll DM (1992) Thermal comfort indexes. Current uses and abuses. *Nat. Weather Digest* 17, 4: 33–38

De Dear RJ and Brager G (1998) Developing an adaptive model of thermal comfort and preference. *ASHRAE Trans.* 104, 1a: 145–167

De Dear R and Pickup J (2000) An outdoor thermal environment index (OUT_SET*) – Part II – Applications. In: R. De Dear, J. Kalma, T. Oke, A. Auliciems (eds.), Biometeorology and Urban Climatology at the Turn of the Millennium. *Selected Papers from the Conference ICB-ICUC'99 (Sydney, 8–12 Nov. 1999). WMO, Geneva, WCASP-*50: 258–290

De Dear RJ and Brager GS (2002) Thermal comfort in naturally ventilated buildings: revisions to ASHRAE Standard 55, *Energ. Build.* 34: 549–561

EPA (2006) Excessive Heat Events Guidebook. EPA 430-B-06-005. www.epa.gov/heatislands/about/heatguidebook

Fanger PO (1970) Thermal comfort. Analysis and application in environment engineering. Danish Technical Press, Copenhagen

Fiala D, Lomas KJ and Stohrer M (1999) A computer model of human thermoregulation for a wide range of environmental conditions: The passive system. *J. Appl. Physiol.* 87, 5: 1957–1972

Fiala D, Lomas KJ and Stohrer M (2001) Computer prediction of human thermoregulatory and temperature responses to a wide range of environmental conditions. *Int. J. Biometeorol.* 45: 143–159

Fiala D, Lomas KJ and Stohrer M (2003) First principles modeling of thermal sensation responses in steady-state and transient conditions. *ASHRAE Trans. Res.* 109, Part I: 179–186

Friedrich M, Grätz A, Jendritzky G (2001) Further development of the urban bioclimate model UBIKLIM, taking local wind systems into account. *Met. Z.* 10, 4: 267–272

Gagge AP, Fobelets AP and Berglund PE (1986) A standard predictive index of human response to the thermal environment. *ASHRAE Trans.* 92: 709–731

Givoni B (1976) Man, Climate and Architecture. Applied Science Publishers, London, pp. 483

Grätz A, Jendritzky G, Sievers U (1992) The Urban Bioclimate Model of the Deutscher Wetterdienst. In: K Höschele (ed.) Proceedings of the Symposium on Planning Applications of Urban and Building Climatology in Berlin 14–15 Oct., 1991, *Wiss. Ber. IMK, Karlsruhe*: 96–105

Havenith G (2001) An individual model of human thermoregulation for the simulation of heat stress response. *J. Appl. Physiol.* 90: 1943–1954

Havenith G (2005) Temperature regulation, heat balance and climatic stress. In: W. Kirch, B. Menne, R. Bertollini (eds.), Extreme Weather Events and Public Health Responses. Springer, Heidelberg, pp. 69–80

Hassi J (2005) Cold extremes and impacts on health. In: W. Kirch, B. Menne, R. Bertollini (eds.), Extreme Weather Events and Public Health Responses. Springer, Heidelberg, pp. 59–67

Höppe P (1984) Die Energiebilanz des Menschen. *Wiss. Mitt. Meteorol. Inst. Uni München* 49

Höppe P (1999) The physiological equivalent temperature – a universal index for the biometeorological assessment of the thermal environment. *Int. J. Biometeorol.* 43: 71–75

Horikoshi T, Tsuchikawa T, Kurazumi Y and Matsubara N (1995) Mathematical expression of combined and seperate effect of air temperature, humidity, air velocity and thermal radiation on thermal comfort. *Arch. Complex Environ. Stud.* 7, 3–4: 9–12

Horikoshi T, Einishi M, Tsuchikawa T and Imai H (1997) Geographical distribution and annual fluctuation of thermal environmental indices in Japan. Development of a new thermal environmental index for outdoors and its application. *J. Human-Environ. Syst.* 1, 1: 87–92

Huizenga C, Zhang H and Arens E (2001) A model of human physiology and comfort for assesssing complex thermal environments. *Build. Environ.* 36: 691–699

Humphreys MA (1981) The dependence of comfortable temperatures upon indoor and outdoor climates. In: K. Cena and J.A. Clark (eds.), Bioengineering, Thermal Physiology and Comfort, Amsterdam, Elsevier, pp. 229–250

Jendritzky G, Sönning W and Swantes HJ (1979) Ein objektives Bewertungsverfahren zur Beschreibung des thermischen Milieus in der Stadt- und Landschaftsplanung ("Klima-Michel-Modell"). *Beiträge d. Akad. f. Raumforschung und Landesplanung*, 28, Hannover

Jendritzky G (1990) Bioklimatische Bewertungsgrundlage der Räume am Beispiel von mesoskaligen Bioklimakarten. In: Jendritzky G, Schirmer H, Menz G, Schmidt-Kessen W: Methode zur raumbezogenen Bewertung der thermischen Komponente im Bioklima des Menschen (Fortgeschriebenes Klima-Michel-Modell). Akad Raumforschung Landesplanung, Hannover, Beiträge 114: 7–69

Jendritzky G, Maarouf A, Fiala D and Staiger H (2002) An update on the development of a universal thermal climate index. Proceedings of the 15th Conference on Biometeorological Aerobiology and 16th ICB02, 27 Oct–1 Nov 2002, Kansas City, AMS, pp. 129–133

Kalkstein LS, Barthel CD, Green JS and Nichols MC (1996) A New Spatial Synoptic classification: application to Air Mass Analysis. *Int. J. Climatol.* 16, 983–1004

Kirch W, Menne B, Bertollini R (2006) Extreme Weather Events and Public Health Responses. Springer, Berlin

Konz S, Hwang C, Dhiman B, Duncan J and Masud A (1977) An experimental validation of mathematical simulation of human thermoregulation. *Comput. Biol. Med.* 7: 71–82

Koppe C, Kovats S, Jendritzky G and Menne B (2004) Heat-waves: risks and responses. World Health Organization. Health and Global Environmental Change, Series, No. 2, Copenhagen, Denmark

Koppe C and Jendritzky G (2005) Inclusion of short-term adaptation to thermal stresses in a heat load warning procedure. *Meteorol. Z.* 14, 2: 271–278

Kosatsky T (2005) The 2003 European heat waves. *Euro Surveill* 10, 7: www.eurosurveillance.org

Kovats SR and Jendritzky G (2006) Heat-waves and human health. In: B. Menne and K.L. Ebi (eds.), Climate Change and Adaptation Strategies for Human Health. Steinkopff, Darmstadt, pp. 63–97

Landsberg HE (1972) The assessment of human bioclimate, a limited review of physical parameters. World Meteorological Organization, *Technical Note No. 123, WMO-No. 331*, Geneva

Levine M, Ürge-Vorsatz D, Blok K, Geng L, Harvey D, Lang S, Levermore G, Mongameli Mehlwana A, Mirasgedis S, Novikova A, Rilling J and Yoshino H (2007) Residential and commercial buildings. In: B. Metz, O.R. Davidson, P.R. Bosch, R. Dave, L.A. Meyer (eds.), Climate Change 2007: Mitigation. Contribution of Working Group III to the Fourth Assessment Report of the Intergovernmental Panel on Climate Change, Cambridge University Press, Cambridge, United Kingdom/New York

McConahey E, Haves P and Christ T (2002) The integration of engineering and architecture: A perspective on natural ventilation for the new San Francisco Federal Building. In proceedings of the ACEEE 2002 Summer Study on Energy Efficiency in Buildings Conference

Moran DS, Shitzer A and Pandolf KB (1998) A physiological strain index to evaluate heat stress. *Am. J. Physiol. Regul. Integr. Comp. Physiol.* 275: R129–R134

Moran DS, Castellani JW, O'Brien C, Young AJ and Pandolf KB (1999) Evaluating physiological strain during cold exposure using a new cold strain index. *Am. J. Physiol. Regul. Integr. Comp. Physiol.* 277: R556–R564

Nicol JF and Humphreys MA (2002) Adaptive thermal comfort and sustainable thermal standards for buildings. *Energ. Build.* 34, 6: 563–572

Oke TR (1987) Boundary Layer Climates, Methuen, London

Parsons KC (2003) Human Thermal Environments: The Effects of Hot, Moderate, and Cold Environments on Human Health, Comfort and Performance. Taylor & Francis, London/New York

Richards M and Havenith G (2007) I.B. Mekjavic, S.N. Kounalakis, N.A.S. Taylor (eds.), Progress Towards the Final UTCI Model. Proceedings of the 12th International Conference on Environmental Ergonomics. August 19–24, Piran Slovenia. Ljubljana, Biomed., pp. 521–524

Roaf S, Chrichton D and Nicol F (2005) Adapting Buildings and Cities for Climate Change, Achitectural Press, Oxford

Pickup J and De Dear R (2000) An outdoor thermal comfort index (OUT_SET*) – Part I – the model and its assumptions. In: R. de Dear, J. Kalma, T. Oke, A. Auliciems (eds.), Biometeorology and Urban Climatology at the Turn of the Millenium. Selected Papers from the Conference ICB-*ICUC'99 (Sydney, 8–12 Nov. 1999)*. WMO, Geneva, WCASP-50: 279–283

Schär C and Jendritzky G (2004) Hot news from Summer 2003. News and views. *Nature.* 432, 2 Dec: 559–560

Shitzer A (2006) Wind-chill-equivalent temperatures: regarding the impact due to the variability of the environmental convective heat transfer coefficient. *Int. J. Biomet.* 50, 4: 224–232

Staiger H, Bucher K and Jendritzky G (1997) Gefühlte Temperatur. Die physiologisch gerechte Bewertung von Wärmebelastung und Kältestress beim Aufenthalt im Freien in der Mabzahl Grad Celsius. *Annalen der Meteorologie*, Deutscher Wetterdienst, Offenbach, 33: 100–107

Steadman RG (1984) A Universal Scale of Apparent Temperature. *J. Climate Appl. Meteor.* 23: 1674–1687

Steadman RG (1994) Norms of apparent temperature in Australia. *Aust. Met. Mag.* 43: 1–16

Stolwijk JAJ (1971) A mathematical model of physiological temperature regulation in man. *NASA contractor report, NASA CR-1855*, Washington, DC

Tanabe SI, Kobayashi K, Nakano J, Ozeki Y and Konishi M (2002) Evaluation of thermal comfort using combined multi-node thermoregulation (65MN) and radiation models and computational fluid dynamics (CFD). *Energ. Build.* 34: 637–646

The Eurowinter Group (1997) Cold exposure and winter mortality from ischaemic heart disease, cerebrovascular disease, respiratory disease, and all causes in warm and cold regions of Europe. Keatinge, W.R., Donaldson, G.C. (Coord.). *Lancet* 349: 1341–1346

Tikuisis P and Osczevski RJ (2002) Dynamic model of facial cooling. *J. Appl. Meteor.* 41: 1241–1246

Tikuisis P and Osczevski RJ (2003) Facial cooling during cold air exposure. BAMS July, 927–934

Ward J and White S (2007) Smart Thermostats Trial. CSIRO Report ET/IR 970/R for Sustainability Victoria

WMO (2004) Proceedings of the Meeting of Experts to Develop Guidelines on Heat/Health Warning Systems. WCASP- No. 63, WMO-TD No. 1212

WMO/WHO (2007) Guide for Heat Health Warning Systems HHWSs. In preparation

Wissler EH (1985) Mathematical simulation of human thermal behavior using whole body models. In: A. Shitzer, R.C. Eberhart (eds.), Heat Transfer in Medicine and Biology – Analysis and Applications, Plenum, New York/London, pp. 325–373

Chapter 3
Heat/Health Warning Systems: Development, Implementation, and Intervention Activities

Laurence S. Kalkstein, Scott C. Sheridan, and Adam J. Kalkstein

Abstract There is an increasing awareness that heat is a major killer in many larger urban areas, and many municipalities have taken renewed interest in how they deal with oppressive heat. The implementation of sophisticated heat/health watch warning systems (HHWWS) is becoming more widespread, and these systems are becoming an important mechanism to save lives. One primary consideration in HHWWS development is the knowledge that response to heat varies through time and space. The more elaborate systems consider not only the intensity of heat, but the variability of the summer climate, which is closely related to urban population vulnerability. Thus, thresholds that induce negative health responses vary from one city to another, as well as over the season cycle at any one city. Warning system development involves a clear and consistent nomenclature (e.g. heat advisory, excessive heat warning), coordination between the agency issuing the warning and other stakeholders, public awareness of the system, targeted intervention procedures, and evaluation of effectiveness. This chapter describes these attributes in greater detail.

Over the course of recent decades, significant heat waves (e.g., North America in 1980 and 1995, Europe in 1976 and 2003, East Asia in 2004) have resulted in significant loss of life and exposed considerable weaknesses in the infrastructure of heat wave mitigation plans and human adaptation to oppressive weather (Klinenberg 2002).

In response to these heat events, many municipalities around the world have taken renewed interest in how they deal with the oppressive heat. In this chapter, we discuss the mechanisms for the development and implementation of heat/health

---

L.S. Kalkstein
Center for Climatic Research, University of Delaware, Newark, DE 19716, USA

S.C. Sheridan
Department of Geography, Kent State University, Kent, OH 44242, USA

A.J. Kalkstein
Department of Geography, Arizona State University, Tempe, AZ 85287, USA

watch warning systems (HHWWS), one of the key methods by which heat events are forecast and their effects are mitigated. We begin by describing the details by which thermal stress is evaluated in current HHWWS and the process by which warning criteria are determined. We then discuss the real-time development of HHWWS along with the "message delivery" to the public, heat mitigation strategies, and checking the effectiveness of HHWWS.

3.1 The Evaluation of Thermal Stress

There is robust literature (Kovats and Koppe 2005) associating what is generally termed "oppressive" heat with some negative health consequence. However, the means by which "oppressive" is defined varies widely (Watts and Kalkstein 2004); accordingly, the HHWWS that have been developed across the world in recent years have utilized a diversity of methods. Each of these methods has their respective strengths and weaknesses.

The utilization of a temperature threshold is perhaps the simplest of all methods. However, as outdoor temperature alone is significantly correlated with human mortality during excessive heat events (EHEs), temperature is considered by some to be a fairly reliable indicator. Moreover, the sole utilization of temperature has a further advantage in that it is the most commonly measured of all meteorological variables and thus is available for more locations. A number of nations, including Spain (Ministero de Sanidad y Consumo 2005), France (Pascal et al. 2006), the United Kingdom (UK Department of Health 2005), and Portugal (Paixao and Nogueira 2002), utilize maximum and/or minimum temperature thresholds in determining heat stress (Fig. 3.1).

An extension of the temperature threshold is the utilization of an "apparent temperature" that takes into account humidity (and wind speed in certain cases) as well as temperature. Several different formulations of the apparent temperature exist, including the *Heat Index* (Steadman 1984), used widely in the USA and Australia, and the *Humidex* (Masterton and Richardson 1979), developed in Canada. These indices are especially useful in locations where summer absolute humidity levels can vary widely, hence their widespread use in North America. Thresholds can then be developed as with temperature; the 40.6°C threshold of heat index across much of the USA is a prime example (Watts and Kalkstein 2004).

Another method of assessing meteorological conditions for application to the heat-health issue involves the classification of weather types, or air masses. The philosophy behind this "synoptic" methodology is to classify an entire suite of meteorological variables and thus holistically categorize the atmospheric situation at a given moment for a particular location or region (Yarnal 1993). This categorization when applied to heat is usually based upon surface weather variables, although upper atmospheric variables may also be incorporated. By categorizing the atmosphere into one of several internally homogeneous groups, other factors, such as solar radiation, wind speed, and cloud cover are inherently accounted for. For example, as a building's "heat load", as expressed by solar radiation income, has been associated with variability in human mortality, cloud cover or a some direct measure of solar radiation can be an important inclusion (Koppe and Jendritzky 2005). In synoptic approaches, discrete categories are created rather than a meteorological threshold

3 Heat/Health Warning Systems

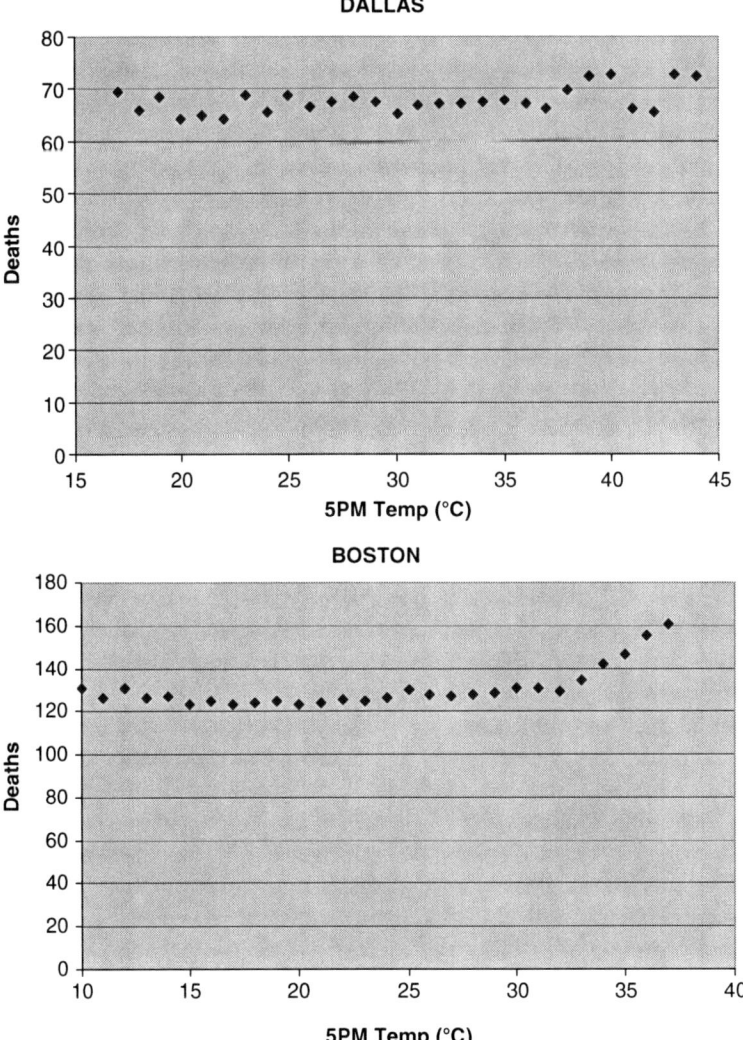

Fig. 3.1 Mean daily mortality in relation to 1700hr temperature for Dallas and Boston, USA

value along the continuum of a continuous variable (e.g., temperature); the result is a determination of "oppressive" synoptic categories that are historically associated with negative health outcomes. The synoptic-based systems generally require meteorological data that is more comprehensive than the temperature- or apparent temperature-based models, including hourly surface data for a number of variables.

A number of systems employ the synoptic methodology. Most notable are around 20 of the newer HHWWS across the USA (Sheridan and Kalkstein 2004), that incorporate the Spatial Synoptic Classification (SSC, Sheridan 2002). Several systems in Italy (Michelozzi and Nogueira 2004), Canada, South Korea, and China (Tan et al. 2003) also utilize the SSC.

A more physiologically based approach by which heat stress is evaluated includes those that are based on modeling the response in the human thermoregulatory system to ambient weather conditions. Rather than rely on proxy indicators, these methods aim to provide a direct assessment based on radiative fluxes to and from a typical human being. In the HeRATE system (Koppe and Jendritzky 2005), the thermal stress of ambient conditions is combined with an evaluation of short-term adaptation in assessing the overall level of heat stress upon the average individual. While thorough, the thermoregulatory system does require the most detailed array of meteorological conditions: in order to correctly model radiative fluxes, detailed information on temperature, humidity, wind, and cloud type and cloud cover at different levels must be assessed. The German HHWWS is the foremost advocate of the thermoregulatory system, and utilizes the HeRATE system as the foundation for its warning system structure (Koppe and Jendritzky 2005).

3.2 Considerations in Evaluating Thermal Stress

Regardless of which procedure above is utilized when devising a HHWWS, several key considerations must be made when correlating meteorological parameters with a human health response. Three of the most important considerations are the spatial variability, temporal variability, and persistence.

One of the primary considerations within the heat-health evaluation is that meteorological conditions in one location do not elicit the same response as they would in another location. There are a number of examples (e.g. Kalkstein et al. 2008; WHO, WMO, and UNEP, 1996) that depict significant differences in the heat/health relationship on the regional or national scale. Those who are accustomed to warmer conditions generally have a higher threshold before becoming stressed; moreover, in regions where the heat is more persistent during the summertime, the mortality response is generally less than in locations where the heat is intermittent (Kalkstein and Davis 1989). These spatial relationships have also changed over time (Davis et al. 2003) as air conditioning has become more commonplace.

Though virtually all HHWWS base forecasts upon local conditions, thereby accounting for local variability in ambient conditions, fewer modify the threshold values to account for local climatology. Many systems, such as the original US National Weather Service, lack regional definitions, and only more recently incorporate them on a basic level (dividing the US into a "northern" and "southern" region, with recommended threshold levels 5°F (3°C) different (NOAA 1995). The number of times different locations will exceed these thresholds varies greatly. The ICARO system in Portugal also utilizes a single threshold of 32°C (Paixao and Nogueira 2002). Most of the newer systems across Europe, including Italy (Michelozzi and Nogueira 2004), Spain (Ministero de Sanidad y Consumo 2005), the United Kingdom (Department of Health 2004), and France (Institute de Veille Sanitaire 2005), incorporate regionally defined thresholds (e.g. France, Fig. 3.2) that vary according to climatology.

The systems with the more elaborate methodology account for spatial variability inherently. For example, HHWWS that utilize the Spatial Synoptic Classification in the US, Canada, Italy, and China identify air masses whose definitions change across space (as well as time), so the spatial component is included (Sheridan and Kalkstein 2004). Similarly, as the HeRATE system evaluates heat stress on a local level, it too defines localized thresholds (Koppe and Jendritzky 2005).

Below the regional scale, an issue of disparity in vulnerability between urban and rural residents also needs to be addressed. In some cases, where thresholds are divided based on regional units, this can be accounted for in the general spatial variability (e.g. see Paris, France in Fig. 3.2). In other cases, where the jurisdiction includes rural and urban areas (as is the case within many US forecast offices), there is little differentiation, although at least one office, Wilmington, Ohio (G. Tipton, 2006, personal communication) uses lower thresholds for urban areas than rural areas, although some recent work (Sheridan and Dolney 2003) suggests that differences in vulnerability from rural to urban areas are minimal.

Just as the heat-health relationship varies spatially, it also varies over the course of the summer season. This *intra-seasonal acclimatization* has been well documented (WHO/WMO/UNEP 1996). Early season heat waves elicit a stronger response than late season heat waves of identical character, as the local population has had a chance to acclimatize to the warmer weather. Additionally, there is a "mortality displacement" effect that is very apparent in many locales shortly after a heat wave has ended; 20–40% of the mortality during an EHE would have occurred shortly afterward had the event not occurred (WHO/WMO/UNEP 1996).

Despite its importance, relatively few systems account for intra-seasonal variability. Nearly all of the systems based on an apparent temperature or temperature threshold do not modify this threshold over the course of the year. Several of

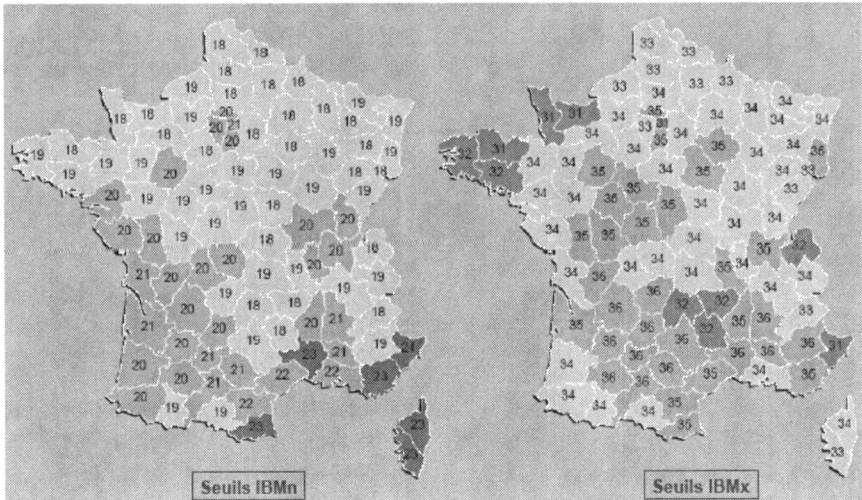

Fig. 3.2 Minimum (*left*) and maximum (*right*) thresholds by division in France's HHWWS (Institute de Veille Sanitaire 2005)

the Italian cities that utilize apparent temperature thresholds are an exception (Michelozzi and Nogueira 2004), with a modifier for time-of-season included in the calculation of heat-related mortality, for example, for Milan (only on days categorized as Moist Tropical Plus by the SSC):

$$MORT = -3.36 + 0.67\ DIS + .36\ T6 - 0.039\ TOS,\ \text{where}$$

$MORT$ = forecast mortality, DIS = day in sequence of offensive weather, $T6$ = 06 h temperature, and TOS = time of season (1 May = 1, 2 May = 2, etc.). Similarly, the HeRATE system as well as all air mass-related systems account for this intra-seasonal acclimatization by altering thresholds throughout the year (Koppe and Jendritzky 2005; Sheridan and Kalkstein 2004; see Fig. 3.3 for example).

The persistence of an EHE is another factor that impacts heat-related mortality in an important manner (Kalkstein 2000). Vulnerability, as expressed by increasing mortality, generally increases through the first several days of an EHE, and then may decrease thereafter. There are two methods by which this can be accounted for. Several systems base their thresholds upon repeat occurrence – for instance, the Swiss heat warning system is based upon the exceedence of a heat index of 32°C on three consecutive days (MeteoSwiss 2006). In other cases, predictive equations account for the persistence of offensive weather by determining mortality changes as thresholds are exceeded over a longer time interval.

3.3 The Determination of Thresholds

Once a meteorological methodology has been selected for HHWWS development, the next stage is the determination of when to describe weather conditions as being "oppressive". Most modern HHWWS are developed using an *inductive* method,

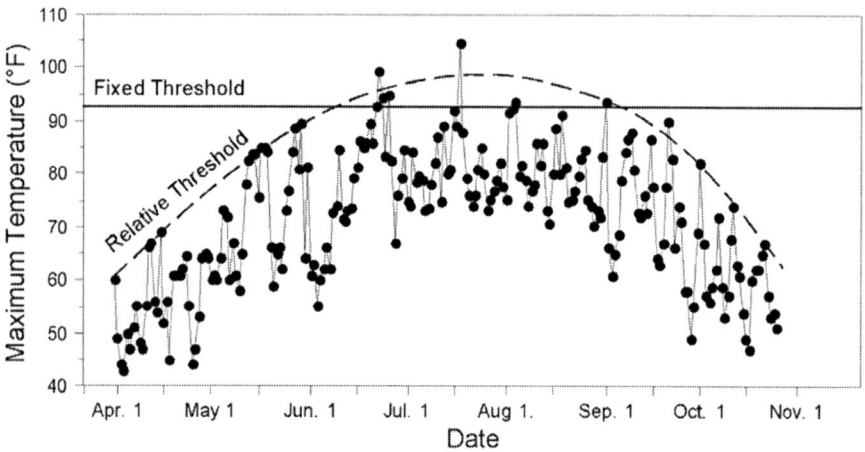

Fig. 3.3 Seasonally adjusted relative temperature threshold compared. Absolute temperature threshold for defining oppressive conditions (B. Davis, August 2005, Personal communication, New Hope EnvironmentalServices)

by which past weather and past health information are analyzed to determine what weather conditions lead to excess mortality. This can be done in either a subjective or statistical fashion.

Regardless, nearly all of the studies that have assessed the heat-health relationship for HHWWS development have utilized mortality data. While this certainly should not imply that the only negative health outcome from excessive heat is death, this dataset is nevertheless the most popular as it records a dichotomous event (either one is dead or alive, unlike hospital admissions, where there are levels of severity), and the records are generally the most readily available and most complete.

Past mortality data are generally obtained for either the total population or the total senior (aged 65 and older) population within a city, region, or metropolitan area (Sheridan and Kalkstein 2004). Mortality of all causes, or almost all causes (sometimes with the exception of accidents) are used in place of just those deaths that are termed by a medical examiner as "heat-related," since this restriction would result in a significant undercount of vulnerability (Sheridan and Kalkstein 2004). Data are usually standardized to account for demographic shifts over time as well as intraseasonal variability in mortality, and may be standardized to allow for comparison across locations.

Whether mortality data are utilized or not, one critical point in setting up a HHWWS is the determination of a threshold at which a heat warning is called, which is based on operational considerations (Smoyer-Tomic and Rainham 2001). If the threshold is set too low, too many heat warnings may be called, and the population may suffer from warning fatigue and disregard warnings. Conversely, if the criterion is set too high, days that are significantly hazardous may go unheralded.

In many of the synoptic-based HHWWS, the identification of an "oppressive air mass" is made when any air mass is associated historically with a statistically significant mean above the normal, "baseline" mortality (Sheridan and Kalkstein 2004). A rise in mortality of 5–10% or more on average is often significant. In contrast, other systems are developed with a specific mortality increase threshold in mind. France's HHWWS is the best example of this, where the threshold temperature values are associated with mortality rises of 50% in Paris and other urban centers, and 100% (i.e., a doubling) in rural locations (Pascal et al. 2006). Similarly, in Portugal, the ICARO system requires a value of 0.31 (representing a 31% increase above normal) for a warning to be called (Paixao and Nogueira 2002).

In a number of other cases, no clear mortality-related threshold is specified. For Germany's HeRATE system, as it is thermal heat load that is quantified, there are no mortality benchmarks but rather the inherent levels of thermal stress that are utilized to determine the warning (Koppe and Jendritzky 2005). In other systems, such as the Swiss system or the original US National Weather Service system, thresholds are defined with no direct association made to specific mortality increases.

Most HHWWS contain greater than one level of warning, with a set of municipal responses based on the level of negative health outcomes (as discussed further below). Where mortality thresholds are defined, it is often a higher threshold that determines a higher warning; for example, in the Philadelphia HHWWS it is a minimum of four excess deaths forecast that leads to an "Excessive Heat Warning;" the lower "Heat Advisory" is associated with excess mortality forecasts of 1–3 deaths (Sheridan and Kalkstein 2004). Similarly, the higher Level 4 in the ICARO system in Portugal is associated with at least a 93% increase in mortality forecast, compared with the lower Level 3, associated with a 31% increase, as noted above (Paixao and Nogueira 2002). With other HHWWS, it is a matter of duration. For example, in the Toronto HHWWS, no higher-level Extreme Heat Alert is called until at least 1 day after a lower-level Heat Alert has been issued, regardless of the magnitude of the heat (U.S. EPA 2006). In France, the third (lower) level of mobilization is associated with the first day of a heat wave, whereas the fourth (higher) level of mobilization is associated with subsequent days (Institute de Veille Sanitaire 2005).

3.4 Creation of a Warning System

Ultimate HHWWS development and the impact of the system on the community is linked with the quality of the message delivery system to both the public and important stakeholders (Bernard and McGeehin 2004). To accomplish this successfully, many current systems have turned to the internet as a way of increasing the speed of communication among all interested parties (Sheridan and Kalkstein 2004). Ideally, meteorological data must be made available in digital form from a forecasting office for at least the next 1–3 days. These data can then be processed and produce an informational web page as output for all associated stakeholders (Fig. 3.4).

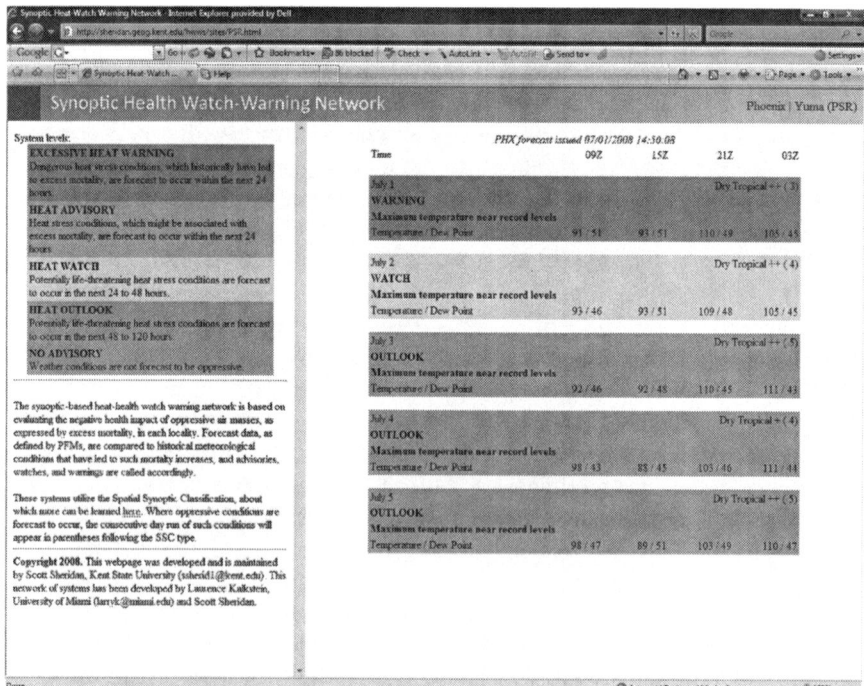

Fig. 3.4 The operational webpage for the Phoenix/Yuma, USA HHWWS

The Philadelphia (USA) HHWWS has served as a prototype for many other systems across the world since its inception (Kalkstein et al. 1996; Sheridan and Kalkstein 2004). The current system begins its daily run by the ingestion of digitized forecast data provided by the local National Weather Service (NWS) office. The computer program determines the SSC air mass type and then categorizes each of the next 7 days based on its potential for leading to a negative health response. These categorizations include an **excessive heat warning** or **heat advisory** for the first day, a **heat watch** if the offensive weather is forecast 2 days out, and an **excessive heat outlook** for days 3–7. Warnings and advisories differ in that the former is forecast to produce significant loss of life, while the latter is not, in spite of the presence of an offensive air mass. If there is an excessive heat warning or heat advisory, the NWS contacts the Philadelphia Department of Health to inform them of their decision. The Department of Health then formulates its intervention plan based on whether a warning or advisory has been issued, and contacts the numerous stakeholders in Philadelphia who must react in some manner when loss of life is expected; this is expanded upon below. Forecasts may be modified by the Philadelphia National Weather Service Office at any time during the day as conditions warrant; the HHWWS software then reassesses the forecast automatically.

It is clear that this process generally requires the collaboration of those in the meteorological field and the public health sector. In a number of cases, the decision on whether or not to call a warning is made on the meteorological end, including many locations in the US (Sheridan and Kalkstein 2004), and Germany (Koppe and Jendritzky 2005). In these cases it is then up to the local health authorities to decide whether to and how to act upon this warning. In other cases, the health authorities bear the primary responsibility for the calling of warnings, including Italy (de'Donato et al. 2005) and Portugal (Calado 2004). Either method can work efficiently, as long as the collaboration between the meteorological and health community within the municipality is close.

It is important that the public understands the message being issued when excessive heat is forecast. Nomenclature varies considerably from one country to the next; in Toronto, "Emergencies" and "Alerts" are issued; in Philadelphia, "Warnings", "Advisories", and "Watches" represent the nomenclature; in Rome, the terminology is "Alarms" and "Advisories". Differing nomenclature among nations is fine, but it is important that the terminology is consistent within countries so the public can understand the consistent message if traveling from one locale to the next. Thus, the media plays a large role in utilizing the proper terminology and in explaining the level of concern to individuals tuning in. Often when a new HHWWS is developed in an urban area, a press conference is scheduled, and the messages to be issued by the HHWWS are explained in detail to the media and the public at large.

3.5 Intervention Measures and Public Outreach

No matter how efficient HHWWS are in estimating the health outcome of an excessive heat event, they cannot be successful in saving lives if proper intervention procedures are not in place. Intervention describes the actions taken by local communities whenever excessive heat warnings are issued by the local or regional weather service or health department. For intervention activities to be successful, there must be close stakeholder interaction between a number of agencies assigned with increasing the well-being of the local population. Some of these include, beyond the local weather service, the department of health, emergency management, local utility companies, institutions that house the elderly, police, civic associations, and church groups. Intervention also implies "getting the message to the people"; even if extensive intervention activities are developed by a particular locale, they are less effective if people are unaware of the existence of an EHE, and the proper response to such an event. Thus, outreach and message delivery are major components to intervention, and sometimes these aspects is ignored.

The intensity of intervention activities varies widely from community to community, region to region, and country to country. Many areas recognize that heat is possibly the major weather-related health issue in their jurisdiction, and these areas tend to have the most elaborate intervention systems. The development of HHWWS in many regions has enhanced awareness and stakeholder collaboration; one good example is Seattle, USA, where prior to the establishment of a HHWWS in 2005, no heat advisories were ever issued by the local National Weather Service office. This cool, marine city did not consider heat to be a major (or even minor) health issue. Today, not only are advisories being issued utilizing a new synoptic-based HHWWS, but the city and surrounding communities have developed a comprehensive intervention plan, fact sheets on how people and agencies should respond to the heat, and recently the area sponsored a highly successful "Partners for Preparedness Conference" attended by the Mayor, a U.S. senator, county health commissioners, utility companies, and of course the developers of the HHWWS for Seattle (National Weather Service 2005; Seattle Partners in Emergency Preparedness 2005).

Although intervention procedures vary across locales, a broad consensus is emerging which describes the most vulnerable segments of the population, and some universal procedures that should be undertaken to lessen the negative health outcomes of excessive heat events. The elderly, very young, homeless, poor, socially isolated, those with mobility restrictions, those on medication, alcoholics, and those engaging in vigorous outside physical activity are most at risk (US EPA 2006). In many communities, these population segments are identified and kept under surveillance to lessen the probability of increased health problems. In addition, the following activities have been broadly accepted as being constructive to lessen the number of heat-related fatalities:

- Establishing and facilitating access to air conditioned public shelters
- Ensuring real-time public access to information on the risks of excessive heat conditions and appropriate responses through broadcast media, web sites, toll-free phone lines, and other means
- Establishing systems to alert public health officials about high risk individuals or those in distress during an excessive heat event (e.g., phone hotlines, high-risk lists)
- Directly assessing and, if needed, intervening on behalf of those at greatest risk (e.g., the homeless, older people, those with known medical conditions)

Beyond these baseline interventions, some communities have developed sophisticated plans to protect their inhabitants from heat-related illnesses. Two of the most elaborate programs are in Philadelphia, USA and Toronto, Canada (Table 3.1), and outreach efforts are not only extensive but costly; the total cost for intervention

Table 3.1 Summary of confirmed EHE notification and response program elements in Philadelphia and Toronto

Program elements	Philadelphia[a]	Toronto[b]
Prediction		
Ensure access to weather forecasts capable of predicting EHE conditions 1–5 days in advance	√	√
Risk assessment		
Coordinate transfer and evaluation of weather forecasts by EHE program personnel	√	√
Develop quantitative estimates of the EHE's potential health impact	√	√
Use of the broader criteria for identifying heat-attributable deaths	√	√
Develop information on high-risk individuals	√	
Develop information on facilities and locations with concentrations of high risk individuals	√	√
Notification and response		
Coordinate public broadcasts of information about the anticipated timing, severity, and duration of EHE conditions and availability and hours of any public cooling centers	√	√
Coordinate public distribution and broadcast of tips on how to stay cool during an EHE and symptoms of excessive heat exposure	√	√
Operate phone lines that provide advice on staying cool and recognizing symptoms of excessive heat exposure, or that can be used to report heat-related health concerns	√	√
Designate public buildings with air-conditioning or specific private buildings as public cooling shelters and provide transportation to those locations.	√	√
Extend hours of operation at community centers with air-conditioning	√	
Arrange for extra staffing of emergency support services	√	
Directly contact and evaluate the environmental conditions and health status of known high-risk individuals and locations likely to have concentrations of these individuals	√	√
Increase outreach efforts to the homeless and establish provisions for their protective removal to cooling shelters	√	√
Suspend utility shut-offs	√	√
Reschedule public events to avoid large outdoor gatherings when possible	√	
Mitigation		
Develop and promote actions to reduce effects of urban heat islands (e.g., increase urban vegetation and albedo of surfaces)	Not evaluated	

[a]NOAA 1995; Kalkstein 2002.
[b]Kalkstein 2002; M. Vittiglio and N. Day, 2005, personal communications, Toronto Public Health.

activities in Philadelphia for summer, 2002 was over $100,000 (Kalkstein 2002). Other locales have no formal heat wave mitigation plan, such as Phoenix and New Orleans, USA. Although both cities have sophisticated heat/health warning systems in operation, these are much less effective in saving lives if they are not pared with the proper intervention procedures.

3.6 Effectiveness of Intervention Activities

One criticism of urban intervention programs is that they do not reach the most vulnerable segments of the population in time to help ameliorate negative health outcomes. Many locales disseminate "passive" heat avoidance advice, which often doesn't reach the intended vulnerable targets, such as homeless and homebound people (Kovats and Ebi, 2006). Some communities, especially in Europe, have registers of vulnerable individuals, but many of these are developed voluntarily by relatives of high-risk people. Thus, intervention programs must include a vigorous dissemination program if they are to be successful.

There have been some evaluations to determine how effective heat intervention outreach has been. During the 2003 heat wave in Portugal, the mass media reached over 90% of the population. TV was the main source of dissemination, followed by radio and newspapers. Less than 5% consulted information on the internet. Less than 2% called the public health emergency line. In summary, it was concluded that the behavior of the people changed during the heat wave and the instructions were closely followed by a large segment of the population (Paixao 2004).

Results obtained from a recent US/Canada study on "getting the message out" were somewhat different (Sheridan 2006). Although there was clear recognition of deadly heat events by the general population, there was considerable confusion involving how people should handle themselves during such an event. Most respondents knew that they should remain hydrated, but few knew that they should not overexert themselves. It appeared that people listened intently to the forecasts indicating dangerous heat, but blocked out the intervention procedures suggested by the local health departments. Additionally, in this study, few people actually modified their behavior or considered themselves highly vulnerable to the negative impacts of excessive heat.

Another issue that may lessen effectiveness in disseminating of heat intervention advice is potential confusion between heat and pollution warnings. In a Toronto and Phoenix evaluation, ozone alerts often coincided with heat events, and some vulnerable individuals chose not to drive to cooling centers because of the pollution alert (Sheridan, in press). Thus, people were deprived the benefit of a cooler surrounding because the pollution alert suggested that driving be limited. One of the future challenges of heat warnings is to determine whether they should be combined with pollution warnings (not recommended by the authors), or whether they should remain separate. If the latter is chosen, it is important that the media does not send "mixed messages" to vulnerable segments of the population.

3.7 Methods to Check Effectiveness of Urban Heat Programs

Heat health watch warning systems are very difficult to evaluate because of the interactive nature of the systems (Kovats and Ebi, in press). Of course, the goal of the systems is to reduce the negative health outcomes in urban areas, particularly heat-related mortality. However, how can you separate mortality reductions attributed to, for example, increased air conditioning penetration, from reductions directly related to HHWWS operation, greater public awareness, and associated intervention activities?

Effectiveness requires evaluation of the system on several different fronts (Kalkstein 2006):

1. How accurate is the actual weather service forecast of a heatwave?
2. Assuming forecast accuracy, how precise is the system in estimating the negative health outcome that is anticipated from the forecast heat event?
3. Is the system, and ancillary intervention activities, actually saving lives?

Forecast accuracy is vital if the system is to be a useful tool to stakeholders. Errors in forecasting come in two forms: false positives and false negatives. A false positive arises if the forecast calls for an excessive heat event and no event materializes. The result would be loss of money to the community, since an advisory or warning would be called when it would not be necessary. There would also be a loss in public confidence, since "crying wolf" too often renders the population skeptical to the overall message that is attempted to be delivered. A false negative occurs if the forecast does not anticipate an excessive heat event and one actually occurs. This is a worse outcome than a false positive, because the local populous is not aware that a dangerous extreme weather event will occur, the system will not call an advisory or warning, and greater numbers of lives will be lost. The newest systems in the U.S. now forecast 5 days out, and care must be taken not to overreact when adverse heat conditions are predicted to happen that far in advance (Sheridan and Kalkstein 2004).

There have been several manuscripts published that have assessed the accuracy of the HHWWS itself in estimating heat-related mortality. The synoptic-based Philadelphia system seemed to accurately estimate heat-related deaths during the hot summer of 1995 (Kalkstein et al. 1996), and the more recent generation detected 21 of 22 hot days when heat-related mortality occurred in 2005 (Szatkowski 2006). Evaluations of the Italian system have shown mixed, but generally positive, results. There was an underestimation of deaths in Rome for the intensely hot summer of 2003 (Michelozzi et al. 2004), but more positive results have been obtained when the Rome system was redeveloped after that oppressive heat event (de'Donato et al. 2005). Obviously it is imperative that HHWWS show some sense of accuracy in estimating negative heat-related health outcomes, as a number of systems tie their intervention activities to these estimates.

Determining if the systems are saving lives is a tricky business, and only a few studies to date have looked into this issue. Probably the best-known study was performed by Ebi et al. (2004) during a 4 year period after the inauguration of the Philadelphia heat/health system in 1995. The evaluation determined that, during the summers of 1995–1998, the system saved 117 total lives. It was concluded that, for each day the National Weather Service office in Philadelphia called a warning, an average of 2.6 lives were saved on the warning day and on each of the following 3 days. The study also estimated that the net benefits of system operation totalled over US$400 million, based upon a conservative standard of US$4 million per statistical life.

Clearly other studies of this type are necessary, but in most cases the newest generation of heat/health systems have not been running sufficiently long to collect the requisite data on lives saved. Nevertheless, it is clear that an accurate system, based on sound biometeorological science, coupled with a quality intervention program and an efficient public delivery system can contribute mightily to lives saved during extremely hot weather.

3.8 Conclusions

Awareness that heat is the major weather-related killer in most large urban areas has increased considerably (Kalkstein et al., 2008), and it is apparent that quality heat/health programs can save lives (Kalkstein 2000). There is no single system or methodology that is "best" when it comes to meteorological methodology or efficiency of operation. Rather, each municipality must choose between an array of possible approaches to deal most effectively with heat outcomes. The chosen approaches are dependent upon the type and frequency of meteorological variables forecast by the local weather service office, the political composition of the urban area, the stakeholder collaboration that exists in the region, and the ultimate intervention plan that each locale must develop. However, there is one unifying theme that is important to emphasize: *systems should be based on finding thresholds that lead to negative health outcomes*. Arbitrary, absolute thresholds are much less efficient, and each urban populous responds differently to extreme heat.

The public is becoming more aware that there are many tools available to them to combat extreme heat/health problems. With the inauguration of each new HHWWS in the United States and Canada, there is an associated press conference that is organized by the local National Weather Service office, the Department of Health, and several other stakeholders. Invitees include the media, politicians, civic leaders, fire and police departments, and other relevant stakeholders. These press conferences are important; they are widely broadcast and familiarize the public with the dangers of excessive heat.

However, there is still evidence that, although the public generally knows when an EHE is occurring, they either feel that they are not vulnerable to the impacts or they are unaware of how to deal with the situation. One study indicates that relatively few people modify their behavior when an excessive heat warning is called (Sheridan 2006). A similar study, which was developed by interviewing over 200 people in Phoenix, USA, a particularly hot city, indicated that people were aware of the issuance of excessive heat warnings (86%), but a much smaller proportion actually modified their behavior during dangerously hot weather (50%; Kalkstein 2006). Thus, even with the most sophisticated HHWWS available, the systems will be less effective if the general populous is unaware of how to respond. This appears to be the "weak link" in system operation and implementation.

The rapid spread of quality HHWWS around the world is a very welcome development, recognized by local and national weather and health officials as well as the World Meteorological Organization and World Health Organization. The links, from system development to public response, are becoming stronger as awareness increases, but there is still considerable work to be done to minimize the vulnerability of the general population to the vagaries of heat.

References

Bernard S.M. and M.A. McGeehin, 2004: Municipal heat wave response plans. *American Journal of Public Health* 94:1520–1521.

Calado, R., 2004: Heat-waves in Portugal. In: World Health Organisation. Regional Office for Europe. Extreme weather and climate events and public health responses. Report on a WHO meeting in Bratislava, Slovakia 9–10 February 2004. Copenhagen, Denmark.

Davis, R.E., P.C. Knappenberger, W.M. Novicoff, and P.J. Michaels, 2003: Decadal changes in summer mortality in U.S. cities. *International Journal of Biometeorology* 47(3):166–175.

de'Donato, F., P. Michelozzi, L. Kalkstein, M. D'Ovidio, U. Krichmayer, and G. Accetta, 2005: The Italian project for prevention of heat-health effects during summer, findings from 2005. Proceedings of the 17th International Congress of Biometeorology ICB 2005, 5–9 September in Garmisch. Annalen der Meteorologie 41, Vol. 1, Deutscher Wetterdienst, Offenbach, pp. 287–290.

Department of Health, 2005: *Heatwave – Plan for England – Protecting health and reducing harm from extreme heat and heatwaves*. London: Department of Health.

Ebi, K.L., T.J. Teisberg, L.S. Kalkstein, L. Robinson, and R.F. Weiher, 2004: Heat watch/warning systems save lives: estimated costs and benefits for Philadelphia 1995–1998. *Bulletin of the American Meteorological Society* 85:1067–1074.

Institute de Veille Sanitaire, 2005: *Système d'alerte canicule et santé 2005* (Sacs 2005). Paris: Institute de Veille Sanitaire.

Kalkstein, L.S., 2000: Saving lives during extreme summer weather. *British Medical Journal* 321: 650–651.

Kalkstein, L.S., 2002: Description of our heat/health watch warning systems: their nature and extent and their required resources. Final report for U.S. EPA, Contract number 68-W-02-027.

Kalkstein, L.S., 2006: Heat and health: methodological considerations for warning system development. Paper given to NOAA/NWS Weather Service Forecasting Offices.

Kalkstein, L.S. and R.E. Davis, 1989: Weather and human mortality: an evaluation of demographic and interregional responses in the United States. *Annals of the Association of American Geographers* 79(1):44–64.

Kalkstein, L.S., P.F. Jamason, J.S. Greene, J. Libby and L. Robinson, 1996: The Philadelphia hot weather-health watch/warning system: Development and application, Summer 1995. *Bulletin of the American Meteorological Society* 77(7):1519–1528.

Kalkstein, L.S., J.S. Greene, D. Mills, A. Perrin, J. Samenow, and J-C. Cohen, 2008: Analog European Heat Waves for U.S. Cities to Analyze Impacts on Heat-Related Mortality. *Bulletin of the American Meteorological Society* 89:75–86.

Klinenberg, E., 2002: *Heat Wave: A Social Autopsy of Disaster in Chicago*. Chicago: University of Chicago Press.

Koppe, C., G. Jendritzky, 2005: Inclusion of short term adaptation to thermal stresses in a heat load warning procedure. *Meteorologische Zeitschrift* 14(2):271–278.

Kovats, R.S. and C. Koppe, 2005: Heat waves: past and future impacts on health. In: Ebi, K.L., Smith, J.B., Burton, I. (eds): *Integration of Public Health Wit Adaptation to Climate Change – Lessons Learned and New Directions*. Taylor & Francis, Leiden, The Netherlands, pp. 136–160.

Kovats, R.S. and K.L. Ebi, 2006: Heatwaves and public health in Europe. *European Journal of Public Health* 16:529–599.

Masterton, J.M. and F.A. Richardson, 1979: Humidex, a method of quantifying human discomfort due to excessive heat and humidity, CLI 1–79, Environment Canada, 45 pp.
MeteoSwiss, 2006: Heat. http://www.meteoschweiz.ch/web/en/weather/health/heat.html Accessed 4 April 2006.
Michelozzi, P. and P.J. Nogueira, 2004: A national system for the prevention of heat health effects in Italy. In: World Health Organisation. Regional Office for Europe. Extreme weather and climate events and public health responses. Report on a WHO meeting in Bratislava, Slovakia 9–10 February 2004. Copenhagen, Denmark.
Michelozzi, P., F. de'Donato, G. Accetta, F. Forastiere, M. d'Ovidio, C. Perucci, and L. Kalkstein, 2004: Impact of Heat Waves on Mortality – Rome, June 1–August 15, 2003. *Journal of the American Medical Association* 291:2537–2538.
Ministero de Sanidad y Consumo, 2005: El Gobbierno se anticipa a una posible ola de calor con la puesta en marcha de un plan de prevención. http://www.msc.es/gabinetePrensa/notaPrensa/desarrolloNotaPrensa.jsp?id=14.
National Oceanic and Atmospheric Administration, 1995: National disaster survey report: July, 1995 heat wave. Silver Spring, MD: U.S. Department of Commerce.
National Oceanic and Atmospheric Administration, National Weather Service, 2005: Seattle Heat/Health Warning System Fact Sheet, 3 pp.
Paixao, E., 2004: Behaviours of Portuguese population during the August 2003 heat wave. Paper given at the ICARÓ Conference, 7 May 2004, Lisbon.
Paixao, E., and P.J. Nogueira, 2002: Estudo da onda de calor de Julho de 1991 em Portugal, efeitos na mortalidade: relatório científico. Lisbon, Observatório Nacional de Saúde.
Pascal, M., K. Laaidi, M. Ledrans, E. Baffert, C. Caserio-Schönemann, A. Le Tertre, J. Manach, S. Medina, J. Rudant, and P. Empereur-Bissonnet, 2006: France's heat health watch warning system. *International Journal of Biometeorology* 50:144–153.
Seattle Partners in Emergency Preparedness, 2005: Conference Brochure, National Weather Service, 10 pp.
Sheridan, S.C., 2002: The re-development of a weather type classification scheme for North America. *International Journal of Climatology* 22:51–68.
Sheridan, S.C., 2006: Municipal response and public perception of heat/health watch warning systems. Final report for U.S. EPA: Contract Number XA-83105001–0.
Sheridan, S.C., 2007: A survey of public perception and response to heat warnings across four North American cities: an evaluation of municipal effectiveness. *International Journal of Biometeorology* 52:3–15.
Sheridan, S.C. and T.J. Dolney, 2003: Heat, mortality, and level of urbanization: Measuring vulnerability across Ohio, USA. *Climate Research* 24:255–266.
Sheridan S.C. and L.S. Kalkstein, 2004: Progress in heat watch warning system technology. *Bulletin of the American Meteorological Society* 85:1931–1941.
Smoyer-Tomic K.E. and D.G.C. Rainham, 2001: Beating the heat: development and evaluation of a Canadian hot weather health response plan. *Environmental Health Perspectives* 109:1241–1247.
Steadman, R.G., 1984: A universal scale of apparent temperature. *Journal of Climate and Applied Meteorology* 23:1674–1687.
Szatkowski, G., 2006: The effectiveness of the Philadelphia heat/health system. Paper given at the annual meeting of the American Meteorological Society, Atlanta.
Tan, J., L.S. Kalkstein, J. Huang, S. Lin, H. Yin, and D. Shao, 2003: An operational heat/health warning system in Shanghai. *International Journal of Biometeorology* 48:157–162.
U.K. Department of Health, 2005: Heatwave – plan for England: Reducing harm from extreme heat and heatwaves. London: Department of Health.
U.S. EPA, 2006: Excessive heat events guidebook. Boulder, CO: Stratus Consulting.
Watts, J.D. and L.S. Kalkstein, 2004: The development of a warm-weather relative stress index for environmental applications. *Journal of Applied Meteorology* 43:503–513.
WHO, WMO, and UNEP, 1996: Climate change and human health. Geneva: WHO.
Yarnal, B., 1993: *Synoptic Climatology in Environmental Analysis*. Bellhaven Press, London, 195 pp.

Chapter 4
Malaria Early Warning Systems

Kristie L. Ebi

Abstract Malaria is the most important vectorborne disease in the world; it is also preventable. Climate patterns and weather events play a role in determining the incidence and geographic range of malaria, including through changes in human behavior, effects on the pathogen (*Plasmodium*), and effects on the malaria vector (*Anopheles*). Better understanding of the associations between malaria and environmental variables has lead to increased interest in developing early warning systems that alert public health and vector control personnel about developing conditions that are associated with epidemics, combined with the ability to implement appropriate and effective interventions to reduce the number of expected cases. A key challenge has been the inability to predict when and where outbreaks will occur far enough in advance that timely interventions can be implemented. Advances in several areas could increase the sensitivity and specificity of malaria early warning systems, including improving long-range forecasting, monitoring of environmental variables, and case surveillance. Better understanding is needed of how to incorporate uncertainties when setting thresholds for early warning systems. In most areas where malaria epidemics occur, additional capacity is needed to effectively respond to a warning that an epidemic is predicted to occur. Despite the uncertainties and constraints, there is significant promise in using climatic and environmental variables to help regions prepare for and effectively respond to malaria epidemics. Projected climate change suggests increased malaria risks, emphasizing the need for more effective warning systems.

K.L. Ebi
ESS, LLC, 5249 Tancreti Lane, Alexandria, VA 22304, USA
e-mail: krisebi@essllc.com

4.1 Introduction

Malaria is the most important vectorborne disease in the world; it is also a preventable disease. It places significant burdens on families and communities, particularly in Africa where 80% of all cases and 90% of mortality occur (Bremen 2001; D'Alessandro and Buttiens 2001). In sub-Saharan Africa, malaria remains the most common parasitic disease and is the main cause of morbidity and mortality among children less than 5 years of age and among pregnant women (WHO 2004). Up to 3 million deaths from the direct effects of malaria occur annually in Africa, more than 75% of them in children (Bremen et al. 2004). This estimate could double if the indirect effects of malaria (including malaria-related anemia, hypoglycemia, respiratory distress and low birthweight) are included when defining the burden of malaria (Breman 2001). The 1990 Global Burden of Disease study estimated that malaria accounted for approximately 10.8% of years of life lost across all sub-Saharan Africa (Murray et al. 1996). The importance of the burden of malaria is reflected in it being chosen as one of the Millennium Development Goals. The Global Strategic Plan for Roll Back Malaria 2005–2015 stated that "six out of eight Millennium Development Goals can only be reached with effective malaria control in place" (http://www.rollbackmalaria.org/forumV/docs/gsp_en.pdf).

Malaria has proved difficult to control (Githeko and Shiff 2005). Despite considerable control efforts, including larviciding, insecticide residual spraying, chemoprophylaxis for particularly vulnerable groups (i.e. pregnant women and children), and effective case management, there has been a global resurgence of epidemic malaria over the past 2 decades, causing significant morbidity and mortality. Reasons suggested for the resurgence include failure of malaria control programs, population redistribution and growth, changes in land use, increasing prevalence of drug and pesticide resistance, degradation of public health infrastructure, and climate variability and change (Githeko and Ndegwa 2001; Greenwood and Mutabingwa 2002).

At the Abuja summit of African leaders in 2000, targets were set for improving malaria epidemic detection and response with the aim of detecting 60% of epidemics within 2 weeks of onset and responding to 60% of epidemics within 2 weeks of detection (WHO/UNICEF 2005). However, the implementation of these targets within current health systems may be difficult. In many epidemic-prone regions, current surveillance is not able to provide timely detection of the onset of epidemics, and response has been limited by human and financial constraints. This has increased interest in using remotely sensed indicators of conditions conducive to an outbreak.

Climate patterns and weather events play a role in determining the incidence and geographic range of malaria worldwide, including through changes in human behavior, effects on the pathogen (*Plasmodium*), and effects on the malaria vector (*Anopheles*). Climate patterns are a primary determinant of whether the conditions in a particular location are suitable for stable *Plasmodium falciparum* malaria transmission (Craig et al. 1999). Numerous laboratory and field studies have documented that a change in temperature may lengthen or shorten the season during which mosquitoes or parasites can survive; and that changes in precipitation or temperature may result in conditions during the season of transmission that are conducive to increased or decreased parasite and vector populations. At warmer temperatures, adult female mosquitoes feed more frequently and digest blood more rapidly, and the *Plasmodium* parasite matures more rapidly within the female mosquitoes. Temperature also affects the duration of the aquatic stage. Therefore, small changes in precipitation or temperature may allow transmission in previously inhospitable altitudes or ecosystems. Most malaria epidemics follow abnormal weather conditions, often in combination with other causes (Abeku 2007).

Better understanding of the associations between malaria and environmental variables has lead to increased interest in developing early warning systems that alert public health and vector control personnel about developing conditions that are associated with epidemics, combined with the ability to implement appropriate and effective interventions to reduce the number of expected cases. A key challenge has been the inability to predict when and where outbreaks will occur far enough in advance that timely interventions can be implemented to control malaria and its consequences. The ability to do so would have significant benefits for the control and prevention of malaria. For example, in Ethiopia, an annual alert implemented during a fixed week just before the beginning of the high transmission season can prevent 28.4% of malaria cases (Teklehaimanot et al. 2004a–c). However, an early warning system based on temperature, rainfall, and other variables can prevent an equivalent percentage of cases with only 0.5 alerts per year, thus saving lives using fewer public health resources. Incorporating advances in climate forecasting with a better understanding of the relationships among weather variables, climate anomalies, malaria epidemics, and of particularly vulnerable sub-populations can be the basis of effective early warning systems.

This chapter reviews malaria transmission dynamics, summarizes key issues with developing early warning systems for infectious diseases, reviews developments in malaria early warning systems, and then summarizes recent projections of the burden of malaria under different climate change scenarios to indicate the possible extent of future needs for early warning systems.

4.2 Malaria

Malaria transmission can be categorized as stable and unstable. High transmission levels with little inter-annual variation characterize regions with stable transmission. Regular exposure to malaria leads to a level of population immunity that means that epidemics are less likely. Where malaria is transmitted stably, the probability of dying from an untreated case of malaria is approximately 2–3% (Kiszewski and Teklehaimanot 2004). In these areas, people are infected repeatedly throughout their lives and there is considerable morbidity and mortality during early childhood. Because those who survive childhood have some immunity against malaria, many adult infections are associated with low morbidity.

Unstable malaria is characterized by transmission levels that vary annually. Population immunity is low, so when malaria transmission occurs, explosive epidemics can occur that cause significant morbidity and mortality in both children and adults. Case fatality rates can be up to ten-times greater during an epidemic (Kiszewski and Teklehaimanot 2004). Although the average number of people infected in epidemic-prone regions may be relatively few compared with where transmission is stable, the impact per case can be far greater due to higher rates of severe disease and mortality. Older, more productive members of the community can be severely affected, amplifying the impacts of the epidemic on families and society. Although there are concerns with the accuracy of malaria data, it is estimated that epidemic malaria causes 12–25% of estimated annual worldwide malaria deaths, including up to 50% of the estimated annual malaria mortality in people less than 15 years of age (Worrall et al. 2004). In Africa it has been estimated that the population at risk of epidemic malaria is nearly 125 million, with 12.4 million annual cases of epidemic malaria (Worrall et al. 2004). Epidemic malaria occurs most often in areas where environmental conditions are marginal for the mosquito vector and parasite development, including warm, semi-arid zones, tropical mountainous areas, and regions where previous levels of control are beginning to fail. Many epidemics in highland areas are superimposed over normal seasonal increases in malaria incidence, a phenomenon that makes early detection difficult (Abeku 2007).

The severity of malaria is a function of interactions among the malaria parasite, the mosquito vector, the host, and the environment. In Africa, the pattern of malaria transmission varies regionally, depending on climate, biogeography, ecology, and anthropogenic activities (i.e. irrigation) (Mabaso et al. 2007). The vectors that carry malaria require specific habitats, with surface water for reproduction, favorable temperatures, and humidity for adult mosquito survival. The development rates of both the vector and parasite are temperature dependent. Malaria vectors lay their eggs in irrigation canals, wetlands, and small pools of water, as long as the water is clean, not too shaded, and relatively still. In many semi-arid areas, these sites are only widely available with the onset of seasonal rains, unless there is dry season irrigation.

Conditions more permissive for malaria transmission appear during climate anomalies (Kiszewski and Teklehaimanot 2004). Epidemic malaria is a serious problem in semi-arid and highland areas (above 2,000 m) in eastern and southern parts of Africa. Reports suggest that the incidence of malaria in the East African highlands has increased since the end of the 1970s (Githeko and Shiff 2005). Analysis of temperature data from 1950 to 2002 for four high-altitude sites found evidence for a significant warming trend (Pascual et al. 2006). The possible biological significance of this trend was assessed using a model of the population dynamics of the mosquito vectors for malaria and concluded that the observed temperature changes could significantly accelerate the mosquito life cycle, particularly the development rate of larvae and adult survival. Therefore, even a small increase in temperature may result in a significant increase in the number of available malaria vectors. Because the probability that an *Anopheles* mosquito will transmit malaria with each bite is low, the number of available competent vectors is an important determinant of an epidemic.

Climate patterns are important determinants of the geographic distribution and incidence of malaria, but other factors are also critical. The many determinants of malaria rarely act in isolation; these determinants form a web of interconnected influences, often with positive feedbacks between malaria transmission and other drivers (Janssen and Martens 1997; Chan et al. 1999; Lindsay and Martens 1998). The non-climatic socioeconomic and biological drivers that have a direct impact on malaria include drug and pesticide resistance, deterioration of health care and public health infrastructure (including vector control efforts), demographic change, and changes in land use patterns. Further, the presence of protein-calorie undernutrition and micronutrient deficiencies, particularly zinc and vitamin A, contribute substantially to the malaria burden (Caulfield et al. 2004).

Early warning systems can save lives when they intensify vector control activities and increase the availability of insecticide-treated bednets, anti-malarial drugs, etc. Inter-epidemic assessments also are critical for improving the responsiveness and effectiveness of interventions.

4.3 Components of an Early Warning System

The principal components of an early warning system include forecasting when and where an epidemic is likely to occur, predicting the number of cases of malaria that could occur (or whether a pre-defined threshold will be exceeded), implementing an effective and timely response plan, and ongoing evaluation of the system and its components (Ebi and Schmier 2005). The distinction between prediction and early warning is important: early warning is prediction but not all predictions are early warnings. An effective and efficient early warning system for malaria should reduce vulnerability to current climate, as well as be designed for easy modification to take into account continuing climate change. The relative importance of forecasting timing and predicting the number of cases will depend on the control decisions that will be taken and the degree of interannual variation of the disease. In some instances, forecasting the timing of a likely epidemic is sufficient. For example, for diseases which are absent from the human population for long periods followed by explosive epidemics, early detection and/or forecasts of the probability of an epidemic may be more important than predictions of epidemic size (Kuhn et al. 2005).

The system should be developed with all relevant stakeholders to ensure that the issues of greatest concern are identified and addressed, thus increasing the likelihood of success of the system. Stakeholders include the agencies and/or organizations that will fund the development and operation of the system, the groups who will be expected to take action, and those likely to be affected. Including those previously affected may provide local knowledge about responses and their effectiveness. Figure 4.1 describes a framework for developing early warning systems for vectorborne diseases (Kuhn et al. 2005).

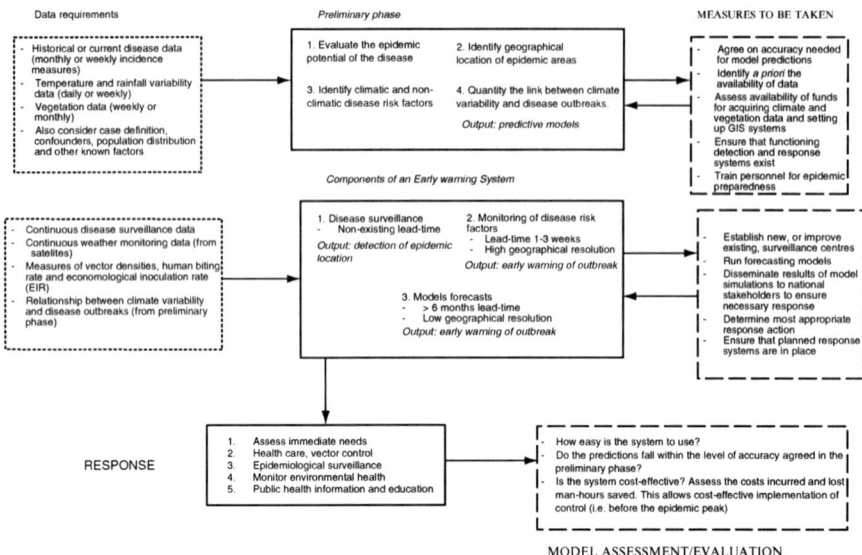

Fig. 4.1 Components of an early warning system for infectious diseases (Kuhn et al. 2005)

4.3.1 Developing Disease Prediction Models

Multiple disciplines are required to develop accurate, population- and location-specific effective and efficient early warning systems (Woodruff 2005). Biometeorology contributes to the development of models to predict possible health burdens associated with changes in weather patterns, through understanding of key aspects of weather/malaria associations. The two principal approaches to modeling associations between weather variables and malaria morbidity and mortality are statistical and biological. Statistical models are derived from direct correlations between the predictor variables such as climate and the outcome of interest. Biological models attempt to represent the processes by which climate affects the population dynamics of pathogens and vectors. The specific approach used will depend on the model purpose.

The usefulness of the model depends on the ability to obtain reliable and up-to-date information on both the health outcome and the factors critical to disease incidence in time for effective responses to be implemented. Various indicators can be used to evaluate different aspects of the accuracy of a system, including overall accuracy (measure of increased predictive accuracy over what should have been expected), model sensitivity (proportion of the epidemics correctly predicted), specificity (proportion of non-epidemic periods correctly predicted), positive predictive value (proportion of predictions of an epidemic that were correct), and negative predictive value (proportion of predictions of a non-epidemic that were correct).

Both long- and short-range forecasts can be used as the basis of an early warning system. Long-range forecasts rely on atmospheric and oceanic phenomena such as the El Nino Southern Oscillation (ENSO) that indicate trends of higher or lower than normal temperature and precipitation 6 months or more in advance. Short-range forecasts focus on monthly or weekly weather patterns. Inputs can come from direct monitoring of risk factors such as rainfall measurements in 1 month to predict the probability of an outbreak of a vectorborne disease in the subsequent months, or forecasting of these risk factors (such as seasonal climate forecasts) (Kuhn et al. 2005). Model choice will depend on the trade-offs between accuracy (generally maximized by using direct observations of risk factors) and lead-time.

Malaria prediction models need to consider not just weather, but also other risk factors that could affect the potential for an outbreak. Because the effects of climate on disease transmission are influenced by the topography and hydrology of each situation, a generalized early warning approach may not be feasible for a large geographic area (Kiszewski and Teklehaimanot 2004).

4.3.2 Developing Epidemic Response Plans

The impacts of a malaria epidemic are determined by the vulnerability of the affected population (i.e. prevalence of immunity, malnutrition, drug resistance, and other factors), changes in transmission risk due to factors such as unusual rainfall, and the ability of health care facilities to detect and effectively respond to an epidemic early.

Key to the control of vector-borne diseases such as malaria is surveillance and response. Surveillance involves the systematic collection of information on health determinants and outcomes necessary to determine the occurrence and geographic range of health outcomes, and the analysis, interpretation, and distribution of this information to relevant actors for timely and effective control activities. Surveillance programs are designed to provide early intelligence on whether a health outcome is increasing in incidence or geographic range. Routine collection and analysis on a weekly basis of surveillance data, depending on the shape of the epidemic transmission curve, can significantly improve implementation of appropriate interventions, thus increasing the chance of effective disease control.

A challenge in many countries with high malaria burdens is that surveillance programs require collecting accurate and timely data. Substantially less than 10% of malaria cases are reported officially (Breman et al. 2007), with most cases self diagnosed and treated at home. Of diseases brought to medical attention, most fevers are presumed to be malaria, without verification of the diagnosis. This misclassification of malaria leads to considerable uncertainty in analyses of the associations between environmental variables and morbidity and mortality. However, presumptive treatment of malaria also has benefits, including promptness of therapy (and, therefore, hopefully reduced risks that the case will progress to severe malaria), reduced costs in time and money, lack of expenditure on diagnostic facilities and skilled staff, and equity in availability of treatment (Koram and Molyneux 2007).

A response plan needs to incorporate components for implementation and for evaluation of effectiveness. The implementation component should, at a minimum, include modules addressing where the response plan will be implemented; when interventions will be implemented, including thresholds for action; what interventions will be implemented; how the response plan will be implemented; and to whom the interventions will be communicated (Ebi and Schmier 2005).

WHO developed a framework for malaria early warning systems (MEWS) that includes four components: vulnerability mapping, seasonal climate forecasting; environmental monitoring; and sentinel case surveillance (WHO 2004). Mapping of areas prone to epidemics is done through historical analysis alone or in combination with climate/environmental suitability for malaria transmission. To support development of an operational MEWS, the International Research Institute for Climate and Society (IRI) created an online resource for information about the climatic conditions that are associated with malaria transmission (Grover-Kopec et al. 2006; http://iridl.ldeo.columbia.edu/maproom/.health/.regional/.africa/.malaria/). Based on the published literature, climatic conditions are considered to be suitable for transmission when the monthly precipitation accumulation is at least 80 mm, the monthly mean temperature is between 18°C and 32°C, and the monthly relative humidity is at least 60%. Because explosive epidemics may occur in regions where rainfall is the constraining variable, usually with a lag of several weeks, the MEWS interface displays the most recent dekadal (about 10-day) rainfall estimates and allows users to generate four time-series graphs that provide an analysis of recent rainfall with respect to that of recent seasons and a short-term historical average to identify epidemic-prone regions.

4.4 Developments in Malaria Early Warning Systems

The first operationally useful early warning system for epidemic malaria was implemented in India (Kiszewski and Teklehaimanot 2004). The system combined human and meteorologic factors to provide about a one month lead-time of epidemic risk. Rainfall alone accounted for about 45% of the variation in malaria transmission.

A number of recent publications describe approaches for forecasting malaria epidemics using climate data; not all of these approaches have been used to develop early warning systems (Table 4.1). Most models aim to predict outbreaks of *Plasmodium falciparum* malaria in Africa. Although different approaches have been used, those that are operational aim to predict epidemics with a lead-time of several months, thus providing ample time and opportunity to prepare. Most publications reporting the successful development of a malaria early warning system were based on a few years of data. Unfortunately, the systems typically don't have follow-up publications demonstrating their effectiveness over time.

Table 4.1 Recent models predicting epidemic malaria using climate information

Country/region	Case information	Variables included	Lags	Model	Alert trigger	Outcome	Result	Reference
East Africa	Malaria data from outpatients at seven highland sites in Ethiopia, Kenya, and Uganda, data from 1978–1999 (depends on clinic)	Monthly maximum and minimum temperature, rainfall; standardized anomalies compared with 1961–1990 baseline	Monthly lags during transmission season	Stepwise regression analysis using autoregressive model including previous number of malaria outpatients, seasonality, and climate variability	Average monthly malaria cases in past 5 years + 2 times SD	Monthly malaria outpatients	Monthly rainfall and maximum and minimum temperature significantly correlated with monthly malaria incidence with a lag of 1–2 months and 2–5 months, respectively; model accounted for 65–81% of the total variance	Zhou et al. 2004
Ethiopia/10 hot (altitude < 1,700 m ASL) and cold districts	Microscopically confirmed *P. falciparum* cases, 1990–2000	Weekly maximum and minimum temperature, and rainfall; week of time; year; autoregressive term	4–10 weeks for temperature variables; 4–12 weeks for rainfall. Autoregressive term = moving average of number of	Poisson regression with lagged weather factors, an autoregressive term, time trend, and indicator variables for week	Defined threshold exceeded by the predicted number of cases for two consecutive weeks	Potentially prevented cases	Overall pattern of cases predicted within 5% of the detection system, with the degree of prediction varying by district. Underestimated the	Teklehaimanot et al. 2004a; Teklehaimanot et al. 2004c

(continued)

Table 4.1 (continued)

Country/region	Case information	Variables included	Lags	Model	Alert trigger	Outcome	Result	Reference
			cases 4, 5, and 6 weeks before	of the year; 4th degree polynomial distributed lag model			height of the observed peaks. Prediction better in cold districts.	
Highlands of Kenya and Uganda	Weekly clinical malaria cases reported in 20 sentinel sites	Earth observation data for maximum and minimum land surface temperature (spatial resolution of 5 km); dekadal (10-day) rainfall estimates and normalized difference vegetation index (NDVI)	Model includes rainfall two and three months earlier, mean minimum temperature of the previous month, and case incidence from the previous month	Biologic model of relationships between meteorological variables and morbidity	An epidemic was flagged if weekly incidence exceeds both the week-specific mean plus one standard deviation and the overall mean plus one standard deviation	Epidemic alert	Model refined based on results of HIMAL (Highland Malaria) project	Abeku et al. 2004
Highlands of western Kenya	Monthly district total outpatient visits for malaria, 1997–2002	Rainfall estimates from Africa Data Dissemination Services, expressed as rainfall totals per dekad, 1995–2001	Dekadal rainfall totals	Association between rainfall and monthly outpatient admissions for malaria	WHO quartile, Cullen, and cumulative sum epidemic detection techniques	Epidemic alert for the 2002 malaria season	The temporal resolution of monthly outpatient admissions was not sufficient for early warning of epidemics,	Hay et al. 2003

					but was useful for quantifying the scale of the emergency. The rainfall estimates provided a warning 4 weeks ahead of the malaria surge.	
Tanzania	Peripheral blood smear examinations from the Ndolage hospital in the Kagera region, 1990–1999	Daily data on rain, humidity, and minimum and maximum temperature from the closest meteorological station; data were aggregated into monthly values; seasonal forecasts from EU DEMETER	Multiple linear regression of normalized climatic variables and malaria incidence for each rainy season	Log malaria incidence during the two rainy seasons	Malaria incidence during the first season (October–March) were positively correlated with total rainfall (August–January). However, the correlation was non-significant when an epidemic in 1998 was excluded. In the second	Jones et al. 2007

(continued)

Table 4.1 (continued)

Country/region	Case information	Variables included	Lags	Model	Alert trigger	Outcome	Result	Reference
							season (April–September), high malaria incidence was associated with increased rainfall during the season and with maximum temperature during the first rainy season.	
Botswana	Laboratory confirmed malaria cases, 1982–2003	Seasonal rainfall averages (December through February) from CMAP; sea surface temperature from Nino3.4 region; whether data are from before or after 1997 policy interventions; year; interaction variables	Seasonal time scale	Linear and quadratic stepwise regression with log malaria incidence; quadratic relationship had the best fit		Log malaria incidence per 1,000 population	About 85% of model variance explained by quadratic relationship between CMAP rainfall and log malaria incidence; provides warning one month in advance of seasonal peak	Thomson et al. 2005

Botswana	Laboratory confirmed malaria cases, 1982–2002	Seasonal rainfall averages (November through February) from DEMETER multi-model ensemble forecasts; sea surface temperature from Nino3.4 region; whether data are from before or after 1997 policy interventions; year; interaction variables	Seasonal time scale	Quadratic regression with log malaria incidence (see Thomson et al. 2005)	Probabilistic predictions for very high, high, low, and very low quartiles	Log malaria incidence per 1,000 population	Refinement of previously reported model; increased lead time up to 4 months over warnings issued with observed precipitation (5 months in advance of seasonal peak)	Thomson et al. 2006
Zimbabwe	Model tested on clinical data from Hwange, 1995–1998	Daily maximum temperature offset to roughly represent mean temperature and total dekadal rainfall (sum of previous 10-days rainfall) from ERA-40		Mathematical – biological model of the life cycle of *Anopheles gambiae*		Temporal and spatial distribution of the prevalence of *falciparum* infections	Numerical evaluations show a good first-order approximation to the prevalence of *falciparum* infections; model captures both the seasonality	Hoshen and Morse 2004

(continued)

Table 4.1 (continued)

Country/region	Case information	Variables included	Lags	Model	Alert trigger	Outcome	Result	Reference
		weather reanalysis data, 1960–2000					and inter-annual variability	
Zimbabwe	Malaria cases reported in the Hwange district to the Ministry of Health and Child Welfare, 1994–1998	Monthly minimum and maximum temperature and cumulative rainfall, 1993–1997	Four months	Model was eveloped using decadal minimum and maximum temperature and daily rainfall from the Western highlands of Kenya for 1993–1997; once the model was developed, it was rerun with monthly Zimbabwe meteorological data		Predicted monthly malaria case data	Model tends to over-predict case numbers, even after adjusting for case reporting, and predicts the magnitude of monthly malaria ($r^2 = 0.8806$)	Worrall et al. 2007
Zimbabwe	Annual clinical malaria case data for children under the	Mean annual values of rainfall, vapor pressure, and minimum, maximum, and		Bayesian negative binomial regression with		Between- and within year variation in monthly malaria cases	High annual malaria incidence coincided with high rainfall and	Mabaso et al. 2006

Location	Data		Method	Outcome	Results	Reference
		age of five reported in 58 districts covering the entire country, 1988–1999	mean temperature, and NDVI, 1998–1999	Markov Chain Monte Carlo simulation to estimate model parameters		relatively warm temperatures; temperature was important only in the presence of sufficient rainfall	
KwaZulu – Natal, South Africa	Active and passive surveillance of all confirmed malaria cases since 1970 for the province and since 1981 for districts	Monthly mean daily maximum and mean temperature and monthly rainfall	Mean monthly temperature of the previous season (January–October) and total rainfall during the summer months (November–March)	Linear regression of climate variables and malaria case data	Variance in seasonal changes in case numbers and total changes in case numbers	Weather variables explained 50% of the variance in seasonal changes in case numbers. No climate variables were significantly associated with total number of cases.	Craig et al. 2004
Colombia	Monthly *vivax* and *falciparum* case data from the provincial health service	Sea surface temperature in Nino3.4 region, mean, minimum, and maximum temperature rainfall, dew	0–5 months	Seasonal cross-correlations and spatial and temporal power	Estimate of the linear association between climate conditions	During normal years, endemic malaria associated with Tmean,	Poveda et al. 2001

(continued)

Table 4.1 (continued)

Country/region	Case information	Variables included	Lags	Model	Alert trigger	Outcome	Result	Reference
	for 16 towns, 1978–1995 along Pacific coast and 1990–1997 along Cauca river floodplain	point, Cauca river discharge		spectral analyses; comparison of normal and El Nino years		and malaria incidence	rainfall, dew point, and river discharges. During El Nino events, the timing of outbreaks does not change, but the number of cases intensifies	

Seasonal weather forecasts need to be accurate enough and at a scale fine enough to describe important microclimatic conditions. Further, the system needs to be sufficiently predictive of the presence or absence of malaria epidemics in years with unusual weather patterns. Hay et al. (2003) assessed whether a combination of seasonal climate forecasts, monitoring of meteorologic conditions, and early detection of cases could have helped prevent the 2002 malaria emergency in highland regions of western Kenya. Seasonal climate forecasts did not anticipate the heavy rainfall. On a shorter time scale, rainfall data gave timely and reliable early warnings. Normal rainfall conditions in two regions led to typical outbreaks, while exceptional rainfall in two other regions led to epidemics. Routine health information and management systems did not give timely warning of the epidemic.

Similar conclusions were reached in studies in Tanzania and Eritrea. Jones et al. (2007) analyzed weather data and peripheral blood smear examinations, and found that malaria incidence was positively correlated with rainfall during the first rainy season (October–March). During the second rainy season (April–September), malaria incidence was correlated with high rainfall that season and high maximum temperature during the first rainy season. However, excluding the 1998 epidemic from the analysis, there was no statistical association between excessive rainfall and malaria incidence in the first season; maximum temperature alone was associated with increased malaria. The authors concluded that the underlying relationship between rainfall and malaria in this district may be too complicated to analyze using regression analysis. Over the period 1990–1999, DEMETER seasonal temperature forecasts were able to predict high malaria incidence during the first season, despite the low statistical significance of the temperature relationship. DEMETER rainfall forecasts for the second season would have been unable to predict the rainfall anomalies in 1993 (low rainfall) and 1994 (high rainfall); both were associated with high malaria incidence. DEMETER rainfall forecasts were predictive for intermediate (non-extreme) years. If DEMETER had been used operationally, the seasonal forecast for 1996 would have resulted in a false alarm. In 1999, the forecast and observed temperature anomalies were positive, but near normal malaria conditions were recorded.

Eritrea has a successful malaria control program for endemic malaria, and is susceptible to epidemics with high morbidity and mortality (Ceccato et al. 2007). Relationships were investigated between monthly clinical malaria incidence in 58 districts and monthly climate data and seasonal forecasts. Although correlations were good between malaria anomalies and measured rainfall (lagged by 2 months), the weather stations did not have sufficient coverage to be widely useful. Satellite derived rainfall was moderately correlated with malaria incidence anomalies with a lead-time of 2–3 months, as was remotely sensed vegetation (NDVI). However, the seasonal forecasting skill was low for the June/July/August rains, except for the Eastern border.

Other research is furthering understanding of disease transmission dynamics that hold the promise of developing more sensitive and specific early warning systems. For example, Kiszewski et al. (2004) developed a global index representing the contribution of regionally dominant vector mosquitoes to the force of malaria transmission. The index incorporates published estimates of the proportion of blood meals taken from human hosts, daily survival of the vector, and duration of the transmission season and of extrinsic incubation. The results found that biologic characteristics of the diverse vector species interact with climate to explain much of the regional variation in the intensity of transmission.

4.5 Projections of the Future Incidence and Geographic Range of Malaria

There has been a great deal of interest in modeling how the incidence and geographic range of malaria could change under different climate change projections. Malaria is a complex disease to model, so no model completely describes the relationships among environmental variables, vector, and parasite. Table 4.2 summarizes recent studies that projected the impacts of climate change on the incidence and geographic range of malaria, including the climate scenarios used. Most of the studies focused on sub-Saharan Africa. The results suggest that climate change will be associated with geographic expansions of the areas suitable for stable *falciparum* malaria in some regions and with contractions in others. Most models do not consider how vector and disease control programs could affect malaria's future incidence and geographic range. For those studies that did, the main approaches used were inclusion of current "control capacity" in the observed climate-health function (Rogers and Randolph 2000) and categorization of the model output by adaptive capacity, thereby separating the effects of climate change from the effects of improvements in public health (van Lieshout et al. 2004).

Few models projected the impacts of climate change on malaria outside Africa. An assessment in Portugal projected an increase in the number of days per year suitable for malaria transmission; however, the risk of actual transmission would be low or negligible if no infected vectors are present (Casimiro and Calheiros 2002). Some central Asian areas are projected to be at increased risk of malaria, and areas in central America and around the Amazon are projected to have reductions in transmission due to decreases in rainfall (van Lieshout et al. 2004). An assessment in India projected shifts in the geographic range and duration of the transmission window for *falciparum* and *vivax* malaria (Bhattacharya et al. 2006). An assessment in Australia based on climatic suitability for the main anopheline vectors projected a likely southward expansion of habitat, although the future risk of endemicity would remain low due to the capacity to respond (McMichael et al. 2003).

The World Health Organization conducted a regional and global comparative risk assessment to project the total health burden attributed to climate change, including malaria, between 2000 and 2030 and how much of this burden could be avoided by stabilizing greenhouse gas emissions at different concentrations (stabilization at CO_2-equivalents of 550 ppm or at 750 ppm) (Campbell-Lendrum et al. 2003; McMichael et al. 2004). Large changes were projected in the risk of *falciparum* malaria in countries at the edge of the current distribution, with relative changes

Table 4.2 Projections of the impacts of climate change on the incidence and geographic range of malaria (Confalonieri et al. 2007)

Region	Metric	Model	Climate scenario, with time slices	Population projections and other assumptions	Main results	Reference
Global and regional	Population at risk in areas where climate conditions are suitable for malaria transmission	Biological model, calibrated from laboratory and field data, for *falciparum* malaria	HadCM3, driven by A1FI, A2, B1, and B2 SRES scenarios. 2020s, 2050s, 2080s	SRES population scenarios; current malaria control status used as an indicator of adaptive capacity	Estimates of the additional population at risk for >1 month transmission range from >220 million (A1FI) to >400 million (A2) when climate and population growth are included. The global estimates are severely reduced if transmission risk for more than 3 consecutive months per year is considered, with a net reduction in the global population at risk under the A2 and B1 scenarios.	van Lieshout et al. 2004
Africa	Person-months at risk for stable falciparum transmission	MARA/ARMA model of climate suitability for stable falciparum transmission	HadCM3, driven by A1FI, A2a, and B1 SRES scenarios. 2020s, 2050s, 2080s	Estimates based on 1995 population	By 2100, 16–28% increase in person-months of exposure across all scenarios, including a 5–7% increase in (mainly altitudinal) distribution, with limited latitudinal expansion. Countries with large areas that are close to the climatic thresholds for transmission show large potential increases across all scenarios.	Tanser et al. 2003

(continued)

Table 4.2 (continued)

Region	Metric	Model	Climate scenario, with time slices	Population projections and other assumptions	Main results	Reference
Africa	Map of climate suitability for stable falciparum transmission (minimum 4 months suitable per year)	MARA/ARMA model of climate suitability for stable falciparum transmission	HadCM2 medium-high ensemble mean. 2020s, 2050s, 2080s	Climate factors only (monthly mean and minimum temperature, and monthly precipitation)	Decreased transmission in 2020s in south east Africa. By 2050s and 2080s, localized increases in highland and upland areas, and decreases around Sahel and south central Africa.	Thomas et al. 2004
Zimbabwe, Africa	Climate suitability for transmission	MARA/ARMA model of climate suitability for stable falciparum transmission	16 climate projections from COSMIC. Climate sensitivities of 1.4°C and 4.5°C; equivalent CO_2 of 350 and 750 ppmv. 2100	None	Highlands become more suitable for transmission. The lowlands and regions with low precipitation show varying degrees of change, depending on climate sensitivity, emission scenario, and GCM.	Ebi et al. 2005
Portugal	Percent days per year with favourable temperature for disease transmission	Transmission risk based on published thresholds	PROMES for 2040s and HadRM2 for 2090s	Some assumptions about vector distribution and/or introduction	Significant increase in the number of days suitable for survival of malaria vectors; however, if no infected vectors are present, then the risk is very low for *vivax* and negligible for *falciparum* malaria	Casimiro and Calheiros 2002

Region	Outcome	Method	Climate scenario	Other assumptions	Main findings	Reference
Australia	Geographic area suitable/unsuitable for maintenance of vector	Empirical – statistical model (CLIMEX) based on current distribution, relative abundance, and seasonal phenology of main malaria vector	CSIROMk2 and ECHAM4 driven by B1, A1B, and A1FI emission scenarios. 2020, 2050	Assumes adaptive capacity; used Australian population projections	"Malaria receptive zone" expands southward to include some regional towns by 2050s. Absolute risk of reintroduction very low.	McMichael et al. 2003
India, all states	Climate suitability for *falciparum* and *vivax* malaria transmission	Temperature transmission windows based on observed associations between temperature and malaria cases	HadRM2 driven by IS92a emission scenario	None	By 2050s, geographic range projected to shift away from central regions toward south western and northern states. The duration of the transmission window is likely to widen in northern and western states and shorten in southern states.	Bhattacharya et al. 2006

Table 4.3 Range of estimates for the relative risks of malaria attributable to climate change in 2030 from the global burden of disease study by exposure scenarios (McMichael et al. 2004)

Region	Relative risks		
	Unmitigated emissions	S750[c]	S550[c]
Africa region	1.00–1.17	1.00–1.11	1.00–1.09
Eastern Mediterranean region	1.00–1.43	1.00–1.27	1.00–1.09
Latin America and Caribbean region	1.00–1.28	1.00–1.18	1.00–1.15
South-East Asia region	1.00–1.02	1.00–1.01	1.00–1.01
Western Pacific region[a]	1.00–1.83	1.00–1.53	1.00–1.43
Developed countries[b]	1.00–1.27	1.00–1.33	1.00–1.52

[a] Without developed countries.
[b] And Cuba.
[c] Stabilization at CO_2-equivalents of 550 ppm or at 750 ppm

much smaller in areas that are currently highly endemic for malaria; Table 4.3 provides relative risk estimates. The upper estimate of the projected potential increase in malaria risk is large in some regions (e.g. 1.83 for the Western Pacific region) and small in others (e.g. 1.02 for the South-East Asian region). The study concluded that if the understanding of broad relationships between climate and disease was realistic, then unmitigated climate change was likely to cause significant health impacts through at least 2030 (Campbell-Lendrum et al. 2003).

Although there are considerable uncertainties with projections of the possible impacts of climate change on the geographic range and incidence of malaria, many studies project increases in malaria along the edges of the present distribution. If the geographic range extends, then large-scale epidemics would likely result when the disease is introduced into non-immune populations. Table 4.2 provides an indication of where additional early warning systems are likely to be needed in coming decades.

4.6 Discussion

Although considerable progress has been made in using environmental variables to predict malaria epidemics, improvements are needed in several areas. Controlling malaria in epidemic-prone areas is a different challenge than in epidemic settings; strategies need to be targeted in time and space (Cox and Abeku 2007). Limitations in long-range forecasting, monitoring of environmental variables, and case surveillance are challenging development of accurate and timely malaria early warning systems.

An assessment of the validity of seasonal climate forecasts for malaria transmission in Africa suggested that model systems forecast climate anomalies correctly in only 6–7 out of 10 years (Blench 1999). That means that weather and health data will continue to need to be collected until the skill of seasonal climate forecasts improves significantly.

Another difficulty associated with the use of climate data in monitoring health impacts lies in linking climate and health data at a suitable high resolution; weather stations are often widely spaced and may not describe either local variation in climate (e.g. temperature in city centers vs. nearby rural areas) or microclimatic conditions in specific important environments (e.g. resting sites of adult mosquitoes). In order to develop early warning systems, it may be necessary to measure weather variables in important microclimates, and to test whether they provide close correlation with health outcomes. When data are not available, there are a variety of approaches for estimating weather variables based on downscaling from regional and global models. However, these are not yet available at high spatial resolutions.

The sensitivity and specificity of a predictive model have implications for the design of interventions. False positives (issuing a warning when none was required) and false negatives (not issuing a warning when one was needed) have consequences, not only in terms of morbidity and mortality, but also in terms of public willingness to rely on subsequent warnings. Incorporating understanding of these uncertainties and their associated costs into the design of an early warning system can improve its effectiveness. In all cases, model performance should always be validated using external data (Cox and Abeku 2007).

In most areas where malaria epidemics occur, additional capacity is needed to effectively respond to the warning that an epidemic is predicted to occur. Health systems are often weak, with significant improvements needed in the capacity for monitoring, surveillance, planning, preparedness, and preventing outbreaks of malaria (DaSilva et al. 2007).

Despite the many uncertainties and constraints, recent research suggests significant promise in using climatic and environmental variables to help regions prepare for and effectively respond to malaria epidemics. Kuhn et al. (2005) reviewed the evidence for a range of epidemic-prone infectious diseases and concluded that the following candidate infectious diseases for development of early warning systems: cholera, malaria, meningococcal meningitis, dengue fever, yellow fever, Japanese ad St. Louis encephalitis, Rift Valley fever, leishmaniasis, African trypanosomiasis, West Nile virus, Murray Valley encephalitis and Ross River virus, and influenza. Climate change projections suggest increased risks for epidemics with changing weather patterns, making the need for more effective warning systems imperative.

References

Abeku TA (2007) Response to malaria epidemics in Africa. *Emerging Infectious Diseases* **13**:681–686.

Abeku TA, Hay SI, Ochola S, Langi P, Beard B, deVlas SJ et al. (2004) Malaria epidemic early warning and detection in African highlands. *Trends in Parasitology* **20**:400–405.

Bhattacharya S, Sharma C, Dhiman RC, Mitra AP (2006) Climate change and malaria in India. *Current Science* **90**:369–375.

Blench R (1999) Seasonal climatic forecasting: who can use it and how should it be disseminated? *Nat Resource Persp* **47**:1–4 (http://www.odi.org.uk/nrp/47.html).

Breman J (2001) The ears of a hippopotamus: anifestation, determinants and estimates of the burden. *American Journal of Tropical Medicine and Hygiene* **64**:1–11.

Breman JG, Alilio MS, Mills A (2004) Conquering the intolerable burden of malaria: what's new, what's needed: a summary. *American Journal of Tropical Medicine and Hygiene* **71**(Suppl 2):1–15.

Breman JG, Alilio MS, White NJ (2007) Defining and defeating the intolerable burden of malaria. III. Progress and perspectives. *American Journal of Tropical Medicine and Hygiene* **77**(Suppl 6):vi–xi.

Campbell-Lendrum DH, Corvalan CF, Pruss-Ustun A (2003) How much disease could climate change cause? In: *Climate Change and Human Health: Risks and Responses.* A.J. McMichael, D. Campbell-Lendrum, C.F. Corvalan, K.L. Ebi, A. Githeko, J.D. Scheraga, A. Woodward (eds.), Geneva: WHO/WMO/UNEP.

Casimiro E, Calheiros J (2002) Human health. In: *Climate Change in Portugal: Scenarios, Impacts and Adaptation Measures – SIAM Project.* F. Santos, K. Forbes, R. Moita (eds.), Gradiva, Lisbon, pp. 241–300.

Caulfield L, Richard SA, Black R (2004) Undernutrition as an underlying cause of malaria morbidity and mortality. *American Journal of Tropical Medicine and Hygiene* **71**(Suppl 2):55–63.

Ceccato P, Ghebremeskel T, Jaiteh M, Graves PM, Levy M, Ghebreselassie S, Ogbamariam A, Barnston AG, Bell M, del Corral J, Connor SJ, Fesseha I, Brantly EP, Thomson MC (2007) Malaria stratification, climate, and epidemic early warning in Eritrea. *American Journal of Tropical Medicine and Hygiene* **77**(Suppl 6):61–68.

Chan NY, Smith F, Wilson TF, Ebi KL, Smith AE (1999) An integrated assessment framework for climate change and infectious diseases. *Environmental Health Perspect*ives **107**:329–337.

Cox J, Abeku TA (2007) Early warning systems for malaria in Africa: from blueprint to practice. *Trends in Parasitology* **23**:243–246.

Craig MH, Snow RW, le Sueur D (1999) A climate-based distribution model of malaria transmission in sub-Saharan Africa. *Parasitology Today* **15**:105–111.

Craig MH, Kleinschmidt I, Nawn JB, Le Sueur D, Sharp B (2004) Exploring 30 years of malaria case data in KwaZulu-Natal, South Africa. Part I. The impact of climatic factors. *Tropical Medicine and International Health* **9**:1247.

D'Alessandro U, Buttiens H (2001) History and importance of antimalaria drug resistance. *Tropical Medicine and International Health* **6**:845–848.

DaSilva J, Connor SJ, Mason SJ, Thomson MC (2007) Response to Cox and Abeku: early warning systems for malaria in Africa: from blueprint to practice. *Trends in Parasitology* **23**:246–247.

Ebi KL, Schmier JK (2005) A stitch in time: improving public health early warning systems for extreme weather events. *Epidemiologic Reviews* **27**:115–121.

Ebi KL, Hartman J, McConnell JK, Chan N, Weyant J (2005) Climate suitability for stable malaria transmission in Zimbabwe under different climate change scenarios. *Climatic Change* **73**:375–393.

Githeko AK, Ndegwa W (2001) Predicting malaria epidemics using climate data in Kenyan highlands: a tool for decision makers. *Global Change and Human Health* **2**:54–63.

Githeko AK, Shiff C (2005) The history of malaria control in Africa: lessons learned and future perspectives. In: *Integration of Public Health with Adaptation to Climate Change: Lessons Learned and New Directions.* K.L. Ebi, J. Smith, I. Burton (eds.), Taylor & Francis, London. pp. 114–135.

Greenwood B, Mutabingwa T (2002) Malaria in 2002. *Nature* **415**:67–672.

Grover-Kopec EK, Benno Blumenthal M, Ceccato P, Dinku T, Omumbo JA, Connor SJ (2006) Web-based climate information resources for malaria control in Africa. *Malaria Journal* **5**:38, doi:10.1186/1475-2875-5-38.

Hay SI, Were EC, Renshaw M, Noor AM, Ochola SA, Olusanmi I, Alipui N, Snow RW (2003) Forecasting, warning, and detection of malaria epidemics: a case study. *Lancet* **361**:1705–1706.

Hoshen MB, Morse AP (2004) A weather-driven model of malaria transmission. *Malaria Journal* **3**:32.

Janssen M, Martens P (1997) Modelling malaria as a complex adaptive system. *Artificial Life* **3**:213–236.

Jones AE, Uddenfeldt Wort U, Morse AP, Hastings IM, Gagnon AS (2007) Climate predictions of El Nino malaria epidemics in north-west Tanzania. *Malaria Journal* **6**:162, doi:10.1186/1475-2875-6-162.

Kiszewski AE, Teklehaimanot A (2004) A review of the clinical and epidemiologic burdens of epidemic malaria. *American Journal of Tropical Medicine and Hygiene* **71**(Suppl 2):128–135.

Kiszewski A, Mellinger A, Spielman A, Malaney P, Ehrlich Sachs S, Sachs J (2004) A global index representing the stability of malaria transmission. *American Journal of Tropical Medicine and Hygiene* **70**:486–498.

Koram KA, Molyneux ME (2007) When is "malaria" malaria? The different burdens of malaria infection, malaria disease, and malaria-like illnesses. *American Journal of Tropical Medicine and Hygiene* **77**(Suppl 6):1–5.

Kovats RS, Bouma MJ, Hajat S, Worrall E, Haines A (2003) El Nino and health. *Lancet* **263**:1481–1489.

Lindsay SW, Martens P (1998) Malaria in the African highlands: past, present and future. *Bulletin of the WHO* **76**:33–45.

Mabaso MH, Craig M, Ross A, Smith T (2007) Environmental predictors of the seasonality of malaria transmission in Africa: the challenge. *American Journal of Tropical Medicine and Hygiene* **76**:33–38.

Mabaso MLH, Vounatsou P, Midzi S, DaSilva J, Smith T (2006) Spatio-temporal analysis of the role of climate in inter-annual variation of malaria incidence in Zimbabwe. *International Journal of Health Geographics* **5**:20, doi:10/1186/1476–072X-5-20.

McMichael AJ, Woodruff R, Whetton P, Hennessy K, Nicholls N, Hales S, Woodward A, Kjellstrom T (2003) *Human Health and Climate Change in Oceania: Risk Assessment 2002*. Department of Health and Ageing, Canberra, 128 pp.

McMichael AJ, Campbell-Lendrum D, Kovats S, Edwards S, Wilkinson P, Wilson T, Nicholls R, Hales S, Tanser F, LeSueur D, Schlesinger M, Andronova N (2004) Global climate change. In: *Comparative Quantification of Health Risks: Global and Regional Burden of Disease due to Selected Major Risk Factors*. M. Ezzati, A. Lopez, A. Rodgers and C. Murray (eds.), World Health Organization, Geneva, pp. 1543–1649.

Murray CJ, Lopez AD (1996) The Global Burden of Disease: a comprehensive assessment of mortality and disability from diseases, injuries and risks factors in 1990 and projected to 2020. Cambridge, MA, Harvard School of Public Health (Global Burden of Disease and Injury Series, Vol. 1).

Pascual M, Ahumada JA, Chabes LF, Rodo X, Bouma M (2006) Malaria resurgence in the East African highlands: temperature trends revisited. *Proceeding of the National Academy of Sciences* **103**(15):5829–5834

Poveda G, Rojas W, Quiñones ML, Vélez ID, Mantilla RI, Ruiz D, Zuluaga JS, Rua GL (2001) Coupling between Annual and ENSO Timescales in the Malaria-Climate Association in Colombia. *Environmental Health Perspectives* **109**(5):489–493.

Rogers DJ, Randolph SE (2000) The global spread of malaria in a future, warmer world. *Science* **289**:1763–1769.

Tanser FC, Sharp B, le Sueur D (2003) Potential effect of climate change on malaria transmission in Africa. *Lancet* **362**:1792–1798.

Teklehaimanot HD, Lipsitch M, Teklehaimanot A, Schwartz J (2004a) Weather-based prediction of *Plasmodium falciparum* malaria in epidemic-prone regions of Ethiopia. I. Patterns of lagged weather effects reflect biological mechanisms. *Malaria Journal* **3**:41, doi:10.1186/1475-2875-3-41.

Teklehaimanot HD, Schwartz J, Teklehaimanot A, Litpsitch M (2004b) Alert threshold algorithms and malaria epidemic detection. *Emerging Infectious Diseases* **10**:1220–1226.

Teklehaimanot HD, Schwartz J, Teklehaimanot A, Lipsitch M (2004c) Weather-based prediction of Plasmodium falciparum malaria in epidemic-prone regions of Ethiopia. II. Weather-based prediction systems perform comparably to early detection systems in identifying times for interventions. *Malaria Journal* **3**:44, doi:10.1186/1475-2875-3-44.

Thomas CJ, Davies G, Dunn CE (2004) Mixed picture for changes in stable malaria distribution with future climate in Africa. *Trends in Parasitology* **20**:216–220

Thomson MC, Mason SJ, Phindela T, Connor SJ (2005) Use of rainfall and sea surface temperature monitoring for malaria early warning in Botswana. *American Journal of Tropical Medicine and Hygiene* **73**:214–221.

Thomson MC, Doblas-Reyes FJ, Mason SJ, Hagedorn R, Connor SJ, Phindela T, Morse AP, Palmer TN (2006) Malaria early warnings based on seasonal climate forecasts from multi-model ensembles. *Nature* **439**:576–579.

Van Lieshout M, Kovats RS, Livermore MTJ, Martens P (2004) Climate change and malaria: analysis of the SRES climate and socio-economic scenarios. *Global Environmental Change* **14**:87–99.

WHO (2004) *Malaria Epidemics: Forecasting, Prevention, Early Warning and Control – From Policy to Practice.* World Health Organization, Geneva.

WHO-UNICEF (2005) World Malaria Report 2005 (WHO/HTM/MAL/2005.1102). World Health Organization – United Nations Children's Fund.

Worrall E, Rietveld A, Delacollette C (2004) The burden of malaria epidemics and cost-effectiveness of interventions in epidemic situations in Africa. *American Journal of Tropical Medicine and Hygiene* **71**(Suppl 2):136–140.

Worrall E, Connor SJ, Thomson MC (2007) A model to simulate the impact of timing, coverage and transmission intensity on the effectiveness of indoor residual spraying (IRS) for malaria control. *Tropical Medicine and International Health* **12**:75–88.

Woodruff RE (2005) Early warning systems: Ross River virus disease in Australia. In: *Integration of Public Health with Adaptation to Climate Change: Lessons Learned and New Directions.* K.L. Ebi, J. Smith, I. Burton (eds.), Taylor & Francis, London, pp. 91–113.

Zhou G, Minakawa N, Githeko AK, Yan G (2004) Association between climate variability and malaria epidemics in the East African highlands. *Proceeding of the National Academy of Sciences* **101**(8):2375–2380.

Chapter 5
Pollen, Allergies and Adaptation

Mikhail Sofiev, Jean Bousquet, Tapio Linkosalo, Hanna Ranta, Auli Rantio-Lehtimaki, Pilvi Siljamo, Erkka Valovirta, and Athanasios Damialis

Abstract This chapter is dedicated to the problem of plant-induced human allergy and to a specific way of adaptation to it via short-to-mid-term forecasts of atmospheric pollen concentrations and following pre-emptive and preparatory measures. It starts from the introduction to the subject, then considers the main forms of human plant-related allergy. Then, basics of pollination ecology are introduced and the mechanisms and possible models for pollination are presented. Apart from the standard local-scale effect of pollination, the chapter considers the long-distance transport of pollen and outlines the methodology for its quantitative evaluation

M. Sofiev
Finnish Meteorological Institute, Finland
e-mail: mikhail.sofiev@fmi.fi

J. Bouszuet
Hôpital Arnaud de Villeneuve, France
e-mail: jean.bousquet@orange.fr

T. Linkosalo
University of Helsinki, Finland
e-mail: tapio.linkosalo@helsinki.fi

H. Ranta
University of Turku, Finland
e-mail: hanranta@utu.fi

A. Rantio-Lehtimaki
University of Turku, Finland
e-mail: ahrantio@utu.fi

P. Siljamo
Finnish Meteorological Institute, Finland
e-mail: pilvi.siljamo@fmi.fi

E. Valovirta
Turku Allergy Centre, Finland
e-mail: erkka.valovirta@terveystalo.com

A. Damialis
Aristotle University, Greece
e-mail: dthanos@bio.auth.gr

and forecasting. The final section is dedicated to possible adaptation measures in changing climate.

5.1 Introduction

Allergy is among the major diseases of the world affecting about 500 million people. These diseases are present in all countries, they pose a major burden to the society, and their prevalence is increasing in nearly all regions, with the fastest rate reported in developing countries. Even allergic diseases considered as non-severe, such as allergic rhinitis, impair the social and professional lives of the patients or their caregivers. Moreover, patients with allergic diseases have reduced learning capabilities and impaired school performance. Finally, the corresponding economic impact is substantial (Bousquet et al. 2006).

Allergic diseases result from a complex interaction between genes, allergens and co-factors, which vary between regions. Allergens are antigens reacting with specific IgE (immunoglobulin E) antibodies and mainly released in a form of molecular-weight fine aerosols from a wide range of mites, animals, insects, plants, or fungi. They are usually distinguished as indoor (mites, some molds, animal dander, insects) and outdoor allergens (pollens and some molds). Exposure to allergens is a trigger for symptoms in sensitised individuals with asthma and allergic rhino-conjunctivitis.

There are regional specifics of the sensitivity to allergens. For example, in Africa, allergic diseases are more common in urban rather than in rural areas possibly because parasites protect from atopic diseases. In most other regions the situation is the opposite suggesting non-allergenic co-factors in development of sensitization and symptoms.

Pollen is among the first identified and most important triggers of allergic asthma and rhino-conjunctivitis, and pollination depends on climatic and meteorological variables. Greater concentrations of carbon dioxide and higher temperatures may increase the amount of released pollen, change the geographical distribution of pollinating plants and induce longer pollen seasons. Pollen allergenicity can also increase due to a combined effect of pollen and air pollution (Laaidi 2001; WHO 2003).

According to the mode of the pollen transport, one can distinguish anemophilous and entomophilous plants.

The pollen grains of anemophilous plants are usually aerodynamic and disperse with wind. They represent a major problem for sensitised patients as they often are emitted in large quantities, may travel long distances and may affect individuals who are far from the pollen source. It is, however, those who are nearest the emission area or directly downwind from it who generally show the most severe symptoms. Severity, affected territory and duration of the pollinating season are mainly driven by actual weather conditions as well as by the larger-scale characteristics of the regional climate. Therefore, changes in environment and climate may in near future considerably affect loads of airborne allergens via changes in vegetation structure, magnitude of flowering and atmospheric transport. Understanding the mechanisms driving the development of pollen seasons and means of their short- and long-term forecasting for taking pre-emptive measures is of great importance to public health at present and in the predicted future conditions.

As shown in several studies (Erdtman 1937; Keynan et al. 1991; Rantio-Lehtimäki 1994; Campbell et al. 1999; Corden et al 2002; Latalowa et al 2002; Hjelmroos 1992; Damialis et al. 2005; Lorenzo et al. 2006), the pollen load in air consists of two main fractions: grains emitted from regional sources of pollen (nearby forests, grasslands, etc.) and grains released from remote sources and transported to the region with moving air masses. Forecasting the local and regional pollen emission can be based on various phenological models and observations, as well as on aerobiological monitoring. The long-range transport phenomenon poses different complicated problems, not considered until recently. Now, the long-range transport of both grains themselves and the allergenic material, which can be released during the atmospheric dispersion (Motta et al. 2006; Majd et al. 2004), is a quickly growing area of research. Large-scale pollen transport can also have environmental impacts well beyond the problem of human allergy. For instance, highlights the possibility of quick distribution of genetic material over large territories, across climate zones and vegetation regions. This can catalyze the changes caused by the altering climate and make them faster and more widespread.

The pollen of entomophilous plants is carried by insects. Attracted by the usually colourful and perfumed flowers they carry the pollen grains from one flower to the female parts of another. The pollen grains are sticky and adhere to one another, often forming heavy lumps, and to the antennae or other parts of the insects. Little pollen is liberated into the atmosphere and sensitisation thus usually requires direct contact between the subject and the pollen source.

Certain entomophilous plants, such as dandelions and spiraeas, produce large amounts of pollen which accidentally gets into the air. This is common when the flowers are pollinated by insects that consume a lot of pollen grains themselves, and when there is an open pollen presentation. Others are ambophilous, which means that they are adapted to wind as well as to insect dispersal, a system that allows for flexibility. Examples are some species of plantain, e.g. *Plantago lanceolata*, and manna ash, *Fraxinus ornus*. Some of these ambophilous species may be among the plants that cause allergy symptoms on a regular basis.

The pollen causing most allergies are found among:

- Grasses that are universally distributed. The grasses pollinate at the end of spring and the beginning of summer, but, in some places such as Southern California or Florida, they are spread throughout the year. Bermuda grass (*Cynodon dactylon*) and Bahia grass (*Paspalum notatum*) do not usually cross-react with other grasses (Davies et al. 2005).
- Weeds such as the Compositeae plants: mugwort (*Artemisia*) and ragweed (*Ambrosia*) (D'Amato et al. 1998; Solomon 2001), *Parietaria*, not only in the

Mediterranean area (D'Amato et al. 1998), *Chenopodium* and *Salsola* in some desert areas (Al-Dowaisan et al. 2004). Weeds such as ragweed flower at the end of summer and the beginning of autumn. *Parietaria* often pollinates over a long period of time (March–November) and is considered as a perennial pollen.

- And trees: birch (*Betula*), other Betulaceae (Lewis and Imber 1975; Eriksson et al. 1984), Oleaceae including the ash (*Fraxinus*) and olive tree (*Olea europea*) (D'Amato et al. 1998), the oak (*Quercus*), the plane tree (*Platanus*) (Varela et al. 1997) and Cupressaceae including the cypress tree (*Cupressus*) (Charpin et al. 2005), junipers (*Juniperus*) (Iacovacci et al. 1998), thuyas (Guerin et al. 1996), the Japanese cedar (*Cryptomeria japonica*) (Ganbo et al. 1995) and the mountain cedar (*Juniperus ashei*) (Ramirez 1984; Bucholtz et al. 1985).

Trees generally pollinate at the end of winter and the beginning of spring. However, the length, duration and intensity of the pollinating period often vary from 1 year to the next sometimes making the diagnosis difficult. Moreover, the change in temperature in Northern Europe has caused earlier birch pollen seasons (Emberlin et al. 2002). Multiple pollen seasons in polysensitized patients are important to consider.

The grass season is early to late summer, whilst weeds such as *Ambrosia* flower at the end of summer and beginning of autumn. *Parietaria* often pollinates over a long period of time (March–November) and is considered a perennial pollen. In warm and humid climate also grass pollens can be found all year round.

The size of the pollen varies from 10 to 100 µm on average. This explains why pollen itself is deposited in the nostrils and, more particularly, the eyes and also why most pollen-allergic patients suffer from rhino-conjunctivitis. However, pollen allergens can be borne on submicronic particles (Solomon et al. 1983; Suphioglu et al. 1992) and induce and/or contribute to the persistence of rhinitis and asthma. This is particularly the case of asthma attacks occurring during thunderstorms (Anto and Sunyer 1997; Bauman 1996; Bellomo et al. 1992; Knox 1993; Venables et al. 1997).

Cross reactivities between pollen types, due to the presence of homologous allergens with common epitopes, are now better understood using molecular biology techniques (Scheiner et al. 1997; Fedorov et al. 1997; Ipsen and Lowenstein 1997; Mothes et al. 2004). However, it is unclear as to whether all *in vitro* cross-reactivities observed between pollens are clinically relevant (Pham and Baldo 1995). Major cross reactivities include pollens of the Gramineae family (Freidhoff et al. 1986; Hiller et al. 1997; Mourad et al. 1988) except for Bermuda and Bahia grasses (Matthiesen et al. 1991; Lovborg et al. 1998) and Bahia grass (Phillips et al. 1989), the Oleacea family (Bousquet et al. 1985; Baldo et al. 1992; Batanero et al. 1996), the Betuleacea family (Hirschwehr et al. 1992; Mari et al. 2003) and the Cupressaceae family (Pham et al. 1994) but not those of the Urticaceae family (Corbi et al. 1985; Bousquet et al. 1986). Moreover, there is clinically little cross-reactivity between ragweed and other members of the Compositeae family (Leiferman et al. 1976; Fernandez et al. 1993; Hirschwehr et al. 1998). For the grass family, cross-reactivity is often extensive within subfamilies. Most of the common grasses within the temperate areas belong to Pooideae, and cross-react to a large extent. Bermuda grass (*Cynodon dactylon*) and Bahia grass (*Paspalum notatum*) belong to Chloridoideae.and cross-reactions between these species and Pooideae species are less common. Many of the reactions may in this way be explained by the close phylogenetic relationship between the pollen-producing plants. But in some cases, they may be the result of so called pan-allergens, which from a evolutionary point of view very conservative protein families. One example is profiling, present in all eukaryots, another is a family of pathogenesis-related proteins that occur in many angiosperms, and to whom the most important Betula allergen belongs. The occurrence of pan-allergens explain why distantly related plants such as banana (Musa) and melon (Cucurbita), or birch (Betula) and celeriac (Apium,) may be involved in cross-reactions.

5.2 Human Allergy, Sensitivity to Natural Pollen and Pre-emptive Measures Based on Forecasting of Pollen Atmospheric Concentrations

Pollen grains typically induce rhinitis, conjunctivitis and asthma. However, they may more rarely be involved in contact urticaria, anaphylactic symptoms or exceptionally in diseases such as nephritic syndrome.

5.2.1 Rhinitis and Conjunctivitis

Allergic rhinitis is a symptomatic disorder of the nose induced by an IgE-mediated inflammation of the nasal membranes in response to allergen exposure. Symptoms of rhinitis include rhinorrhea, nasal obstruction, nasal itching and sneezing which will be cured spontaneously or with treatment. The severity of allergic rhinitis can be classified as "mild" or "moderate-severe" (Fig. 5.1) based on symptoms and quality of life parameters.

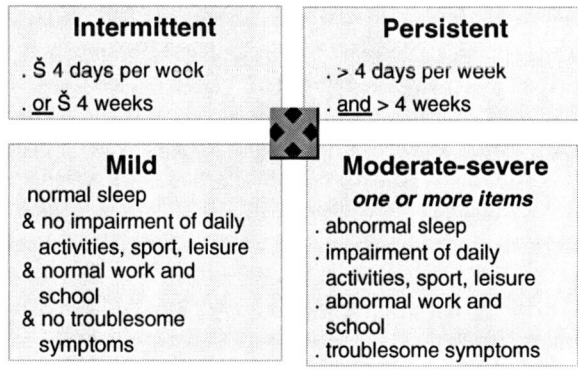

Fig. 5.1 Classification of allergic rhinitis

Previously, allergic rhinitis was subdivided based on the time of exposure into seasonal, perennial and occupational diseases (9–11). Perennial allergic rhinitis is most frequently caused by indoor allergens such as dust mites, moulds, insects (cockroaches) and animal danders. Seasonal allergic rhinitis is related to a wide variety of outdoor allergens such as pollen or moulds. However, this subdivision is not entirely satisfactory since:

- It is often difficult to differentiate between seasonal and perennial symptoms.
- The exposure to some pollen allergens is long-standing.
- The exposure to some perennial allergens is not similar over the year.
- The majority of patients are now sensitised to pollen and perennial allergens.

Thus, a major change in the subdivision of allergic rhinitis has been proposed by ARIA with the terms "intermittent" and "persistent" (Fig. 5.1). However, in the present document, the terms "seasonal" and "perennial" are still used for the description of published studies.

Eye symptoms are common in patients suffering from allergic rhinitis, however, they do not exist in all patients. In some cases, eye symptoms may predominate over rhinitis, hence the presence of conjunctivitis should always be considered. On the other hand, conjunctivitis is not always induced by allergy triggers.

The high prevalence of allergic rhinitis and its effect on quality of life have led to its being classified as a major chronic respiratory disease. It is reported to affect 10–40% of the global population and its prevalence is increasing both in children and adults. There are approximately 350 million people suffering from allergic rhinitis. The prevalence in European Union varies from 17% to 28% in different countries (Bauchau & Durham, 2001).

Allergic rhinitis can significantly reduce quality of life, impairing sleep and adversely affecting leisure, social life, school performance, and work productivity (McMenamin 1994). The direct and indirect financial costs of allergic rhinitis including sick leave, school and work absenteeism and loss of productivity are substantial as well.

Asthma is a chronic inflammatory disorder of the airways usually associated with variable airflow obstruction, which is often reversible, spontaneously or under treatment, and airway hyper-responsiveness (Global strategy 2002). Allergen sensitisation is an important risk factor for asthma and is also often associated with rhinitis (an inflammation of nose) (Bousquet et al. 2001). Typical symptoms of asthma are wheezing, shortness of breath, chest tightness and cough. Acute exacerbations may lead to even life-threatening asthma attacks. Asthma impairs school and work performance and social life. Physical quality of life is impaired by bronchial symptoms, whereas social life is also impaired by rhinitis co-morbidity.

Asthma affects both children and adults. Using a conservative definition, it is estimated that as many as 300 million people of all ages and all ethnic backgrounds suffer from asthma, which is also the most common chronic disease in childhood. Two large multinational studies have assessed the prevalence of asthma around the world: the European Community Respiratory Health Survey (ECRHS) in adults (EC Survey 1996) and the International Study of Asthma and Allergies in Childhood (ISAAC 1998) (Fig. 5.2).

Trends in asthma prevalence vary between countries. For the past 40 years, the prevalence of asthma has increased in all countries in parallel with that of allergy.

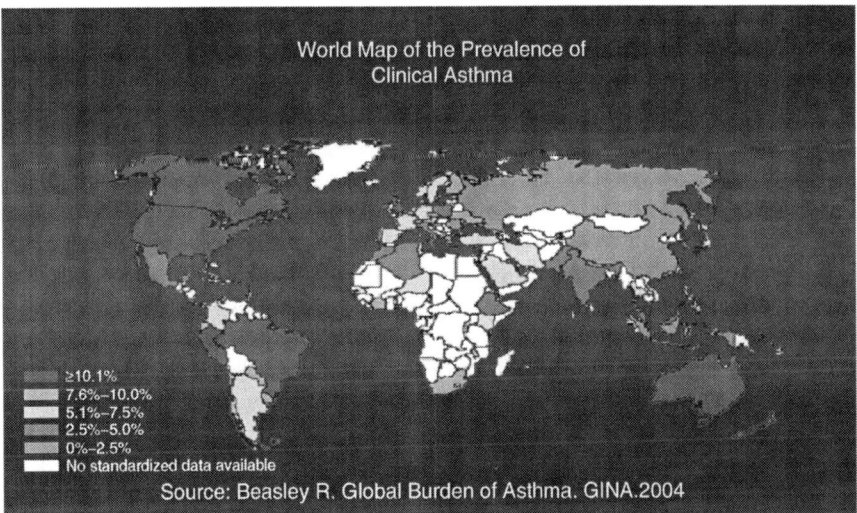

Fig. 5.2 A map of world prevalence of asthma

Asthma still increases worldwide as communities adopt modern lifestyles and become urbanized. With a projected increase of proportion of the world population living in urban areas, there is likely to be a marked increase in the number of people with asthma worldwide over the next two decades. It is estimated that there may be an additional 100 million people with asthma by 2025. However, the prevalence of childhood asthma and allergy may decrease in some countries with a high prevalence of the disease and the increase of the asthma epidemic possibly coming to an end.

The links between rhinitis and asthma are not always straightforward. Epidemiologic studies have consistently shown that asthma and rhinitis often co-exist in the same patients. In epidemiologic studies, over 70% of asthmatics have concomitant rhinitis. However, only 15–40% of rhinitis patients have clinically demonstrable asthma. Patients with severe persistent rhinitis more often suffer from asthma than those with intermittent disease. Both allergic and non-allergic rhinitis can be associated with asthma. Although differences exist between rhinitis and asthma, upper and lower airways may be considered as a unique entity influenced by a common and probably evolving inflammatory process, which may be sustained and amplified by intertwined mechanisms.

The elements of a successful allergy treatment are:

- Patient education and training
- Avoidance of allergens which induce symptoms to sensitized person whenever it is practically possible
- Pharmacotherapy
- Allergen-specific immunotherapy by subcutaneous injections or by sublingual route with drops or tablets

For all of them, timely information on forthcoming pollen episodes and their severity would allow for tuned measures with higher efficiency, lower costs and reduced adverse effects.

The patient-physician relationship and communication should be based on the needs of the patient. A patient-centered and structured consultation has shown to be effective, as well as individualized and guided self-management plans. Patients are able to modify their asthma treatment according to their symptoms and severity of the disease. The main goal of allergy treatment and management is to interfere with the underlying allergic inflammation. With early introduction of effective medications after a proper diagnosis of the allergic disorder inflammation can be controlled. Patients tend to regard themselves as "allergic to pollen" in general; for physicians, however, understanding the individual trigger mechanism can be an important guide to the effective treatment and control of the disease.

5.2.2 Pollen Counts and Symptoms

Trees tend to trigger allergic reactions of patient with asthma and allergic rhinoconjunctivitis in early spring while grass contribution is mostly visible during late spring and early summer; weeds can trigger symptoms from spring to fall. Each plant pollinates at approximately the same time each year. However, during the last decades, a significant trend for earlier flowering in many species on the Northern hemisphere has been observed (Menzel et al. 2006). Weather conditions strongly affect concentration of pollen in the air and therefore the prevalence of allergic symptoms. The airborne pollen quantification is a fundamental piece of information for validation and interpretation of scientific clinical trials for the efficiency of medication in pollen related-diseases. It is possible to significantly improve the treatment and its efficiency by using the quantitative pollen information as one of the guidelines. The specific roles of aerobiological observations and modelling as the sources of this information can be summarized in the following main items:

- For health-care authorities, planning protective measures against pollen-related diseases
- For allergists, who must be aware of pollen situation in order to educate their patients
- For patients, planning and adapting their medication before and during the pollen season, planning their holidays, physical training, traveling, education and career, as well as domestic habits, such as garden and plants in the house
- For pharmacists, preparatory measures for coming pollen season and for being able to give information and advises to their customers
- For medical industry, planning the manufacturing capacities of medications
- For allergists and medical scientists, in the clinical trials, for selecting the allergen extracts for diagnosis and treatment, and for evaluating the effects of the treatment.
- For schools and universities, planning the schedule for examinations

- For hospitals and emergency rooms, preparing to outbreaks of acute exacerbations of allergic diseases in case of upcoming episodes of heavy pollen loads
- For allergists, when treating patients with immunotherapy, when information about the ambient allergen load is necessary

A wish-list from health-care professionals to aerobiologists and meteorologists would then include:

- Detailed timing of pollen season in different regions (patients live in a certain area and the more local the pollen-related information is the better it is for the patient).
- Estimated duration of individual pollen (trees, grasses, weeds) seasons.
- Estimated strength of the season and expected pollen counts.
- Timely warnings about major changes in pollen counts.
- Easy access to global pollen information with the native language of the patient. This is crucial to allergic people when planning their activity and traveling.

5.3 Pollination Ecology

5.3.1 Basics of Flowering Phenology, Processes Determining Flowering Timing and Intensity

Considering the seasonal development of allergenic plants, a special attention must be dedicated to early-flowering tree species and herbaceous taxa. For trees, the genera in Fagales (*Betula, Alnus, Carya, Corylus, Carpinus, Juglans, Quercus*) and the genus mulberry (Moraceae; *Morus*) together with sugi (Taxodiaceae: *Cryptomeria japonica*) in the boreal and temperate zones and olive (Oleaceae: *Olea*) in the Mediterranean climate are among the most common agents of seasonal allergies. Important herbaceous taxa include one of the largest plant family, the grasses (Poaceae) with about 640 genera and over 10,000 species (Mabbelrey 1987), widely distributed weeds, such as mugwort (*Artemisia*), ragweed (*Ambrosia*), pigweed (*Amaranthus*) and goose foot (*Chenopodium*), as well as pellitory (*Parietaria*) growing predominantly in the Mediterranean climate (www.pollen.com/Pollen.com.asp, www.polleninfo.org).

For many anemophilous plants, effective pollination and reproduction requires adaptations that make plants release pollen simultaneously. Photoperiod, especially the length of the diurnal dark period, has an important part in regulating the development of herbaceous plants (Flood and Halloran 1982; Hay 1990; Mahoney and Kegode 2004). Temperature is also important for the growth rate of herbaceous plants and the discrimination of seasons (Yanovsky and Kay 2003). For trees, a complicated interplay of changes in the light environment and temperature conditions controls the timing of events related to flowering and leaf bud burst. Dormancy, the state of arrested growth in autumn and winter, is released by exposure to chilling temperatures while the following bud development is driven by increasing temperature in spring. There is also recent evidence of importance of light environment for the onset of the bud development (Heide 1993a, b; Partanen et al. 1998, 2001, 2005; Linkosalo and Lechowicz 2006).

Flowering of wind-pollinated trees for most broadleaf species takes place before the leaf bud burst. The advantage is obvious: leaves would hinder the movement of pollen within the tree while the flowering before the bud burst enables more efficient spread of pollen. The timing of leaf bud burst poses a crucial optimization problem between maximizing the period of active photosynthetic production and minimising the risk of damage due to spring frost (Chabot and Hicks 1982; Kikuzawa 1989; Reich et al. 1992). As the flowering precedes the leaf bud burst, its time window is quite narrow and requires proper regional synchronization between the plants. High correlation between these phenological events within and between many boreal tree species seems to suggest that the trees utilise similar mechanisms to control the timing of the spring phenological events. Yet the timing of flowering between species varies somewhat. For example, in Southern Finland the mean date of birch flowering is in mid-May, while aspen flowers about two, and alder up to 4 weeks earlier (Linkosalo 2000).

For herbaceous species, the studies of common ragweed (*Ambrosia artemisiifolia*) and two mugwort species suggest a juvenile phase, during which it is not sensitive to photoperiod, but after that the light for about 14 h a day is needed for flower bud development and seed production (Deen et al. 1998; Sheldon and Hewson 1960; Edmonds 1979; Ferreira et al. 1995; Mahoney and Kegode 2004). In northern Europe, the pollen season of mugwort starts earlier in northern locations where the length of darkness grows more rapidly in late summer (Ekebom et al. 1997).

For all wind-pollinated plants, the weather parameters have significant impact on daily and diurnal patterns of pollen release. The release of mature grains is largely a physical process. Dry conditions, sometimes just a temporary decrease in relative air humidity, are needed for anther walls to dry out and anthers to split, thus allowing the pollen to release. A warm night accelerates the release while it can be almost inhibited by a cold one, especially in herbaceous species (Edmonds 1979; Sheldon and Hewson 1960; Solomon 2002).

After the pollen liberation, air flows agitate the source plant and/or lift the pollen free from the bounder layer surrounding the surface on which it rests. A few species liberate the grains actively, for example, mulberries and nettles catapult their pollen into the air (Solomon 2002). During the pollination period pollen dispersal is favored by warm and windy conditions, low relative humidity and absence of precipitation. Rain causes a break in pollination, removes pollen from the air and may, if continuing for several days or in combination with night frosts, even spoil catkins thus decreasing the total pollen yield. In general, any strong deviation of external factors from the regional climatologic norms (e.g. temperatures or rainfall above or below normal level) during the critical points of developmental phase may result in lowered growth and flower production, or abortion of flower buds (Sharp and Chisman 1961). In a long run, the pollination seasons of species may be able to become adapted to the new conditions through natural selection (Flood and Halloran 1982; Hay 1990).

5 Pollen, Allergies and Adaptation

The phenological phases and their key features can be quantified and simulated by means of models. The basis for phenological modelling was laid more than 270 years ago, when French scientist Réaumur (1735) first discovered, while trying to find out why cereals ripen at different times in different years, that the heat sum accumulated from April until the ripening date always reaches about the same value. This model is essentially similar to the so-called Thermal Time or Growing Degree Day model, that is still the most common model describing various phenological events. As Réaumur found out, most developmental events proceed at rate depending on ambient air temperature. Several experimental studies have since verified this finding, including the classical work by Sarvas (1972), who found that the relationship between ambient temperature and rate of progress of meiosis in pollen mother cells is constant for several tree species (Fig. 5.3a). The relationship follows a sigmoidal curve, and is rather similar to the linear dependence suggested by the Thermal Time models (Fig. 5.3b).

Sarvas claimed that the same relationship applies to the flower development of several boreal tree species. Even further, the good performance of similar models in describing both flowering and leaf bud burst of various species seems to suggest that the boreal trees utilise rather similar control mechanisms to drive all their spring phenological events. More support for this claim was given by Linkosalo (1999), who found using data from historical phenological time series, that the correlation between flowering and leaf bud burst of *Betula* is 0.97, while the two events take place on the average 1.1 days apart. Similar figures for the flowering of *Betula* and *Populus* are 0.83 and 22.7 days. The lower correlation in the latter case is due to the longer time difference between the two events.

Leaf and flower buds of boreal and temperate trees develop already during the previous summer. To keep the buds viable until the next spring, they need to reach a frost-resistant state in the autumn, keep it throughout the winter and avoid premature onset of development during that time. Various studies starting from the classical work of Coville (1920) have shown that buds fall dormant in the autumn – a state where the development is hindered even if the environmental conditions, most of all air temperature, are feasible. Dormancy is released by exposure to chilling temperatures, to certain changes in the light environment, or to a combination of the two (e.g. Sarvas 1974; Heide 1993a, b; Myking and Heide 1995; Leinonen 1996; Linkosalo and Lechowich 2006). Sarvas (1974) found the optimal temperature for

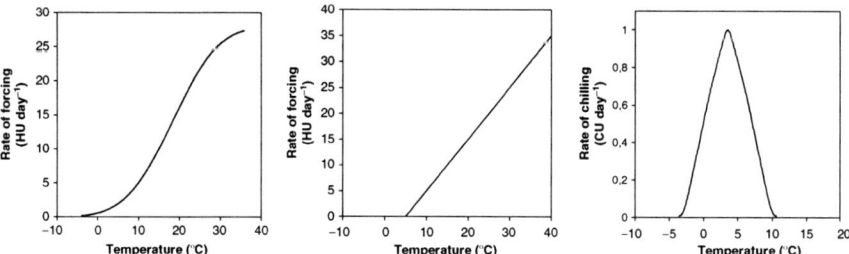

Fig. 5.3 The rate of ontogenetic bud development (forcing) as a function of air temperature according to Sarvas (1972) (**a**) and according to Thermal Time model (**b**). The rate of dormancy development (chilling) as a function of temperature according to Sarvas (1974) (**c**)

dormancy release to be around 3°C (Fig. 5.3c). He formulated a model to describe the sequence of chilling followed by the bud development. Thus the model is called "Sequential" (Hunter and Lechowicz 1992; Hänninen 1995).

Curiously, some more recent work (e.g. Häkkinen et al. 1998; Hannerz 1999; Linkosalo et al. 2000; Hänninen et al. 2006) have found that the standard Thermal Time model, where the bud development is assumed starting from a fixed date in spring, fits empirical data better than the Sequential model. It seems that even though chilling, as described by Sarvas, is required, it is not a sufficient condition for the dormancy release, but an additional environmental forcing is needed. Indeed, some studies on *Betula* indicate that light environment affects the dormancy release (Heide 1993a, b; Linkosalo and Lechowicz 2006). These results are for leaf bud burst, but the above-mentioned high correlation of leaf bud burst and flowering suggests that the events are controlled by same mechanisms, and thus the use of the same models is justified.

In the temperate climatic zone the role of chilling in dormancy release seems to have much more importance, as chilling temperatures are not as prevailing as in the boreal zone, so that the sufficient amount of chilling can be reached rather late in winter. Yet also Thermal Time model has been used with good success to describe the bud development of temperate trees (e.g. Kramer 1994; Chuine et al. 1999).

The reasons causing the year to year variation of the flowering intensity are less known. Yet the intensity may fluctuate from year to year by more than an order of magnitude. The variation is irregular but synchronised over large geographic areas (Koenig and Knops 2000). Accumulated heat sum of previous summer seems to correlate positively with flowering intensity, which is also physiologically reasonable as the flower buds are developed during the previous summer. However, even after taking this into account the variability of the results still remains large. There are several possible reasons for this. Firstly, environmental stress of non-meteorological origin can affect the flowering intensity. Secondly, the resource allocation within the tree varies from year to year, for example the allocation to the ripening seeds that affects the development of catkins and forces less pollen production in years following the strong flowering (Tuomi et al. 1982; Masaka & Maguchi, 2001). Thirdly, the pollination success depends heavily on stochastic environmental features, such as wind, rain and air humidity during the flowering (Peternel et al. 2005). These are hard to take into account in the phenological model and therefore they are often ignored despite potentially considerable impact on the pollen release. Finally, models describing the pollination intensity are typically based on empirical data collected with pollen traps, which are unable to distinguish between local pollen and pollen originating from more distant regions (Ranta et al. 2005).

5.3.2 Local- and Regional-Scale Aspects of Pollen Seasons

Climate is a major determinant of regional seasons of allergenic pollen. In boreal and temperate vegetation zones cold period limits the pollination schedule of plants to certain months, while in warmer climate allergenic pollen may, in addition to periodical peaks, be present in the air all year round (Fang et al. 2001; Hurtado and Alson 1990; Murray et al. 2002; Savitsky and Kobzar 1996). Arid climate may also confine the pollen season to a few months (Halwgay 1994). More specificities arise from edaphic factors, land-use and cultivation, which all affect the distributions and productivity of allergy plants. Extensive geological formations, such as mountains, have their own unique vegetation and pollen flora. Also urban settlements with their characteristic temperature, land-use, high CO_2 concentration and light environment are specific habitats for allergy plants (Ziska et al. 2004).

In boreal and temperate zones (after Allaby 2002) the local pollen season is created by three groups of wind-pollinated plants with different seasonality of pollination: (1) early-flowering trees and shrubs, (2) annual and perennial herbaceous plants pollinating during late summer/early fall near the end of the growing season, (3) herbaceous plants emitting pollen during several months or throughout the growing season (Edmonds 1979).

The first group – trees and shrubs flowering at the beginning of growing season or even in winter – start the season of allergenic pollen. Many tree genera with circumpolar distribution, such as birch (*Betula*), alder (*Alnus*) hazel (*Corylus*), poplar (*Populus*), oaks (*Quercus*) and mulberry (*Morus*), belong to this group. Hickory and Walnut are important spring-time allergenic plants in North America. Taxa belonging to Cupressaceae release pollen already during winter months in the Mediterranean region and in the Southern Great Plains, USA (*Juniperus*), as does the most important allergy plant in Japan, sugi (*Cryptomeria*). Olive (*Olea*) is an important allergy plants in springtime in the Mediterranean basin.

The duration and magnitude of pollen liberation of different species vary, but at a local scale a tree population may, under favourable conditions, release pollen during a relatively short period of time, resulting in sharp peak of high concentrations of airborne pollen. Because many species in this group represent the common regional forest, ornamental or cultivated trees, pollen concentrations may be very high during these peaks (Fig. 5.4).

The second group – annual and perennial herbaceous plants pollinating during late summer and autumn – contain two main allergenic taxa: ragweed (*Ambrosia*) and mugwort (*Artemisia*). Ragweed is the etiological agent in half of pollinosis cases in the US and its significance is increasing in many areas. Most ragweed species originate from North America. During last 80 years they have distributed over Australia and Japan, western parts of South America and Europe, limiting to south Scandinavia and Asian parts of Russia (Makra et al. 2005). Outside their natural habitats (Hall 1994), both ragweed and mugwort are weeds, and readily invade habitats recently disturbed by man, for example wastelands, abandoned fields and road sides, which brings them near the human population.

At a local scale ragweed and mugwort release pollen during several weeks of time per year (Fig. 5.4). Counts of airborne pollen are usually lower than those of common forest trees, but, for example, in Hungary where ragweed is abundant, daily average of 1,000 grains m^{-3} has been exceeded several times (Makra et al. 2005). It must be also noted that the concentrations of mugwort pollen may

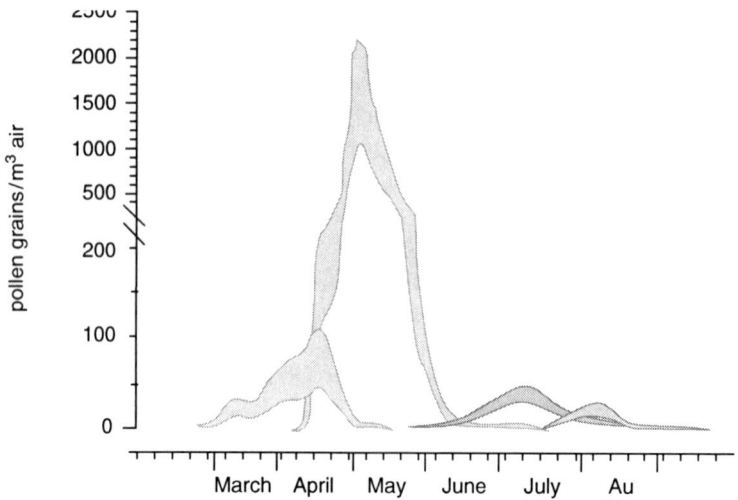

Fig. 5.4 The seasonal development of airborne pollen counts (95% probability) in South Finland, boreal zone of Europe; yellow = alder, green = birch, red = grasses and blue = mugwort

be underestimated at ground level, since the pollen grains of this species are heavy and do not readily rise to the position of the elevated pollen traps.

The third group is somewhat artificial and includes taxonomic groups, such as grasses (Poaceae), goosefoot (*Chenopodium*) and pigweed (*Amaranthus*), with many cross reacting species flowering in different times during the growing season. Human activities greatly affect the distributions and abundance of many of these species: important allergic genera as fescue (*Festuca*), rye grass (*Lolium*), blue grass (*Poa*), Bermuda grass (*Cynodon*), bent (*Agrostis*) are pasture grasses and/or cultivated as forage or lawn. Many of them, like orchard grass (*Dactylis glomerata*) and timothy (*Phleum pratense*) have been introduced from Europe and naturalised in North America and other continents.

Other examples of allergenic taxa releasing pollen during the whole growing season are pellitory (*Parietaria*) predominantly growing in the Mediterranean climate but found up to the UK, and cosmopolitans goosefoot (*Chenopodium*) and pigweed (*Amaranthus*). The latter are noxious weeds and invade recently disturbed soils. Both genera contain several cross-allergenic species flowering during different times of the growing season (Edmonds 1979; Fang et al. 2001; Hall 1994; Hurtado and Alson 1990). Pellitory-of-the-wall (*Parietaria judaica*) predominantly grows close to human habitats, in wall fissures and disturbed soils. Individual plants consist of shoots arising from common root-stock. Although most shoots flower during spring, some individuals have flowering shoots nearly a year round, which makes the pollen season practically continuous in some regions (Guardia and Belmonte 2004).

Many grass species are cross reacting, so even if individual species flower during a certain time period (specific for the particular taxa), the overall pollen season of grasses lasts several months (Zanotti and Puppi 2000). In East Australia and other southern hemisphere regions with similar climate such season lasts over 200 days (Green et al. 2004), while in temperate climate the period of high pollen concentrations is normally confined to a couple of months (WHO 2003). During the main season, counts over 1,000 grains m^{-3} of airborne grass pollen can be exceeded in some areas, UK for example, in peak season (Emberlin et al. 1999).

Pollen allergens are present not only in grains themselves but practically in all parts of the plant, especially when the pollination time approaches. The allergens are found, for instance, in small-sized (1–4 μm) particles called orbicules (El-Ghazaly et al. 1995). The orbicules (or Ubisch bodies) are sporopollenin granules stored together with pollen in anther lobes and often released before the liberation of the pollen itself. These small-size particles contain the same allergens as pollen and may penetrate deep into human airways, down to the alveoli triggering the allergic reactions (El-Ghazaly et al. 1995). Concentration of such small-size allergenic particles in the air can increase up to 2 weeks before the main pollen period, which may explain the early allergic reactions occurring a few days prior to flowering (Matikainen and Rantio-Lehtimäki 1998). Allergens are also present e.g. in birch leaves and in the sticky exudates surrounding bursting leaf buds and thus may cause problems to allergic people when branches are used as decoration indoors. The presence of allergens in grass leaves may also be a cause of allergy symptoms that may occur in connection with lawn mowing, although this does not seem to be thoroughly investigated.

Many countries have national networks of pollen monitoring stations. Currently available forecasts of pollinating season are mainly based on local observations together with information about the weather, growing seasons and typical flowering times of the local plants. Many regional or site-specific forecasting models based on regression of empirical pollen data sets against weather variables have been published in recent years (WHO 2003, numerous local regression models for trees, grasses, etc.). In particular, in Austria the local predictions for the onset of pollen seasons of alder, hazel and birch are based on thermal phenological models (http://www.polleninfo.org). The National Pollen Research Unit, UK, provides regional pollen forecasts for several timescales and characteristics of the pollen season. The information on season start, expected severity and duration is released a month in advance. Medium range (5 days ahead) and short-range forecasts are used to predict day-to-day variation of concentrations once the season started. The techniques are based on regression of empirical pollen data against weather variables, as well as on local vegetation information (Emberlin et al. 1999; Adams-Groom et al. 2002). EU project A.S.T.H.M.A was an attempt to create regional service of short-term forecasts of pollen concentrations and health risks (http://www.enviport.com). In Denmark, a statistical model based on empirical pollen data and weather parameters is used to estimate the pollen amount of birch (*Betula*), grasses (*Poaceae*) and mugwort (*Artemisia*) for up to 48 h ahead (Janne Sommer, Danish Asthma and Allergy Association, personal communication). In Tulsa (Oklahoma, USA) a trajectory model is being used to predict the regional scale transport of *Juniperus ashei* pollen (http://pollen.utulsa.edu).

5.4 Impact of Long-Range Pollen Transport to Allergenic Episodes: Forecasting Possibilities

The above-outlined main mechanisms that control the plant behavior at a specific place lead to significantly different regional phenological calendars dependent on specifics of regional climate and vegetation. Some key features of these processes can be simulated using semi-empirical models, such as Thermal-Time, thus allowing the forecasting of pollen seasons using meteorological forecasts and local plant observations. The current section will discuss the process that challenges such regionalization and provides large-scale links between different parts of continents: a long-range transport of genetic and allergenic material released during the flowering.

A typical shape of observed pollen concentrations in air during the spring season is shown in Fig. 5.4. One can distinguish the main peak representing the local flowering season with well-seen start and end times, which is surrounded by smaller-scale increases of counts a few days before and after the main rise of concentrations. In some places these "tails" turn into two peaks comparable with the main one. These early and late blows of pollen often originate from long-distance transport. The evidence of such phenomenon is quickly growing and obtained at nearly all climatic zones (Corden et al. 2002; Latalowa et al. 2002; Hjelmroos 1992; Damialis et al. 2004; Rantio-Lehtimäki 1994). However, from the point of view of atmospheric science and air pollution, pollen could not be expected to be dispersed farther than a few tens of kilometres in the atmosphere: the grains are too big (tens of micrometres in diameter) and thus ought to be deposited much too fast to be a large-scale pollutant. According to atmospheric science, a so-called Junge size spectrum of aerosol with long air lifetime is confined between 0.1 and 1 µm.

A more detailed theoretical consideration (Sofiev et al. 2006) showed, however, that at least some types of pollen are susceptible for distribution with air masses for a distance of several hundreds or thousands kilometers, which is well sufficient for transport between climatic zones. These conclusions can be extended by suggesting that the allergenic material can be emitted from the pollen grain during its active period. There are also suggestions that anthropogenic pollution can escalate such process (Majd et al. 2004; Motta et al. 2006). As stated in the previous section, this material consists of organic particles with characteristic size close to the "classical" fine-particle range 0.10–1 µm, which are known to travel over large scales.

The four most-evident consequences of the long-range pollen transport are: (i) earlier start of pollen seasons in northern regions where the local flowering is late in comparison with (remote) neighbors (Fig. 5.5); (ii) a delay of the end of the season in central and southern regions where the local flowering ends early (Fig. 5.6); (iii) an increase of the actual concentrations during the main season, especially in regions with weak or moderate flowering intensity (this is not observable directly, and requires detailed modelling to highlight); (iv) a fast (in a scale of days) transport of genetic material across the continents, climate and vegetation zones.

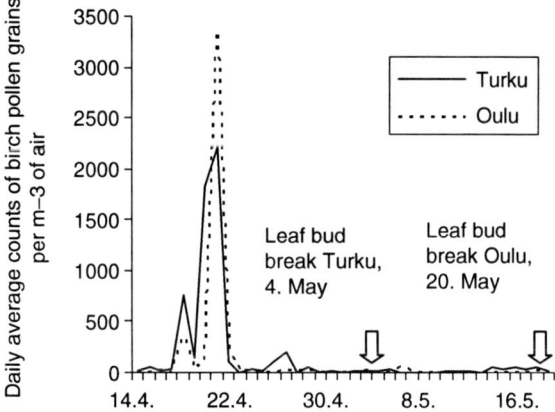

Fig. 5.5 An example of a strong early-spring pollen episode in central Finland caused by the long-range transport from central Europe in 1999 (Sofiev et al. 2006)

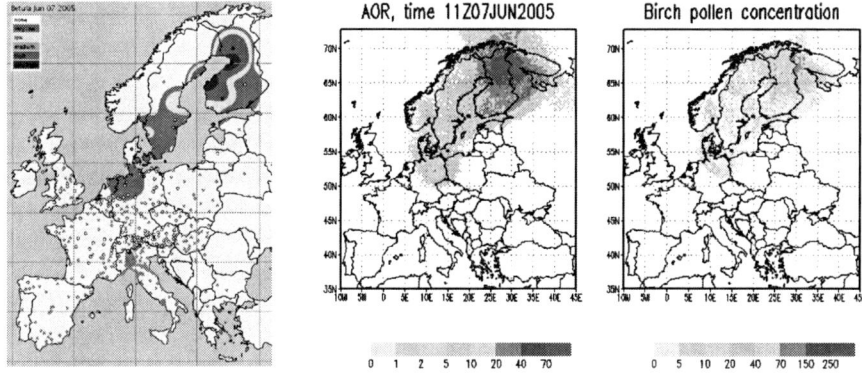

Fig. 5.6 An example of north-to-south transport of pollen from Finnish Lapland to Denmark after the end of the season in the south. *Left panel*: observed by European Aeroallergen Network sites (courtesy of S. Jaeger), *right-hand panel*: probability (Area Of Risk) and pollen concentrations modelled by SILAM model, http://pollen.fmi.fi

An analysis of the phenomenon and its forecasting requires the involvement of atmospheric dispersion models in combination with tools for the prediction of pollen emissions over large territories, such as continents. A principal scheme of such a system is shown in Fig. 5.7.

From the point of view of atmospheric science, pollen is a specific aerosol and simulation of its transport does not pose principal difficulties, providing that existing dispersion models are applicable. These models assume that the transported species follow the path of surrounding air, including small turbulent eddies, and do not pose any feedback to the atmospheric flows. As shown by Sofiev et al. (2006), these assumptions are fulfilled for birch pollen (a comparatively small grain), but are not necessarily correct for larger particles because the neglected terms

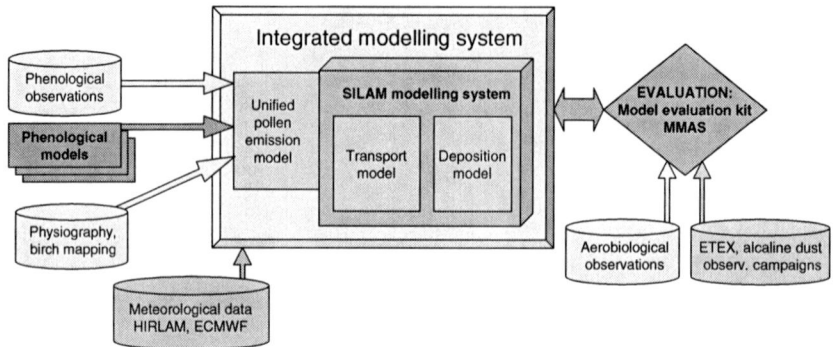

Fig. 5.7 A scheme of analysis and forecasting system for evaluation of pollen long-range transport (being developed in Finland, http://pollen.fmi.fi)

in the transport and deposition equations might no longer be small. The inertial penetration of such grains through low-turbulence layer can be taken into account via a classical parameterization described by Seinfeld and Pandis (2006), chapter 19.4.2. However, it would improve only the estimation of dry deposition, still assuming that the grain follows the main air flows far from the surface. Additional complexity can be introduced by including the aerodynamic shape of some grains. The presence of wings or strong asymmetry may lead to detachment of the grain from surrounding air. For such pollen, the existing dispersion models will manifest large errors in deposition intensity and, possibly, miscompute the diffusion. Therefore, the usage of existing models for each specific taxon requires the explicit applicability checking.

It is known that pollen grains loose and gain water depending on air humidity and temperature, but the corresponding processes are quite poorly known and so far none of working systems takes these into account.

The most important and the most difficult challenge, however, is the construction of the pollen emission model. As seen from Fig. 5.5 and also pointed by e.g. Damialis (2004), Rantio-Lehtimäki (1994), Sofiev et al. (2006) and Siljamo et al. (2004), the long-range transport episode usually lasts for a couple of days and entirely depends on meteorological conditions. Flowering start time varies between the neighboring climatic zones for just a few days. Therefore, in order to catch the long-range transport episode the emission model has to forecast the emission timing with accuracy of 1–2 days – homogeneous over large areas. Unfortunately, even local-scale phenological models have the standard deviations of predicted flowering time about 4–5 days and therefore cannot serve as a sole source of emission information for deterministic pollen forecasts.

There can be two ways to cope with the problem of high uncertainty of pollen emission: (i) to use probabilistic forecasts via ensemble simulations or straightforwardly computed probability distributions, and (ii) to utilize additional information via data assimilation mechanisms to adjust the phenological models "on-the-fly".

Three most-evident data sources for the assimilation are the near-real-time aerobiological and phenological observations and satellite images.

Probabilistic large-scale simulations and forecasting have been used in Finland for several years with positive outcome (see http://pollen.fmi.fi, Sofiev et al. 2006). The model estimated both absolute pollen concentrations over Europe and the territories affected by each of six major birch forest areas (probability distributions). The concentration estimate relies on an European-scale phenological emission model, while the second one does not use it at all – just a map of birch forests. The combination of these two characteristics appeared to be very useful for the allergy forecasting as it (i) shows a risk of getting pollen from major remote sources, (ii) indicates the forests affecting the receptor region, and (iii) estimates an absolute concentration of pollen grains in air using some emission model with known formulations and accuracy. The next level of sophistication of this methodology is to build a probabilistic emission model that would include the above-mentioned uncertainty as a feature of the flux probability distribution.

Adjustment of the emission module using independent observations are proved to be useful but also has certain limitations. In particular, near-real-time phenological observations currently do not exist. Many countries have very dense networks (Fig. 5.8) but the data collection is manual, as well as the processing and compilation

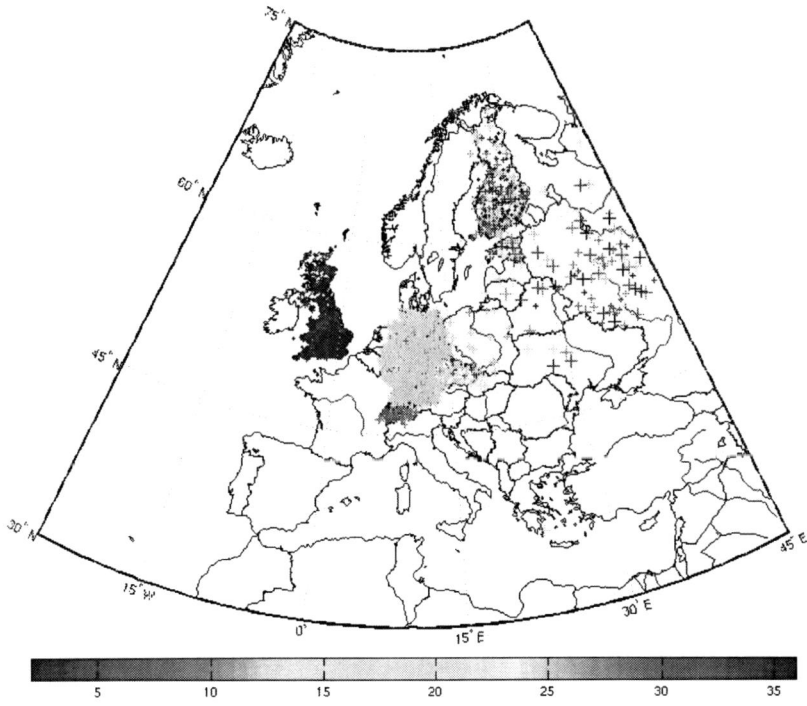

Fig. 5.8 Combined map of 15 national phenological networks in Europe. Colour denotes the number of observed years by each station

of national databases. There is no European-wide regular phenological database; the one presented in Fig. 5.8 was compiled for birch at Finnish Meteorological Institute and is owed to courtesy of the national phenological networks listed in Table 5.1.

Pollen counts are the only data partly available with a delay of less than 2 days (http://www.polleninfo.org). However, such observations do not provide information about the pollen source and thus cannot distinguish between the local and long-range transported pollen. Therefore, this information can be used for two purposes: now-casting the pollen concentrations in the area surrounding the trap and forecasting the counts downwind. Size of the area where a specific pollen trap is representative varies strongly depending on local conditions and microclimate. Usually, however, it extends over a few tens of km – for daily pollen concentrations.

A promising set of information emerges from the growing satellite fleet. Several modern methodologies for evaluating the state of vegetation with very high spatial and temporal resolutions (250 m and 15 min, respectively) are expected or already

Table 5.1 National phenological networks contributed to the database presented in Fig. 5.8

Country	No. of stations	Years available at least for some stations	Species	Data handler/ provider
Belarus	5	1967–1998, 2002–2005	Betula	Minsk aerobiology group
Czech Republic	206	1955–2004, few older	Betula pendula	Czech Hydrometeorological Inst.
Estonia	19	1947–2003	Betula	EMHI
Finland	586	1997–2005, 1800s–2004	Betula pendula, Betula pubescens	Finnish Forest Research Institute METLA
Germany	2,119	1985–2004	Betula pendula	DWD
Latvia	2	1958–1993	Betula	University of Tartu
Lithuania	3	1962–1996	Betula	University of Tartu
Norway	1	1927–2004	Betula pubescens	Biofork Nord
Poland	20	1980–1992, 2005	Betula pendula	Institute of Meteorology and water management
Russia	89	1951–1998	Betula	University of Tartu, Moscow State University
Slovakia	4	1986–2004	Betula pendula	Slovak Hydromet. Inst.
Spain	4	2002–2003	Betula pendula, Betula alba	University of Vigo
Switzerland	138	1996–2004	Betula pendula	Meteoswiss
Ukraine	5	1951–1998	Betula	Moscow State University
United Kingdom	3,414	1999–2004	Betula pendula	UK Phenological Network
Total	6,615			

available for users (http://modis.gsfc.nasa.gov, http://www.esa.int/msg/pag4.html, http://www.eumetsat.int). These products allow various approaches for evaluation of phenological stages, such as NDVI (Normalized-Difference Vegetation Index), micro-channel based evaluation of leaf area indices, etc. (Manninen et al. 2006; Høgda et al. 2002). However, all these tools have a significant limitation: the flowering itself is not visible from the space. The observable set of parameters describes the status of leaves, their size, area, color, etc. For some species the moment of e.g. bud burst correlates well with the flowering (with a possible shift for a few days), while for the others this dependence is either weak or overshadowed by other species opening their leaves earlier (the satellite monitors evidently see the unfolding of leaves of the earliest taxa). Therefore, the satellite-based correction of predicted flowering timing, being otherwise highly accurate and timely, is available not for all plants and not at all for herbaceous plants.

To conclude the list of challenges on the way of pollen large-scale forecasting, it is worth recalling that absolute amount of pollen released during flowering is largely determined by the previous-year conditions during the active vegetation period and, to a less extent, by winter conditions. This is an entirely new feature for atmospheric dispersion modelling: usually such models have to "remember" the situation for just a few hours to the past. Pollen forecasting model has to keep (albeit, in an aggregated form) the information over the past year.

The above-outlined complexity resulted in existence of only few modelling systems approaching the pollen forecast problem. Two of them working at local scale are mentioned above – the A.S.T.H.M.A system in S. Europe and University of Tulsa in the USA. To our knowledge, the only system approaching the European scale exists in Finland (see example of its results in Fig. 5.6). Similar models are being developed in Denmark, France, the US and other countries.

The usefulness of such systems for the purpose of the human adaptation to the pollen seasons strongly depends on the forecast horizon: a warning made few hours before the episode may be too late for preparatory actions. Based on existing statistics of application meteorological models, the high-quality detailed forecast is usually possible for a period of about 2 days, while the main trends can be computed almost a week in advance. Providing that the pollen emission is predicted accurately enough, the allergenic forecasting can be made for a similar period.

5.5 Perspectives of Changing Climate: Most-Evident and Probable Impact and Means of Adaptation

As shown in the previous sections, the climate is a major determinant for the phenology, distribution and productivity of plants, as well as for the distribution of the released pollen. With climate change, the margins of species ranges, their relative abundances and boundaries may start to shift (Walther 2003), which will result in changes in local allergenic pollen spectra. For example Finland will probably experience an annual mean temperature increase of 3–5°C during the next

100 years (Carter et al. 2005) and the CO_2 concentration increase of 0.5–2% per decade (Vingarzan 2004). Forest resources are expected to increase throughout the country, and at the same time, the proportion of coniferous trees will be reduced with an increased dominance of birches (Kellomäki et al. 2001). Likewise, eastern cottonwood (*Populus deltoides*) that is a fast-growing broad-leaved tree, able to respond rapidly to changing environment, is expected to benefit from rising CO_2 concentration (Gielen and Ceulemans 2001).

A meta-analysis of reports on 79 crop and wild plant species indicated that CO_2 enrichment resulted in 19% more flowers, 18% more fruits and 16% more seeds (Jablonski et al. 2002). Studies with several species have indicated that increase in ambient growth temperature may accelerate onset of flowering (Blázquez et al. 2003). Stinson and Bazzaz (2006) observed that elevated CO_2 stimulated significantly the stand-level reproduction of common ragweed and minimized the differences in the reproductive output of large and small plants.

A number of recent changes in pollen patterns have been reported regarding onset (earlier) and duration (longer) of the pollen season (Clot 2001, 2003; Emberlin et al. 1997, 2002; Frenguelli et al. 2002; Galán et al. 2005; Inoue et al. 2002; Newnham 1999; Tedeschini et al. 2006), and also the airborne pollen load (Bortenschlager and Bortenschlager 2005; Corden and Millington 1999; Frei 1998; Frei and Leuschner 2000; Jäger et al. 1996; Rasmussen 2002; Spieksma et al. 2003). Most of these changes have been attributed to the increased levels of air pollutants or the higher temperatures associated with global warming, while some others have been attributed to urbanisation and land use change (Voltolini et al. 2000). Nevertheless, the ever increasing eutrophication of the planet due to increased CO_2 or nitrogen concentrations and its effect on pollen production (Wayne et al. 2002; Townsend et al. 2003) should also be taken into consideration.

Several reports suggest that the synchronous increase in pollen abundance or pollen season length and in respiratory allergy severity or prevalence are to some extent interdependent, the latter being potentially the result of the first (Ault 2004; Clot 2003; Frei 1998; Frei and Leuschner 2000; Voltolini et al. 2000; Wayne et al. 2002; Ziska et al. 2003; Menzel et al. 2006). The number of taxa and covered area vary between the studies. Several reports have concentrated on a selected taxon (Clot 2001; Corden and Millington 1999; Emberlin et al. 1997, 2002; Frenguelli et al. 2002; Galán et al. 2005; Inoue et al. 2002; Orlandi et al. 2005; Osborne et al. 2000; Rasmussen 2002; Tedeschini et al. 2006). Others considered up to six (Bortenschlager and Bortenschlager 2005; Frei 1998; Frei and Leuschner 2000; Jäger et al. 1996; Voltolini et al. 2000; Menzel et al. 2006), for the territory covering up to a whole of Europe. However, Beggs (2004) has argued for the inclusion of more taxa in pollen studies of each study area so as to quantify how global the observed trends are, or how intense is the regional factor.

Shifts to an earlier onset of the pollen season have been more frequently observed than increases in airborne pollen abundance (Clot 2001, 2003; Emberlin et al. 1997, 2002; Frenguelli et al. 2002; Galán et al. 2005; Inoue et al. 2002; Orlandi et al. 2005; Osborne et al. 2000; Tedeschini et al. 2006). For other pollen variables, the picture is more complicated. Often no trends are observed (Frenguelli et al. 2002; Leuschner et al. 2000) and when they are, they tend not to be consistent across different taxa and regions (Spieksma et al. 2003; Tedeschini et al. 2006; Voltolini et al. 2000). Overall more attention has been paid to pollen distribution patterns than to pollen abundance.

Clot (2003) and Damialis et al. (2007) have examined a wider sector of the regional pollen flora (more than 15 plant taxa) concentrating on the onset and length of pollen season, together with the atmospheric pollen loads. Clot (2003) found mainly earlier onset of pollen season in Neuchatel, Switzerland, whereas Damialis et al. (2007) found mainly increasing trends in atmospheric pollen levels in Thessaloniki, Greece, with no systematic shift in the onset of the pollen season.

Differences in regional responses to climatic change are related to differing geographical location (viz. latitude) or to specific regional characteristics (i.e. local climate). The Mediterranean environment is characterised by an annual alternation of a hot and dry period with a cold and moist one. Species living in this environment are adapted to large changes of environmental factors, among which is air temperature. Under such strongly varying conditions, systems have developed high resilience and species are able to survive environmental fluctuations (Dell et al. 1986). The increasing trends of annual pollen levels in Thessaloniki (Greece) could be interpreted in terms of increased daily atmospheric pollen concentrations deriving from a higher reproductive effort of various plant species in response to a changing environment. This is supported by the fact that the peak daily count also showed an increasing trend for a large number of taxa and that the two (annual load and daily peak of pollen counts) were interrelated (Damialis et al. 2007).

Such results, showing strong upward trends in annual pollen abundance that reflect increased levels of pollen production, but much fewer significant changes in phenological characters, have not been reported by other researchers. However, Menzel (2000) and Menzel and Fabian (1999), who studied flowering phenology of plant species across Europe, concluded that due to regional characteristics and in contrast to what is observed in western and northern Europe, no shifts to an earlier onset of flowering are observed in the Balkan peninsula, where Thessaloniki is located. The fact that Damialis et al. (2007) did observe trends in pollen production patterns suggests existence of a major factor leading to responses that can mask between-species variability in the temporal features of their reproductive output.

Increased reproductive effort, in terms of flower and/or pollen production, in response to higher temperatures is widely documented; this holds true for both animal and wind pollinated species (see, for example, Aerts et al. 2004; Mølgaard and Christensen 1997; Stenström and Jónsdóttir 1997; Wan et al. 2002; Ziska et al. 2003). In the example of Greece, minimum air temperature presented an increase of the order of 1.0°C in 2005, compared to that in 1987. In this timescale, airborne pollen levels have displayed remarkable changes that suggest exponential increase. This is particularly true for Cupressaceae, *Carpinus*, *Quercus*. Nevertheless, a few taxa did not follow this pattern and either displayed decreasing trends (*Populus*) or no trends at all (i.e. *Ambrosia*). Regarding *Populus*, which is the only taxon with a negative trend in airborne pollen levels over the last 2 decades, we could remark that *Populus* species have high demands in water (Kailidis 1991). Increasing air temperatures influence water availability, as they influence evapotranspiration rates. For a most sensitive to water-shortage species, rising temperatures might result into a stress situation that could lead to reduced reproductive effort. In the case of *Ambrosia*, its low participation in the regional vegetation may be responsible for the inability to detect trends (Damialis et al. 2007). In the absence of commensurate changes in the abundance of the respective flora, Damialis et al. (2007) argued that changes of airborne-pollen load over the last 20 years in the area of Thessaloniki could be interpreted in terms of the concurrent temperature increase, without ignoring the eutrophication of the planet as an alternative or additional cause. The study has not revealed the influence of the other strong meteorological driver in the Mediterranean region – precipitation amount – onto the pollen abundance. The main reason for that is believed to be a weak and irregular trend of this parameter during the considered period. A significant decreasing trend, however, if projected for this region in the recent IPCC (2007) report (Intergovernmental Panel on Climate Change, http://www.ipcc.ch).

Climate change may also facilitate the spread and naturalization of some allergy plants in new areas. Common ragweed has its northern limits in the south of Sweden, but because the growing season is usually too short for the seeds to mature, the establishment of flowering plants require recurrent import of contaminated birdseed, for instance (Dahl et al. 1999). The establishment of flowering populations of ragweed in new areas increases also the risk of long-distance transport of its allergenic pollen to new areas (Dahl et al. 2000; Lorenzo et al. 2006). Some processes seem to be gaining the stream: in summer 2005 considerable amounts ragweed pollen originated from Eastern Europe was observed during 4 days in sub-arctic Scandinavia. There is also a possibility that the species not too sensitive to light limitations, such as pellitory-of-the-wall and other Urticaceae-plants, as well as some grass species, may benefit from longer warm season and extend their distribution towards the north resulting in longer pollen period.

As a result, a combination of the environmental changes and atmospheric pollution may in the near future considerably affect loads of airborne allergens via changes in vegetation structure, magnitude of flowering and allergen content of pollen (Singer et al. 2005; WHO 2003; Williams 2005). If such changes will be realised, the only way to cope would be an adaptation to the changing environment and, in particular, to longer and possibly more diverse flowering seasons. Adaptation The adaptation of people is possible via a set of pre-emptive measures of behavioural and medical kinds, which both require timely and detailed forecasts of forthcoming pollinating seasons. As shown by the existing systems, such forecasting is possible and useful. However, it heavily relies on many empirical models and statistical relationships, which may appear to be wrong in the future climate. Therefore, the new systems should be based on more universal parameterizations and include possibility for adjustment of the model setup and state using real-time observations.

Phenological and aerobiological data, reflecting the behaviour of various ecosystems, from regions differing in latitude or altitude and belonging into different climatic zones, have far-reaching importance. They can serve in predicting future patterns, in interpreting fossil pollen patterns, but they can also enable us to anticipate associated human health impacts and take appropriate measures, given that the prevalence and severity of respiratory allergies have been significantly increasing worldwide and several reports link them to increases in the pollen abundance or pollen season length. Taxa like *Carpinus*, *Quercus*, Cupressaceae or *Populus* that show trends might serve as bio-indicators of the expected climatic change in the Mediterranean region and possibly in other climatic zones as well.

References

Aerts R, Cornelissen JHC, Dorrepaal E, van Logtestijn RSP, Callaghan TV (2004) Effects of experimentally imposed climate scenarios on flowering phenology and flower production of subarctic bog species. *Glob Change Biol* **10**:1599–1609.

Adams-Groom B, Emberlin J, Corden J, Millington W, Mullins J (2002) Predicting the start of the birch pollen season at London, Derby and Cardiff, United Kingdom, using a multiple regression model, based on data from 1987 to 1997. *Aerobiologia* **18**:117–123.

Al-Dowaisan A, Fakim N, Khan MR, Arifhodzic N, Panicker R, Hanoon A (2004) Salsola pollen as a predominant cause of respiratory allergies in Kuwait. *Ann Allergy Asthma Immunol* Feb; **92**(2):262–267.

Allaby M (2002) Encyclopedia of Weather and Climate. New York: Facts On File.

Anto JM, Sunyer J (1997) Thunderstorms: a risk factor for asthma attacks [editorial; comment]. *Thorax* **52**(8):669–670.

Ault A (2004) Report blames global warming for rising asthma. *Lancet* **363**:1532.

Baldo BA, Panzani RC, Bass D, Zerboni R (1992) Olive (Olea europea) and privet (Ligustrum vulgare) pollen allergens. Identification and cross-reactivity with grass pollen proteins. *Mol Immunol* **29**(10):1209–1218.

Batanero E, Villalba M, Ledesma A, Puente XS, Rodriguez R (1996) Ole e 3, an olive-tree allergen, belongs to a widespread family of pollen proteins. *Eur J Biochem* **241**(3):772–778.

Bauchau V, Durham S (2001) Prevalence and rate of diagnosis of allergic rhinitis in Europe. *Eur Respir J* **24**:758–764.

Bauman A (1996) Asthma associated with thunderstorms [editorial; comment]. *Bmj*. **312**(7031):590–591.

Beggs PJ (2004) Impacts of climate change on aeroallergens: past and future. *Clin Exp Allergy* **34**:1507–1513.

Bellomo R, Gigliotti P, Treloar A, Holmes P, Suphioglu C, Singh MB et al. (1992) Two consecutive thunderstorm associated epidemics of asthma in the city of Melbourne. The possible role of rye grass pollen. *Med J Aust* **156**(12):834–837.

Blázquez MA, Ahn JH and Weigel D (2003) A thermosensory pathway controlling flowering time in Arabidopsis thaliana. *Nat Genet* **33**:168–171.

Bortenschlager S, Bortenschlager I (2005) Altering airborne pollen concentrations due to the Global Warming. A comparative analysis of airborne pollen records from Innsbruck and Obergurgl (Austria) for the period 1980–2001. *Grana* **44**:172–180.

Bousquet J, van Cauwenberge O, Khaltaev N (2001) Allergic rhinitis and its impact on asthma. *J Allergy Clin Immunol* **108**:S147–334.

Bousquet J, Bieber T, Fokkens W (2006) Themes in allergy. Editorial. *Allergy* **61**:1–2.

Bousquet J, Guerin B, Hewitt B, Lim S, Michel FB (1985) Allergy in the Mediterranean area. III: Cross reactivity among Oleaceae pollens. *Clin Allergy* **15**(5):439–448.

Bousquet J, Hewitt B, Guerin B, Dhivert H, Michel FB (1986) Allergy in the Mediterranean area. II: Cross-allergenicity among Urticaceae pollens (Parietaria and Urtica). *Clin Allergy* **16**(1):57–64.

Bucholtz GA, Lockey RF, Serbousek D (1985) Bald cypress tree (Taxodium distichum) pollen, an allergen. *Ann Allergy* **55**(6):805–810.

Campbell ID, McDonald K, Flannigan MD, Kringayark J (1999) Long-distance transport of pollen into the Arctic. *Nature* **399**:29–30.

Carter TR, Jylhä K, Perrels A, Fronzek S, Kankaanpää S (2005) FINADAPT scenarios for the 21st century: alternative futures for considering adaptation to climate change in Finland. FINADAPT Working Paper 2, Finnish Environment Institute Mimeographs 332, Helsinki, 42 pp.

Chabot BF, Hicks DJ (1982) The ecology of leaf life spans. *Ann Rev Ecol Syst* **13**:229–259.

Charpin D, Calleja M, Lahoz C, Pichot C, Waisel Y (2005) Allergy to cypress pollen. *Allergy* Mar; **60**(3):293–301.

Chuine I, Cour P, Rousseau DD (1999) Selecting models to predict the timing of flowering of temperate trees: implications for three phenology modelling. *Plant Cell Environ* **22**:1–13

Clot B (2001) Airborne birch pollen in Neuchâtel (Switzerland): onset, peak and daily patterns. *Aerobiologia* **17**:25–29.

Clot B (2003) Trends in airborne pollen: An overview of 21 years of data in Neuchâtel (Switzerland). *Aerobiologia* **19**:227–234.

Corbi AL, Cortes C, Bousquet J, Basomba A, Cistero A, Garcia-Selles J et al. (1985) Allergenic cross-reactivity among pollens of Urticaceae. *Int Arch Allergy Appl Immunol* **77**(4):377–383.

Corden J, Millington W (1999) A study of *Quercus* pollen in the Derby area, UK. *Aerobiologia* **15**:29–37.

Corden JM, Stach A, Milligton W (2002) A comparison of *Betula* pollen season at two European sites; Derby, United Kingdom and Poznan, Poland (1995–1999). *Aerobiologia* **18**:54–53.

Coville FV (1920) The influence of cold in stimulating the growth of plants. *J Agric Res* **20**:151–160.

Dahl Å, Strandhede S-O, Wihl J-Å (1999) Ragweed – An allergy risk in Sweden ? *Aerobiologia* **15**:293–297.

D'Amato G, Spieksma FT, Liccardi G, Jager S, Russo M, Kontou-Fili K et al. (1998) Pollen-related allergy in Europe. *Allergy* **53**(6):567–578.

Damialis A, Gioulekas D, Lazopoulou C, Balafoutis C, Vokou D (2005) Transport of airborne pollen into the city of Thessaloniki: the effects of wind direction, speed and persistence. *Int J Biometeorol* **49**:139–145. DOI: 10.1007/s00484-004-0229-z.

Damialis A, Halley JM, Gioulekas D, Vokou D (2007) Long-term trends in atmospheric pollen levels in the city of Thessaloniki, Greece. *Atmosph Environ* **41**:7011–7021. DOI: 10.1016/j.atmosenv.2007.05.009.

Davies JM, Bright ML, Rolland JM, O'Hehir RE (2005) Bahia grass pollen specific IgE is common in seasonal rhinitis patients but has limited cross-reactivity with Ryegrass. *Allergy* Feb; **60**(2):251–255.

Dell B, Hopkins AJM, Lamont BB (1986) Introduction, in: Lieth, H. and Mooney, H.A. (Eds.), *Tasks for Vegetation Science*, vol. 16, Junk Publishers, Dordrecht/Boston, MA/Lancaster, pp. 1–4.

Deen W, Hunt T, Swanton CJ (1998) Influence of temperature, photoperiod, and irradiance and the phenological development of common ragweed (Ambrosia artemisiifolia). *Weed Sci* **46**:555–560.

EC Survey (1996) Variations in the prevalence of respiratory symptoms, self-reported asthma attacks, and use of asthma medication in the European Community Respiratory Health Survey (ECRHS). *Eur Respir J* **9**(4):687–695.

Ekebom A, Nilsson S, Saar M, van Hage-Hamsten, M (1997) A comparative study of airborne pollen concentrations of three allergenic types in Tartu (Estonia), Roma/Gotland/Stockholm (Sweden) 1990–1996. *Grana* **36**:366–372.

Edmonds RL (1979) *Aerobiology. The Ecological Systems Approach.* Dowden, Hutchinson & Ross, Stroudsburg, Pennsylvania.

ElGhazaly G, Takahashi Y, Nilsson S, Grafstrom E, Berggren B (1995) Orbicules in Betula pendula and their possible role in allergy. *Grana* **34**:300–304.

Emberlin J, Mullins J, Corden J, Millington W, Brooke M, Savage M, Jones S (1997) The trend to earlier birch pollen seasons in the UK: A biotic response to changes in weather conditions? *Grana* **36**:29–33.

Emberlin J, Mullins J, Corden J, Millington W, Brooke M, Savage M (1999) Regional variations in grass pollen seasons in the UK. Long term trends and forecast models. *Clin Exp Allergy* **29**:347–357.

Emberlin J, Detandt M, Gehrig R, Jäger S, Nolard N, Rantio-Lehtimäki A (2002) Responses in the start of *Betula* (birch) pollen seasons to recent changes in spring temperatures across Europe. *Int J Biometeorol* **46**:159–170.

Erdtman, G. 1937. Pollen grains recorded from the atmosphere over the Atlantic. *Meddel Göteb Bot Trädg* **12**:186–196.

Eriksson NE, Wihl JA, Arrendal H, Strandhede SO (1984) Tree pollen allergy. II. Sensitization to various tree pollen allergens in Sweden. A multi-centre study. *Allergy* **39**(8):610–617.

Fang R, Xie S, Wei F (2001) Pollen survey and clinical research in Yunnan, China. *Aerobiologia* **17**:165–169.

Fedorov AA, Ball T, Mahoney NM, Valenta R, Almo SC (1997) The molecular basis for allergen cross-reactivity: crystal structure and IgE-epitope mapping of birch pollen profilin. *Structure* **5**(1):33–45.

Fernandez C, Martin-Esteban M, Fiandor A, Pascual C, Lopez Serrano C, Martinez Alzamora F et al. (1993) Analysis of cross-reactivity between sunflower pollen and other pollens of the Compositae family. *J Allergy Clin Immunol* **92**(5):660–667.

Ferreira JFS, Simon JE, Janick J (1995) Developmental studies of Artemisia annua: flowering and artemisin production under greenhouse and field conditions. *Planta Medica* **61**:167–170.

Flood RG, Halloran GM (1982) Flowering behaviour of four annual grass species in relation to temperature and photoperiod. *Ann Bot* **49**:469–475.

Frei T (1998) The effects of climate change in Switzerland 1969–1996 on airborne pollen quantities from hazel, birch and grass. *Grana* **37**:172–179.

Frei T, Leuschner RM (2000) A change from grass pollen induced allergy to tree pollen induced allergy: 30 years of pollen observation in Switzerland. *Aerobiologia* **16**:407–416.

Freidhoff LR, Ehrlich-Kautzky E, Grant JH, Meyers DA, Marsh DG (1986) A study of the human immune response to Lolium perenne (rye) pollen and its components, Lol p I and Lol p II (rye I and rye II). I. Prevalence of reactivity to the allergens and correlations among skin test, IgE antibody, and IgG antibody data. *J Allergy Clin Immunol* **78**(6):1190–1201.

Frenguelli G, Tedeschini E, Veronesi F, Bricchi E (2002) Airborne pine (*Pinus* spp.) pollen in the atmosphere of Perugia (Central Italy): Behaviour of pollination in the two last decades. *Aerobiologia* **18**:223–228.

Galán C, Garcia-Mozo H, Vazquez L, Ruiz L, Diaz de la Guardia C, Trigo MM (2005) Heat requirement for the onset of the *Olea europaea* L. pollen season in several sites in Andalusia and the effect of the expected future climate change. *Int J Biometeorol* **49**:184–188.

Ganbo T, Hisamatsu K, Inoue H, Kitta Y, Nakajima M, Goto R et al. (1995) Detection of specific IgE antibodies to Japanese cypress pollen in patients with nasal allergy: a comparative study with Japanese cedar. *Auris Nasus Larynx* **22**(3):158–164.

Gielen B, Ceulemans (2001) The likely impact of rising atmospheric CO_2 on natural and managed *Populus*: a literature review. *Environ Pollut* **115**:335–358.

Global strategy for asthma management and prevention, www.ginasthma.com, 2002.

Green BJ, Dettmann M, Yli-Panula E, Rutherford S, Simpson R (2004) Atmospheric Poaceae pollen frequencies and associations with meteorological parameters in Brisbane, Australia: a 5-year record, 1994–1999, *Int J Biometeorol* **48**:172–178.

Guardia R, Belmonte J (2004) Phenology and pollen production of *Parietaria judaica* L. in Catalonia (NE Spain). *Grana* **43**:57–64.

Guerin B, Kanny G, Terrasse G, Guyot JL, Moneret-Vautrin DA (1996) Allergic rhinitis to thuja pollen. *Int Arch Allergy Immunol* **110**(1):91–94.

Häkkinen R, Linkosalo T, Hari P (1998) Effects of dormancy and environmental factors on timing of bud burst in *Betula pendula*. *Tree Physiol* **18**:707–712.

Hänninen H, Slaney M, Linder S (2006, submitted) Dormancy release of Norway spruce under climatic warming: Testing ecophysiological models of bud burst with a whole-tree chamber experiment.

Hall SA (1994) Modern pollen influx in tallgrass and shortgrass prairies, southern Great Plains, USA. *Grana* **33**:321–326.

Halwgay MH (1994) Airborne pollen of Kuwait city, Kuwait, 1975–1987. *Grana* **33**:333–339.

Hannerz, M (1999) Evaluation of temperature models for predicting bud burst in Norway spruce. *Can J Forest Res – Revue Canadienne de Recherche Forestiere* **29**(1):9–19.

Hänninen H (1995) Effects of climatic change on trees from cool and temperate regions: an ecophysiological approach to modelling of bud burst phenology. *Can J Bot* **73**(2):183–199.

Hay RKM (1990) Tansley Review No. 26. The influence of photoperiod on the dry-matter production of grasses and cereals. *New Phytol* **116**:233–254.

Heide OM (1993a) Dormancy release in beech buds (*Fagus sylvatica*) requires both chilling and long days. *Physiol Plantarum* **88**:187–191.

Heide OM (1993b) Daylength and thermal responses of bud burst during dormancy release in some northern deciduous trees. *Physiol Plantarum* **88**:531–540.

Hiller KM, Esch RE, Klapper DG (1997) Mapping of an allergenically important determinant of grass group I allergens. *J Allergy Clin Immunol* **100**(3):335–340.

Hirschwehr R, Valenta R, Ebner C, Ferreira F, Sperr WR, Valent P et al. (1992) Identification of common allergenic structures in hazel pollen and hazelnuts: a possible explanation for sensitivity to hazelnuts in patients allergic to tree pollen. *J Allergy Clin Immunol* **90**(6 Pt 1): 927–936.

Hirschwehr R, Heppner C, Spitzauer S, Sperr WR, Valent P, Berger U et al. (1998) Identification of common allergenic structures in mugwort and ragweed pollen. *J Allergy Clin Immunol* **101**(2 Pt 1):196–206.

Hjelmroos M (1992) Long-distance transport of *Betula* pollen grains and allergic symptoms. *Aerobiologia* **8**:231–236.

Høgda KA, Karlsen SR, Solheim I, Tømmervik H, Ramfjord H (2002) The start dates of birch pollen seasons in Fennoscandia studied by NOAA AVHRR NDVI data. *Proceedings of IGARSS*. Toronto, Canada, ISBN 0-7803-7536-X.

Hunter AF, Lechowicz MJ (1992) Predicting the timing of budburst in temperate trees. *J Appl Ecol* **29**:597–604.

Hurtado I, Alson J (1990) Air pollen dispersal in tropical area. *Aerobiologia* **6**:122–127.

Iacovacci P, Afferni C, Barletta B, Tinghino R, Di Felice G, Pini C et al. (1998) Juniperus oxycedrus: a new allergenic pollen from the Cupressaceae family. *J Allergy Clin Immunol* **101**(6 Pt 1):755–761.

Inoue S, Kawashima S, Takahashi Y (2002) Estimating the beginning of Japanese cedar pollen release under global climate change. *Glob Change Biol* **8**:1165–1168.

Ipsen H, Lowenstein H (1997) Basic features of crossreactivity in tree and grass pollen allergy. *Clin Rev Allergy Immunol* **15**(4):389–396.

ISAAC (1998) Worldwide variation in prevalence of symptoms of asthma, allergic rhinoconjunctivitis, and atopic eczema. The International Study of Asthma and Allergies in Childhood (ISAAC) Steering Committee, *Lancet* **351**(9111):1225–1232.

IPCC (2007) Intergovernmental Panel on Climate Change Forth Assessment Report. *IPCC*, Geneva, 2007. Available online from http://www.ipcc.ch.

Jablonski LM, Wang X, Curtis PS (2002) Plant reproduction under elevated CO_2 conditions: a meta-analysis of reports on 79 crop and wild species. *New Phytol* **156**:9–26.

Jäger S, Nilsson S, Berggren B, Pessi AM, Helander M, Ramfjord H (1996) Trends of some airborne tree pollen in the Nordic countries and Austria, 1980–1993. A comparison between Stockholm, Trondheim, Turku and Vienna. *Grana* **35**:171–178.

Kailidis DS (1991) *Pollution of Natural Environment*. K. Hristodoulidis, Thessaloniki, Greece. [in Greek]

Kellomäki S, Rouvinen I, Peltola H, Strandman H, Steinbrecher R (2001) Impact of global warming on the tree species composition of boreal forests in Finland and effects of emissions of isoprenoids. *Glob Change Biol* **7**:531–544.

Keynan N, Waisel Y, Shomerilan A, Goren A, Brener S (1991) Annual variations of airborne pollen in the coastal plain of Israel. *Grana* **30**:477–480.

Kikuzawa K (1989) Ecology and evolution of phenological pattern, leaf longevity and leaf habit. *Evol Trend Plant* **3**(2):105–110.

Knox RB (1993) Grass pollen, thunderstorms and asthma. *Clin Exp Allergy* **23**(5):354–359.

Koenig WD, Knops JMH (2000) Patterns of annual seed production by northern hemisphere trees: A global perspective. *Am Nat* **155**:59–69.

Kramer, K (1994) A modelling analysis of the effects of climatic warming on the probability of spring frost damage to tree species in the Netherlands and Germany. *Plant Cell Environ* **17**:367–377.

Laaidi K (2001) Predicting days of high allergenic risk during Betula pollination using weather types. *Int J Biometeorol* **45**:124–132.

Latalowa M, Miętus M, Uruska A (2002) Seasonal variations in the atmospheric *Betula* pollen count in Gdańsk (southern Baltic coast) in relation to meteorological parameters. *Aerobiologia* **18**: 33–43.

Leiferman KM, Gleich GJ, Jones RT (1976) The cross-reactivity of IgE antibodies with pollen allergens. II. Analyses of various species of ragweed and other fall weed pollens. *J Allergy Clin Immunol* **58**(1 PT. 2):140–148.

Leinonen I (1996) Dependence of dormancy release on temperature in different origins of Pinus sylvestris and Betula pendula seedlings. *Scandinavian J Forest Res* **11**(2):122–128.

Leuschner RM, Christen H, Jordan P, Vonthein R (2000) 30 years of studies of grass pollen in Basel (Switzerland). *Aerobiologia* **16**:381–391.

Lewis WH, Imber WE (1975) Allergy epidemiology in the St. Louis, Missouri, area. III. Trees. *Ann Allergy* **35**(2):113–119.

Linkosalo T (1999) Regularities and patterns in the spring phenology of some boreal trees. *Silva Fennica* **33**(4):237–245.

Linkosalo T (2000) Mutual dependency and patterns of spring phenology of boreal trees. *Can J Forest Res* **30**(5):667–673.

Linkosalo T, Carter TR, Häkkinen R, Hari P (2000) Predicting spring phenology and frost damage risk of Betula spp. Under climatic warming: a comparison of two models. *Tree Physiol* **20**(17):1175–1182.

Linkosalo T, Lechowicz MJ (2006). Twilight far-red treatment advances leaf bud-burst of silver birch (Betula pendula). *Tree Physiol* (in print).

Lorenzo C, Marco M, Paola DM, Alfonso C, Marzia O, Simone O (2006) Long distance transport of ragweed pollen as a potential cause of allergy in central Italy. *Ann Allergy Asthma Immunol* **96**(1):86–91.

Lovborg U, Baker P, Tovey E (1998) A species-specific monoclonal antibody to Cynodon dactylon. *Int Arch Allergy Immunol* **117**(4):220–223.

Mabbelrey DJ (1987) *The Plant Book*, Cambridge University Press, Cambridge.

Mahoney KJ, Kegode GO (2004) Biennial wormwood (Artemisia biennis) biomass allocation and seed production. *Weed Sci* **52**:246–254.

Majd A, Chehregani A, Moin M, Gholami M, Kohno S, Nabe T, Shariatzade MA (2004) The Effects of Air Pollution on Structures, Proteins and Allergenicity of Pollen Grains. *Aerobiologia* **20**:111–118.

Makra L, Juhász M, Béczi R, Borsos E (2005) The history and impacts of airborne *Ambrosia* (Asteraceae) pollen in Hungary. *Grana* **44**:57–64.

Manninen T, Stenberg P, Rautiainen M, Voipio P, Smolander H (2006) Leaf Area Index estimation of Boreal Forest using ENVISAT ASAR. *IEEE Trans Geosci Remote Sensing* (in press).

Mari A, Wallner M, Ferreira F (2003) Fagales pollen sensitization in a birch-free area: a respiratory cohort survey using Fagales pollen extracts and birch recombinant allergens (rBet v 1, rBet v 2, rBet v 4). *Clin Exp Allergy* Oct;**33**(10):1419–1428.

Masaka K, Maguchi S (2001) Modelling the masting behaviour of *Betula platyphylla* var *japonica* using the resource budget model. *Ann Bot* **88**:1049–1055.

Matikainen E, Rantio-Lehtimäki A (1998) Semiquantitative and qualitative analysis of preseasonal airborne birch pollen allergens in different particle sizes – Background information for allergen reports. *Grana* **37**(5):293–297.

Matthiesen F, Schumacher MJ, Lowenstein H (1991) Characterization of the major allergen of Cynodon dactylon (Bermuda grass) pollen, Cyn d I. *J Allergy Clin Immunol* **88**(5):763–774.

McMenamin P (1994) Costs of hay fever in the United States in 1990. *Ann Allergy* **73**:35–39.

Menzel A (2000) Trends in phenological phases in Europe between 1951 and 1996. *Int J Biometeorol* **44**:76–81.

Menzel A, Fabian P (1999) Growing season extended in Europe. *Nature* **397**:659.

Menzel A, Sparks TH, Estrella N, Koch E, Aasa A, Ahas R, Alm-Kubler K, Bissolli P, Braslabska O, Briede A, Chmielewski F-M, Crepinsek Z, Curnel Y, Dahl Å, Defila C, Donnelly A, Filella Y, Jatchak K, Mage F, Mestre A, Nordli Ø, Penuelas J, Pirinen P, Remisova V, Scheifinger H, Striz M, Susnik A, Van Vliet AJH, Wielgolaski F-E, Zach S, Zust (2006) European phenological response to climate change matches the warming pattern. *Glob Change Biol* **12**:1969–1976.

Mølgaard P, Christensen K (1997) Response to experimental warming in a population of *Papaver radicatum* in Greenland. *Glob Change Biol* **3**:116–124.

Mothes N, Westritschnig K, Valenta R (2004) Tree pollen allergens. *Clin Allergy Immunol* **18**:165–184.

Motta AC, Marliere M, Peltre G, Sterenberg PA, Lacroix G (2006) Traffic-related air pollutants induce the release of allergen-containing cytoplasmic granules from grass pollen. *Int Arch Allergy Immunol* **139**:294–298.

Mourad W, Mecheri S, Peltre G, David B, Hebert J (1988) Study of the epitope structure of purified Dac G I and Lol p I, the major allergens of Dactylis glomerata and Lolium perenne pollens, using monoclonal antibodies. *J Immunol* **141**(10):3486–3491.

Murray MG, Sonaglioni MI, Villamil CB (2002) Annual variation of airborne pollen in the city of Bahia Blanca. *Argentina* **41**:183–189.

Myking T, Heide OM (1995) Dormancy release and chilling requirements of buds of latitudinal ecotypes of Betula pendula and Betula pubescens. *Tree physiol* **15**(11):697–704.

Newnham RM (1999) Monitoring biogeographical response to climate change: The potential role of aeropalynology. *Aerobiologia* **15**:87–94.

Orlandi F, Ruga L, Romano B, Fornaciari M (2005) Olive flowering as an indicator of local climatic changes. *Theor Appl Climatol* **81**:169–176.

Osborne CP, Chuine I, Viner D, Woodward FI (2000) Olive phenology as a sensitive indicator of future climatic warming in the Mediterranean. *Plant Cell Environ* **23**:701–710.

Partanen J, Koski V, Hänninen H (1998) Effects of photoperiod and temperature on the timing of bud burst in Norway spruce (*Picea abies*). *Tree Physiol* **18**:811–816.

Partanen J, Leinonen I, Repo T (2001) Effect of accumulated duration of the light period on bud burst in Norway spruce (*Picea abies*) of varying ages. *Silva Fenn* **35**(1):111–117.

Partanen J, Hänninen H, Häkkinen R (2005) Bud burst in Norway spruce (*Picea abies*): preliminary evidence for age-specific rest patterns. *Trees-Struct Funct* **19**:66–72.

Pham NH, Baldo BA (1995) Allergenic relationship between taxonomically diverse pollens. *Clin Exp Allergy* **25**(7):599–606.

Pham NH, Baldo BA, Bass DJ (1994) Cypress pollen allergy. Identification of allergens and cross-reactivity between divergent species. *Clin Exp Allergy* **24**(6):558–565.

Phillips JW, Bucholtz GA, Fernandez-Caldas E, Bukantz SC, Lockey RF (1989) Bahia grass pollen, a significant aeroallergen: evidence for the lack of clinical cross-reactivity with timothy grass pollen [see comments]. *Ann Allergy* **63**(6 Pt 1):503–507.

Peternel R, Srnec L, Hrga I, Hercog P, Culig J (2005) Airborne pollen of *Betula*, *Corylus* and *Alnus* in Zagreb, Croatia. A three-year record. *Grana* **44**:187–191.

Ramirez DA (1984) The natural history of mountain cedar pollinosis. *J Allergy Clin Immunol* **73**(1 Pt 1):88–93.

Ranta H, Oksanen A, Hokkanen T, Bondestam K, Heino S (2005) Masting by *Betula*-species; applying the resource budget model to north European data sets. *Int J Biometeorol* **49**:146–151.

Rantio-Lehtimäki A (1994) Short, medium and long range transported airborne particles in viability and antigenicity analyses. *Aerobiologia* **10**:175–181.

Rasmussen A (2002) The effects of climate change on the birch pollen season in Denmark. *Aerobiologia* **18**:253–265.

Réaumur M (1735) Observations du thermometre, faites à Paris pendant l'annèe MDCCXXXV. *Mem Acad Roy Sci Paris* 737–754.

Reich PB, Walters MB, Ellsworth DS (1992) Leaf life-span in relation to leaf, plant, and stand characteristics among diverse ecosystems. *Ecol Monographs* **62**(3):365–392.

Sarvas R (1972) Investigations on the annual cycle of development of forest trees. Active period. *Comm Inst For Fenn* **76**(3):1–110.

Sarvas R (1974) Investigations on the annual cycle of development of forest trees II. Autumn dormancy and winter dormancy. *Comm Inst For Fenn* **84**:1–101.

Savitsky VD, Kobzar VN (1996) Aerobiology in Russia and neighbouring countries, 1980–1993. A bibliographic review. *Grana* **35**:314–318.

Seinfeld JH, Pandis SN (2006) *Atmospheric Chemistry and Physics. From Air Pollution to Climate Change*. Second edition. Wiley, New York, 1203 pp.

Sharp WM, Chisham HH (1961) Flowering and fruiting in white oaks. I staminate flowering throughout pollen dispersal. *Ecology* **42**:365–372.

Scheiner O, Aberer W, Ebner C, Ferreira F, Hoffmann-Sommergruber K, Hsieh LS et al. (1997) Cross-reacting allergens in tree pollen and pollen-related food allergy: implications for diagnosis of specific IgE. *Int Arch Allergy Immunol* **113**(1–3):105–108.

Sheldon JM, Hewson EW (1960) *Atmospheric Pollution by Aeroallergens*. University of Michigan Research Institute Progress Report 4. University of Michigan, Ann Arbor, MI, 191 p.

Siljamo P., Sofiev M., Ranta H. (2004) An approach to simulation of long-range atmospheric transport of natural allergens: an example of birch pollen. In Air Pollution Modelling and its Applications XVII (eds. C. Borrego, A.-L. Norman), Springer (2007), ISBN-10: 0-387-28255-6, pp. 331–340.

Singer B, Ziska LH, Frenz DA, Gebhard DE, Straka JG (2005) Increasing Amb a 1 content in common ragweed (Abmrosia artemisiifolia) pollen as a function of rising atmospheric CO2 concentration. *Funct Plant Biol* **32**:667–670.

Solomon WR. (2001) Ragweed pollinosis: answers awaiting explanations. *Ann Allergy Asthma Immunol* Feb; **86**(2):141–142.

Solomon WR (2002) Airborne pollen: A brief life. *J Allergy Clin Immunol* **109**:895–900.

Solomon WR, Burge HA, Muilenberg ML (1983) Allergen carriage by atmospheric aerosol. I. Ragweed pollen determinants in smaller micronic fractions. *J Allergy Clin Immunol* **72**(5 Pt 1):443–447.

Sofiev M, Siljamo P, Ranta H, Rantio-Lehtimäki A (2006) Towards numerical forecasting of long-range air transport of birch pollen: theoretical considerations and a feasibility study. *Int J Biometeorol* **50**:392–402.

Spieksma FThM, Corden JM, Detandt M, Millington WM, Nikkels H, Nolard N, Schoenmakers CHH, Wachter R, de Weger LA, Willems R, Emberlin J (2003) Quantitative trends in annual totals of five common airborne pollen types (*Betula*, *Quercus*, Poaceae, *Urtica*, and *Artemisia*), at five pollen-monitoring stations in western Europe. *Aerobiologia* **19**:171–184.

Stenström A, Jónsdóttir IS (1997) Responses of the clonal sedge, *Carex bigelowii*, to two seasons of simulated climate change. *Glob Change Biol* **3**:89–96.

Stinson KA, Bazzaz FA (2006) CO2 enrichment reduces reproductive dominance in competing stands of Ambrosia artemisiifolia (common ragweed). *Oecologia* **147**:155–163.

Suphioglu C, Singh MB, Taylor P, Bellomo R, Holmes P, Puy R et al. (1992) Mechanism of grass-pollen-induced asthma. *Lancet*. **339**(8793):569–572.

Tedeschini E, Rodríguez-Rajo FJ, Caramiello R, Jato V, Frenguelli G (2006) The influence in climate changes in *Platanus* spp. Pollination in Spain and Italy. *Grana* **45**:222–229.

Townsend AR, Howarth RW, Bazzaz FA, Booth MS, Cleveland CC, Collinge SK, Dobson AP, Epstein PR, Holland EA, Keeney DR, Mallin MA, Rogers CA, Wayne PR, Wolfe AH (2003) Human health effects of a changing global nitrogen cycle. *Front Ecol Environ* **1**:240–246.

Tuomi J, Niemelä P, Mannila R (1982) Resource allocation on dwarf shoots of birch (*Betula pendula*): reproduction and leaf growth. *New Phytol* **91**:483–487.

Varela S, Subiza J, Subiza JL, Rodriguez R, Garcia B, Jerez M et al. (1997) Platanus pollen as an important cause of pollinosis. *J Allergy Clin Immunol* **100**(6 Pt 1):748–754.
Venables KM, Allitt U, Collier CG, Emberlin J, Greig JB, Hardaker PJ et al. (1997) Thunderstorm-related asthma – the epidemic of 24/25 June 1994. *Clin Exp Allergy* **27**(7):725–736.
Vingarzan R (2004) A review of surface ozone background levels and trends. *Atmos Environ* **38**:3431–3442.
Voltolini S, Minale P, Troise C, Bignardi D, Modena P, Arobba D, Negrini AC (2000) Trend of herbaceous pollen diffusion and allergic sensitisation in Genoa, Italy. *Aerobiologia* **16**:245–249.
Walther G-R (2003) Plants in warmer world. *Perspect Plant Ecol Evol Syst* **6**:169–185.
Wan S, Yuan T, Bowdish S, Wallace L, Russell SD, Luo Y (2002) Response of an allergenic species, *Ambrosia psilostachya* (Asteraceae), to experimental warming and clipping: implications for public health. *Am J Bot* **89**:1843–1846.
Wayne P, Foster S, Connolly J, Bazzaz F, Epstein P (2002) Production of allergenic pollen by ragweed (*Ambrosia artemisiifolia* L.) is increased in CO_2-enriched atmospheres. *Ann Allergy Asthma Immunol* **88**:279–282.
WHO (2003) *Phenology and Human Health: Allergic Disorders*. Copenhagen, WHO regional office for Europe, 55 p.
Williams R (2005) Climate change blamed for rise in hay fever. *Nature* **434**:1059. www://nature.com/nature/journal/v434/n7037/full/4341059a.htm
Yanovsky MJ, Kay SA (2003) Living by the calendar: How plants know when to flower. *Nature Rev Mol Cell Biol* **4**:265–275.
Zanotti AL, Puppi G (2000) Phenological surveys of allergenic species in the neighbourhood of Bologna (Italy). *Aerobiologia* **16**:190–206.
Ziska LH, Bunce JA, Goins (2004) Characterization of an urban-rural CO2/temperature gradient and associated changes in initial plant productivity during secondary succession. *Oecologia* **139**:454–458.
Ziska LH, Gebhard DE, Frenz DA, Faulkner S, Singer BD, Straka JG (2003) Cities as harbingers of climate change: Common ragweed, urbanization, and public health. *J Allergy Clin Immunol* **111**:290–295.

Chapter 6
Plant Biometeorology and Adaptation

Simone Orlandini, Marco Bindi, and Mark Howden

Abstract Weather conditions play a fundamental role in plant growth and development, due to their direct and indirect influence on physical, chemical and biological processes. It is then crucial the analysis of possible impact due to current and future climate change and variability. With this main objective, this chapter starts with a preliminary analysis of plant responses to weather conditions and then with an evaluation of climate change and variability, deeply discussing the main impacts and stresses of climate change, considering enhanced CO_2, higher temperature and available water. Then responses of different systems are discussed together with available adaptation options and conditions for their effective deployment. Different levels of adaptations were considered, from the short and long term measures, to the farm , regional and national options. The chapter concludes with a discussion of future directions in plant biometeorology and research gaps.

6.1 Plant Responses to Weather Conditions

Weather conditions play a fundamental role in plant growth and development due to their direct and indirect influence on each physical, chemical and biological process, that is regulated by specific requirements and any deviation from these patterns may exert a negative influence (Das et al. 2003). Air and soil temperature, air and soil humidity, solar radiation, wind speed and direction, rainfall, evapotranspiration are the most important variables affecting the vegetative and productive responses of

---

S. Orlandini and M. Bindi
Department of Agronomy and Land Management – University of Florence,
Piazzale delle Cascine 18, 50144 Firenze. Italia
e-mail: simone.orlandini@unifi.it; marco.bindi@unifi.it

M. Howden
Climate Adaptation Flagship, GPO Box 284, Canberra, ACT 2601
e-mail: mark.howden@csiro.au

plants, both in natural and cultivated systems. Others can be of interest in particular cases, such as leaf wetness for the analysis of many plant pathogens (Table 6.1).

To define plant responses to weather conditions, and so also to their change and variability, studies can be based on the application of agroclimatic indices and simulation models. These are basically formal expressions of biological, physical and chemical functions fed with environmental and climatic forcing variables. Models are often the only tool available to study the behaviour of complex plant production systems under a variable and changing climate, and they offer unique insights to understand the frequently non-linear interactions among processes in soil-plant systems. In the last few years an increasing interest in this subject has occurred and a high number of computer applications for biometeorological purposes have been developed. Among these it is possible to emphasise plant growth and development, crop yield quality and quantity estimation, water balance, plant protection against pests, diseases, weed and weather hazards, soil erosion and conservation, etc. (Eitzinger et al. 2008).

The analysis of plant responses requires different sources of data, such as meteorological (temperature, rainfall, relative humidity, leaf wetness, solar radiation, wind direction and speed), physical (CO_2 concentration, soil structure) and biological (observed symptoms, crop monitoring, plant parameters). Meteorological data are generally required with hourly time step for epidemiological models, while daily data are required for the other kinds of simulations (growth and development); some soil erosion models require a shorter time step (minutes). The availability of meteorological information can be improved by further developing the spatial interpolation methods and by a more effective use of weather radar and satellite information in addition to traditional meteorological ground data. Automation of weather observing stations may have different impacts on the availability of some meteorological parameters. For instance, cloud cover is now rarely measured as it requires visual observation, whereas automatic station are frequently equipped with solar radiation sensors.

Table 6.1 Role of weather conditions in plant-pathosystems processes

Variable	Effect
Temperature	Phenological development
Solar radiation	Biomass assimilation and growth
High temperature	Rate of infection
	Threshold of development and survival
Low temperature	Spore and insect conservation
	Lower threshold of development and survival
Leaf wetness	Inoculation
	Survival of organism
Precipitation	Dispersion of spore and insect
	Survival of spores
Relative humidity	Survival of spores
Wind	Dispersion of spore and insect

The use of atmospheric models as a source of meteorological data is also an alternative that is worth considering. Large scale models, such as global circulation ones, are provided with a spatial resolution which is still largely insufficient for local scale purposes. The grid size of global models like the ECMWF-model is at the moment about 60 km. Limited Area Models (LAM) were designed with the aim to provide higher resolution atmospheric fields at the continental and regional scale while retaining global modelling capability. Nowadays LAMs use resolutions of 5–30 km and need some external information, i.e. initial conditions and lower and upper boundary conditions, generally provided by Global Circulation Models (Dalla Marta et al. 2003).

6.2 Preliminary Considerations of Climate Change and Variability

The evaluation of climate change and variability impacts can be performed both on past and future data. There are different ways to analyse past climate: use of historical data, remote sensing and reanalysis. Historical weather data are available from data banks of many Meteorological Services or Research Institutions. Direct use of these data however, is sometimes not possible, due to poor data quality, requiring significant quality control and data patching. Sometimes data validation take a great deal of time, compromising the effective realization of analysis. In case of good quality data, the analysis of trend, variability and extremes can be performed (Bartolini et al. 2008). Finally, reanalysis projects reproduce historical global scale atmospheric circulation over decades by analysing various meteorological observation data (Kalnay et al. 1996).

Remote sensing can be a valid alternative. In particular, earth observation images show the world through a wide-enough frame, so that complete large-scale phenomena can be observed. Earth observation from space has the capacity to provide such global data sets continuously and consistently not only on this level, but also on the national and local levels and the use of warning systems must be based on such data. Some example of vegetation products are: vegetation index, maximum greenness during the growing season, total greenness during the growing season, leaf area index, etc. Some of them are available for more than 25 years, enough to identify climate change trends (Struzik et al. 2008).

To analyse future climate conditions, scenarios can be used to explore changes in a range of climate variables and management responses. They are a simplified description of how the future may develop. Scenarios are neither predictions nor forecasts, instead they derive from projections. Nevertheless climatic effects can differ considerably in different regions. Thus, global average values are not adequate for quantifying the potential hazard at the regional level. To simulate climatic variability and extreme events depending on small-scale effects and influenced substantially by topography and geography, high-resolution downscaling techniques are of crucial importance (Calanca et al. 2008). They are applied both on reanalysis and future scenarios and allow the users to obtain information at mesoscale and local scale, where the impacts on agriculture are more evident and assessable. For this aim, the use of numerical weather models can be of vital importance due to their capacity to simulate meteorological conditions on a small scale, when properly validated.

There is considerable evidence that regional variations in climate, particularly the rise of temperature, have already affected plant systems, increasing the hazard impacts (Table 6.2). Examples of observed changes include lengthening of the growing season, latitudinal and altitudinal shifts of plant range, earlier flowering, outbreak of plant diseases, acceleration in breakdown of organic matter in soils, and insects development. With respect to the latter for instance, during the period between 1964 and 2004 in England, a 1°C increase in temperatures can be associated with a 16-day earlier shift in the first appearance of peach-potato aphid (*Myzus persicae*) with peak flight time of the orange tip butterfly (*Anthocharis cardamines L.*) 6 days earlier (Cannell et al. 1999). More frequent precipitation and more humid conditions favour the spread of diseases. The highest intensity of rainfall reduces the infiltration of water in the soil, decreasing the net available soil water content, contributing to an increase of drought conditions.

At times climate variability can exceed a defined threshold, determining an extreme event. The perceived severity of said event depends on the vulnerability of the natural environment and human society. These events usually cause very extensive or local hazards with high intensity: drought, frost, windstorms, heat waves, cold injury, fire, heavy precipitation, floods, snow, wind and hail.

Table 6.2 Climate variables and their effect on plant production systems processes and some of the observed/predicted impacts

Variation	Effect	Impact
Increase of temperature	Increasing of the length of growing season	Variation of harvest period and modification of quality and quantity of productions
	Variation of range of cultivation areas	More variable production
	Anticipation of bud break	Frost risk
		Modification of pollination periods
	Favourable conditions for pest and disease	Possible introduction of new disease and pest
Increase of high rainfall	Drought during Summer months and flood and land slide risks	Strong damages to vegetation
Increase of Summer temperature	Increase frequency of heat waves	Modification of plant responses and possible impacts on quality and quantity of yield
Increase of climate variability	Rapid modification of weather conditions	Problems with vegetation acclimatisation to high and low temperature
		More variable productions

The importance of extreme events derives from their capacity to cause serious damage in very short time frames. For example, in the late 1990s, a drought that affected the central and the southern parts of Spain caused losses of more than 800 million euro in the cereal, olive oil and livestock sectors (more than 50% of the total value of these fields) (European Environment Agency 2003). The impact of the summer-2003 heat wave and drought on agriculture and forestry (potato, maize, wheat, fodder, poultry) caused losses of 4–5 billion euro in Italy, 1.5 billion euro in Germany, and 4 billion euro in France (COPA-COGECA 2003). In the Philippines, typhoons, floods and droughts caused 82.4% of the total Philippine rice (*Oryza sativa*) losses from 1970 to 1990.

On these grounds it is possible to emphasise that climate variations affect the complex agriculture-climate system, by influencing its main biological, chemical and physical elements. Therefore all the management and planning aspects of vegetation must be considered and adapted to climate change and variability with specific short and long term strategies with respect to crop protection, watering, fertilisation, plant breeding, production, site selection, etc.

The importance of the potential impacts of increased frequency and magnitude of climate extremes as a result of potential climate change has been emphasised in the recent IPCC Fourth Assessment report. This report suggested that such changes may be more important than the projected changes in mean climate (Easterling et al. 2007). Improved understanding of both the impacts of changes in climate extremes and the adaptation options for reducing their impacts is a vital step in determining the course of emission-reduction policies (Howden et al. 2007). Consequently the assessment of meteorological hazard impacts on agricultural systems represents an important goal for researchers in the biometeorology community and discipline. Given that there are many aspects in common between both the driving variables of climate extremes and the physiological impacts that they have on crop systems, there appears to be a case for a more coherent and coordinated approach to this issue across the biometeorology research community including through the more effective linkage of data, methods and techniques. However, the incredibly diverse nature of crop systems across the world and the possible responses to changes in climate extremes means that dealing with this issue is not just the domain of researchers but also that of decision-makers: farmers, extension workers, industry leaders and government policymakers. Importantly, these people and institutions not only are in a position to take action but they also have important information in their own right to contribute to improved understanding of the risks and responses (Howden et al. 2007). Closing the 'relevance gap' between researchers and such decision-makers is an important step (Meinke et al. 2006). Consequently, approaches to risk assessment, design and construction of early warning systems and development of innovative technologies and practices needs to have effective input from users: a move from 'supply-driven' science to 'demand-driven' science. Almost by definition, this will require a much more multi-disciplinary approach to analysis of the issue and an appreciation that progress needs to be made at a range of spatial and temporal scales. For example, there exist many barriers to adaptations to current and future climate changes and fostering change at the farm scale will often be facilitated by improvements in the decision-environment that the farmers operate in (e.g. Easterling et al. 2007). Consequently, while the focus of this paper is on the organ, plant and crop scales at which biometeorology has traditionally addressed, we also address some of the larger scale factors that may allow more effective adaptation. Consequently, in climate terms, just as the past may no longer be a good guide for the future, in terms of the discipline of biometeorology, past scales of research may need expanding to account for future changes in users information needs.

The International community is requesting information on climate change and variability and relevant mitigation and adaptation strategies. Consequently the assessment of meteorological hazard impacts represents a fundamental goal for researchers and scientific societies with objective evaluation of current and future climatic conditions by using, harmonising and integrating all the available data, methods and technologies. Risk assessment, definition of warning systems and addressing specific recommendations and evaluations for policy makers, extension services, and other users are therefore crucial. It is crucial to support the transition from "passive acceptance" of climate change and variability by providing decision-makers with suitable information and know-how in order to make an "active response". Moreover, both biometeorologists and users should realize as soon as possible that the past may no longer be a good guide for the future.

In the next three sections, climate change impacts, adaptation strategies and key research challenges will be described to provide an example on how plant biometeorology may be helpful for studying and addressing this fundamental issue. We will focus on cropping systems from which to draw some key lessons because they are relatively well studied, of critical and increasing importance to food security, and are typically more amenable to adaptation options. However, many of the issues raised here are transferable to other domains of biometeorological research such as grazing systems, forests and natural systems although the decisions and context change.

6.3 Impact of Climate

6.3.1 Biophysical Response

6.3.1.1 Enhanced CO_2

It is well known and demonstrated that plants, when exposed to increased concentrations of CO_2, respond with an increased rate of photosynthesis (Kimball et al. 2002; Ainsworth and Long 2005). This behaviour is particular evident in C_3 plants (plants that use a 3-carbon compound in the first step of photosynthesis) that include most of the small grains, legumes, root crops, cool season grasses, and trees. C_4 plants (that have an additional 4-carbon compound step to capture CO_2), which are tropical grasses such as maize, sorghum, millet, and sugarcane, have a more efficient photosynthetic pathway than the C_3 plants under current CO_2 concentrations. C_4 plants also increase their photosynthesis with elevated atmospheric CO_2 concentration, but less markedly (Kimball et al. 2002; Ainsworth and Long 2005). Moreover, both C_3 and C_4 plants respond to elevated CO_2 by partially closing their stomata. This reduces the loss of water from the interior of the leaf to the atmosphere (transpiration). This reduction in transpiration is not accompanied by any significant loss in photosynthesis. As a result of these combined responses to elevated levels of CO_2 plants have a higher water use efficiency (production per unit of water consumed) that can be quite large (about 40% for both C_3 and C_4 plants with a doubling of CO_2, Kimball et al. 2002; Ainsworth and Long 2005). Whilst, the effect on total water use (evapotranspiration per unit area) is less marked (Rozema 1993).

Recent studies confirm that the effects of elevated CO_2 on plant growth and yield depend on photosynthetic pathway, species, growth stage and management regime, such as water and nitrogen (N) applications (Kimball et al. 2002; Ainsworth and Long 2005). On average across several species and under unstressed conditions, recent data analyses find that, compared to current atmospheric CO_2 concentrations, crop yields increase at 550 ppm CO_2 in the range of 10–20% for C3 crops and 0–10% for C4 crops (Ainsworth et al. 2004; Gifford 2004; Long et al. 2004).

These differential effects of CO_2 on C_3 and C_4 crops may alter plant competitive interactions of very sensitive ecosystems (such as savanna and grassland) (e.g. Stokes et al. 2006)

6.3.1.2 Higher Temperature

The trends towards mean higher temperature already occurring are bringing forward the start of the growing season of many plants at middle to higher latitudes (e.g. IPCC 2007b) resulting in changes in agronomic practices (e.g. Stephens and Lyons 1998). It is likely that global warming will further modify the length of the potential growing season (shorter life cycle of determinate crops such as cereal crops, peas, beans, and oil seed crops., and longer life cycle of indeterminate crops like grass, sugar beet, carrots) if there is no adaptation through change in crops to those with different thermal time requirements (e.g. Easterling et al. 2007). Less severe winters will also allow more productive cultivars of winter annual and perennial crops to be grown. This may be of particular importance for C_4 species as they perform better at higher temperatures. As a result of relative changes between crop types (and even crop to livestock production), there is a growing expectation that crop areas will shift. The shifts are likely to be most pronounced along the current margins for production of specific crops. In warm, low latitude regions, increased temperatures increase respiration, resulting in less than optimal conditions for net growth.

Short periods of high temperature can have significant impacts on crop yield through spikelet sterility in rice (Matsui et al. 1997) or lowering grain number and grain size (Ferris et al. 1999). High temperatures during grain-filling may also reduce grain quality through development of heat shock proteins such as betagliadins, which negatively impact on dough-making (e.g. Blumenthal et al. 1991).

Temperature is also related to potential evaporation through influencing the vapour pressure deficit (atmospheric demand) and also being related to incident solar radiation: two of the three key drivers of evaporation (Monteith 1980).

Vapour pressure deficit is the difference between the potential and the actual moisture content of the atmosphere. The highly non-linear nature of the relationship between temperature and the potential vapour content means that increases in temperatures will increase potential evaporation, all else being equal. This effect will be lessened if night-time temperatures (which are related to vapour pressure, Tanner and Sinclair 1983) increase faster than daytime temperatures but the global climate models do not currently show a consistent response like this (IPCC 2007a). If evaporative demand increases, this will decrease the water use efficiency of crops (yield per mm of water used) and also increase the demand for water (e.g. increased water needs for a given crop such as irrigated rice).

6.3.1.3 Available Water

Plant growth is strongly influenced by the availability of water. Climate change is expected to modify rainfall, evaporation, runoff, and soil moisture storage. Changes in total seasonal precipitation or in its pattern of variability are both important. The occurrence of moisture stress during flowering, pollination, and grain-filling is harmful to most crops (e.g. grain, maize, soybean and wheat). Increased evaporation from the soil and increased transpiration from the plants will cause drought stress; resulting in an increased need for crop varieties with greater drought tolerance.

The demand for water for irrigation is projected to rise in a warmer climate, increasing the competition between agriculture and urban as well as industrial users of water (e.g. Doll 2002). Falling water tables and the resulting increase in the energy needed to pump water will make the practice of irrigation more expensive, and more water will be required per unit area under drier conditions with more greenhouse gas emissions generated. Peak irrigation demands are also likely to rise due to more severe heat waves. Additional investment for dams, reservoirs, canals, wells, pumps, and piping may be needed to develop irrigation networks in new locations or to maintain the existing irrigation systems. Finally, changes in soil water availability will alter soil salinisation processes in both irrigated and dryland cropping systems (e.g. van Ittersum et al. 2003) with either beneficial or problematic outcomes possible depending on the system and climate changes. For example, dryland salinisation risk may be lessened by climate changes which involve lower rainfall (van Ittersum et al. 2003).

6.3.1.4 Climate Variability

Extreme meteorological events, such as spells of high temperature, heavy storms, or droughts, can severely disrupt plant production. Similarly, frequent droughts not only reduce water supplies but also increase the amount of water needed for plant transpiration. Recent studies carried out on specific aspects of increased climate variability (e.g. increased heavy precipitation, increased flood, risks of soil erosion, water and soil salinisation, heat and frost stress during growing seasons) demonstrated that increased climate variability may have greater impacts on agriculture than changes in climate means alone (Howden et al. 2003a; van Ittersum et al. 2003; Monirul and Mirza 2002; Rosenzweig et al. 2002; Nearing et al. 2004; Easterling et al. 2007).

6.3.2 Responses of Crop Systems

6.3.2.1 Cereals and Seed Crops

Cereals, oilseed and protein crops including pulses are generally determinate species (plants with a discrete life cycle which ends when the grain is mature), and the duration to maturity depends on temperature and, in many cases, day length. A temperature increase will therefore shorten the length of the growing period, reducing yields, if management is not altered (Porter and Gawith 1999; Tubiello et al. 2000; Howden 2002), and change the area of cultivation assuming no changes to variety or timing of planting (Olesen et al. 2007). Simple management options to counteract the warming effect are changes that include sowing dates and use of longer season cultivars (Olesen et al. 2000; Tubiello et al. 2000, Howden 2002). Changes in yield will be a function of changes in CO_2, rainfall, temperature and management with potential interactions and synergies between these factors leading to non-linear changes. For example, in mid- to high-latitude, the effects of increased warming effect may be wholly or partly counteracted by the CO_2 fertilisation effect by to an increase of about 2–3°C, whereas at low latitudes the positive effect of CO_2 fertilisation may still exist but with lower temperature increase (up to 1.5–2°C) (Tubiello et al. 2007).

6.3.2.2 Root and Tuber Crops

Potato, as well as other root and tuber crops, is expected to show a large response to rising atmospheric CO_2 due to its large below ground sinks for carbon (Miglietta et al. 1998; Bindi et al. 2005) and phloem loading mechanisms (Komor et al. 1996). On the other hand, warming may reduce the growing season in some species and increase water requirements with consequences for yield (Bindi et al. 2008). Climate change scenario studies performed for Europe using crop models show no consistent changes in mean potato yield (Wolf 2002), but a consistent increase in yield variability is predicted for the whole of Europe, which raises the agricultural risk for this crop. However, available crop management strategies (i.e. advanced planting and the cultivation of earlier varieties) seem likely to be effective in overcoming these changes (Wolf 2002).

Root crops such as sugar beet may be expected to benefit from both the warming and the increase in CO_2 concentrations, as these crops are indeterminate (plants that continue to grow and yield throughout the season) in their development and an extended growing season will increase the duration of growth, provided sufficient water is available (Richter et al. 2006).

6.3.2.3 Horticultural Crops

Horticultural crops include vegetables and ornamental crops, either field-grown or grown under protected conditions. The main effects of a climatic warming anticipated for protected crops are changes in the heating and cooling requirements of the housing; which may feed back to altered greenhouse gas emissions. Most field-grown vegetables are high value crops, which are grown under ample water and nutrient supply. Therefore they are likely to mainly respond to changes in temperature and CO_2. Responses to these factors vary among species, largely depending on the type of yield component and the response of phenological development to temperature change. For determinate crops like onion, warming will reduce the duration of crop growth and hence yield, whereas warming stimulates growth and yield in indeterminate species like carrot (Wheeler et al. 1996; Wurr et al. 1998). For lettuce, temperature has been found to have little influence on yield, whereas yield is stimulated by increasing CO_2 (Pearson et al. 1997).

For many field-grown vegetable crops in Europe, increasing temperature will generally be beneficial, with production expanding out of the presently cultivated areas. A temperature increase will in some areas offer the possibility of a larger span of harvesting dates thus giving a more continuous market supply during a longer period of the year. For cool-season vegetable crops such as cauliflower, large temperature increases may decrease production during the summer period in dry areas due to decreased crop quality (Olesen and Grevsen 1993).

6.3.2.4 Industrial and Biofuel Crops

No literature on the impact of climate change on industrial crops like gums and resins, and medicinal and aromatic plants is available; whereas limited knowledge of climate change impacts on other industrial crops has been recently acquired. Van Duivenbooden et al. (2002) estimated that rainfall reduction associated with climate change could reduce groundnut production in Niger by 11–25%. Varaprasad et al. (2003) also concluded that groundnut yields would decrease under future warmer climates, particularly in regions where present temperatures are near or above optimum despite increased CO_2. Moreover, impacts of climate change and elevated CO_2 on perennial industrial crops will be greater than on annual crops, as both damages (temperature stresses, pest outbreaks, increased damage from climate extremes) and benefits (extension of latitudinal optimal growing ranges) may accumulate with time (Rajagopal et al. 2002).

The large increases in cotton yields due to increases in ambient CO_2 concentration reported in the IPCC 2001 have been strongly rescaled when changes in temperature and precipitation were also included in the assessments (Reddy et al.

2002). Literature still does not exist on the probable impacts of climate change on other fibre crops such as jute and kenaf.

Impacts of climate change on typical biofuel crops such as maize and sorghum are discussed earlier in the cereals and seed crops section. Studies with other biofuel crops such as switchgrass (Panicum virgatum L.), a perennial warm season C_4 crop, have shown yield increases with climate change similar to those of grain crops (Brown et al. 2000). Although there is no information on the impact of climate change on non-food, tropical biofuel crops such as Jatropha and Pongamia, it is likely that their response will be similar to other regional crops.

6.3.2.5 Permanent Crops

Permanent crops are very sensitive to climate change since they usually need several years to reach reproductive maturity and then they remain economically productive for a long time. Several studies have been conducted on high value (economic and environmental) permanent crops like grapevine and olive.

Grapevine is a woody perennial plant, which requires relatively high temperatures for production. A climatic warming will therefore modify suitable areas for its cultivation and season duration (from budburst to harvest). Studies carried out for some the main grapevine producing areas (Europe, California and Australia) show both enlargement of the suitable areas to higher latitudes (Kenny and Harrison 1992; Harrison et al. 2000; Jones et al. 2005) and compressed season durations (e.g. early harvest) (Webb et al. 2007; Moriondo et al. 2008). These changes may have strong repercussions on mean yields. Without CO_2 fertilization or adaptation measures, general yield reductions are expected (e.g. for California losses may range from 0% to >40% depending on the trajectory of climate change, Lobell et al. 2006). Yield losses that may be only marginally reduced when the fertilisation effect of increasing CO_2 concentration is included (Bindi et al. 2001; Moriondo et al. 2008). In both cases, however, yield variability is expected to increase under future climate causing higher economic risks for growers (Bindi et al. 1996; Bindi and Fibbi 2000).

Further the impact on grape yields, climate change may have strong repercussion also on grape quality. Currently, many regions appear to be at or near their optimum growing season temperatures. Under future climate, in these regions producing high quality grapes at the margins of their climatic limits, the ripening of balanced fruit required for existing varieties and wine styles will become progressively more difficult. Whilst, in other regions, predicted climate changes could push some regions into more optimal climatic regimes for the production of current varietals. For example in the main Australian winegrowing regions grape quality may be reduced on average from 7% with lower warming to 39% with higher future warming by the year 2030, and from 9% with lower warming to 76% with higher warming by the year 2050 (Webb et al. 2008). The same for United States where potential premium grape production area in the conterminous United States could decline by up to 81% by the late twenty-first century (White et al. 2006).

Olive is a typical Mediterranean species that is particularly sensitive to low temperature and water shortage, thus the northern and southern limits of cultivation in Europe may be conditioned by climate change. Accordingly, studies performed for the Mediterranean basin showed that the area climatically suitable for olive cultivation could be enlarged due to changes in temperature and precipitation patterns that make some areas of France, Italy, Croatia, and Greece newly suitable for olives (Bindi et al. 1992; Moriondo et al. 2008). However, since olive requires chilling to break dormancy (i.e. vernalization), this expansion of olive cultivation to higher latitude may be limited in some areas due to climate warming (e.g. olive phenology and yield will be affected in the southern part of California due to high temperature, but may expand in northern areas until limited by low winter temperatures) (Gutierrez et al. 2008).

In addition to the limit of cultivation, the time and the length of the main olive's phenological stages may be affected by climatic warming. In particular, the time of flowering may be strongly affected (e.g. from 1 to 3 weeks early flowering by the end of the century in the south of Spain) (Galan et al. 2005).

6.4 Main Adaptive Options Available

The purpose of crop adaptation strategies is to manage potential risks related to climate change over the next decades (i.e. counter negative impacts or take advantage of positive ones). These adaptations can be thought of as being applicable at different temporal and spatial scales, e.g., short term adjustments and long term adaptations, farm-level, regional or national policy level. Some of these adaptations are outlined below.

6.4.1 Short-Term Adjustments

Short-term adjustments to climate change are efforts to optimise production without major system changes. They are autonomous in the sense that no other sectors (e.g. policy, research, etc.) are needed in their development and implementation. Thus, short-term adjustment can be considered as the first defence tools against climate change. A large range of short-term adjustments have been reported for dealing with the effect of climate change, these include:

Changes in planting dates and cultivars. For spring crops, climate warming will allow earlier planting or sowing than at present. Crops planted earlier are more likely to be already matured when extreme high temperatures, such as temperature in the middle of summer, can cause injury. Earlier planting in spring increases the length of the growing season; thus earlier planting using a long season cultivar will increase yield potential, provided moisture is adequate and the risk of heat damage is low. Care will be needed to not increase the risk of damage from a late frost, however, as warming is generally expected to reduce frost risk, this may be manageable. Alternatively, in scenarios with large increases in temperature and significant reductions in rainfall, short-season cultivars may be used to escape heat and drought risks during anthesis to grainfill (van Ittersum et al. 2003). Deeper planting of seeds will also contribute to make seed germination more likely. This approach may also be used for winter crops (i.e. cereals). Early sowing and late-season cultivars may be used to match cold temperature requirements (vernalization) that may be not completely fulfilled during warmer winters and to offset the reduction in the length of the growing season due to warmer climate.

Changes in external inputs. External inputs are used to optimise the production of crops in terms of productivity and profitability. The use of fertilisers is generally adjusted to fit the removal of nutrients by the crop and any losses of nutrients that may occur during or between growing seasons. A change in yield level will therefore, all other things being equal, imply a corresponding change in fertiliser inputs. The projected increases in atmospheric CO_2 concentration will cause a larger nitrogen uptake by the crop, and thus larger fertiliser applications. This may be exacerbated by the tendency for lower grain nitrogen levels with elevated CO_2 concentration particularly in low nitrogen situations (e.g. Sinclair et al. 2000). On the other hand climatic constraints on yields may lead to less demand for fertilisers. Changes in climate may also cause larger (or smaller) losses of nitrogen through leaching or gaseous losses. This may also lead to changes in the demand for fertiliser.

The use of pesticides reflects the occurrence of weeds, pests and diseases. Global warming will in many areas lead to a higher incidence of these problems and thus to a potentially increased use of pesticides. The use of pesticides can, however, be limited through the adoption of integrated pest management systems, which adjust the control measures to the observed problem and also takes a range of influencing factors (including weather) into account.

Current fertiliser and pesticide practices are partly based on models and partly on empirical functions obtained in field experiments. These models and functions are updated regularly with new experimental evidence. This process will probably capture the response of changes in the environment through CO_2 and climate. It is, however, important that agricultural researchers and advisors are aware of the possible impact of global change on use of external inputs, so that older empirical data are used with proper caution.

Practices to conserve moisture. A number of water conserving practices are commonly used to combat drought. These may also be used for reducing climate change impacts (Easterling 1996). Such practices include conservation tillage and irrigation management. *Conservation tillage* is the practice of leaving some or all the previous season's crop residues on the soil surface. This may protect the soil from wind and water erosion and retain moisture by reducing evaporation and increasing infiltration of precipitation into the soil. Conservation tillage may also decrease soil temperature. *Irrigation management* can be used to improve considerably the utilisation of applied water through proper timing of the amount of water distributed. For example with irrigation scheduling practices, water is only applied when the crop needs it. This tunes the proper timing and amount of water to actual field conditions allowing a reduction in water use and cost of production.

Easterling et al. (2007) synthesised the results from adaptation studies on wheat, rice and maize showing that on average these adaptations allowed to avoid damage caused by a temperature increase of up to 1.5–3°C in tropical regions and 4.5–5°C in temperate regions; while further warming than these ranges in either region exceeds adaptive capacity.

6.4.2 Long-Term Adaptations

Many long-term adaptations (planned adaptations), including major structural changes to overcome adversity caused by climate change, have been identified (Howden et al. 2003b; Kurukulasuriya and Rosenthal 2003; Aggarwal et al. 2004; Antle et al. 2004; Easterling et al. 2004).

Changes of land use to respond to the differential crop performance under climate change and to stabilise production. In this case crops with high inter-annual variability in production (e.g. wheat) may be substituted with crops with lower productivity but more stable yields (e.g. pasture).

Crop breeding through the use of both traditional and biotechnology techniques to allow introduction of heat and drought resistant crop varieties. Collections of genetic resources in germ-plasm banks may be screened to find sources of resistance to changing diseases and insects, as well as tolerances to heat and water stress and better compatibility to new agricultural technologies. For example, crop varieties with higher "harvest index" will help maintain irrigation efficiency under conditions of reduced water supplies or enhanced demands. Genetic manipulation may also offer another possibility to adapt to stresses (heat, water, pest and disease, etc.) enhanced by climate change allowing the development of "designer-cultivars" much more rapidly than it is possible today. Species not previously used for agricultural purposes may be identified and others already identified may be quickly used.

Crop substitution for the conservation of soil moisture. Some crops use a lower amount of water, are more water and heat resistant, so that they tolerate dry weather better than others do. For example, sorghum is more tolerant of hot and dry conditions than maize.

New land field techniques (laser-levelling of fields, minimum tillage, chiselling compacted soils, stubble mulching, etc.) or new management strategies (e.g. irrigation scheduling and monitoring soil moisture status) to improve irrigation efficiency in agriculture. Moreover a wide array of techniques (such as inter-cropping, multi-cropping, relay cropping etc.) to improve water use efficiency.

Changes in nutrient management to reflect the modified growth and yield of crops, but also changes in the turn-over of nutrients in soils, including losses. It may thus be necessary to revise standards of soil nitrogen mineralisation and the efficiency of use of animal manures and other organic fertilisers. There is a range of management options that will affect the utilisation of fertilisers and manure, including fertiliser placement and timing, reduced tillage and altered crop rotation management.

Changes in farming systems to remain viable and competitive. Specialised farms, especially dairy farms and arable farms, will probably respond more to climate change than mixed farms. On mixed farms with both livestock and arable production there are more options for change, and thus a larger resilience to change in the environment.

6.4.3 Farm Level Adaptations

There is a large range of farm level options for adapting to climate change. Key adaptations (Howden et al. 2003a) include:

- Further develop risk amelioration approaches (e.g. zero tillage and other minimum disturbance techniques, retaining residue, extending fallows, row spacing, planting density, staggering planting times, controlled traffic, erosion control infrastructure)
- More opportunistic cropping – more effectively taking into account environmental condition (e.g. soil moisture), climate (e.g. seasonal climate forecasting) and market conditions
- Expand routine record keeping of weather, production, degradation, pest and diseases, weed invasion
- Tools/training to access/interpret climate data and analyse alternative management options
- Learning from farmers in currently more marginal areas
- Selection of varieties with appropriate thermal time and vernalisation requirements, heat shock resistance, drought tolerance (i.e. 'Staygreen' varieties), high protein levels, resistance to new pests and diseases and perhaps that set flowers in hot/windy conditions
- Improve seasonal and other climate forecasting and also develop warnings prior to planting of likelihood of very hot days and high erosion potential

Whilst a range of technological and managerial options may exist as indicated above, the adoption of these new practices will require:

- Confidence that climate changes several years or decades into the future can be effectively predicted against a naturally high year to year variability in rainfall that characterises these systems.
- The motivation to change to avoid risks or use opportunities.
- Development of new technologies and demonstration of their benefits.
- Protection against establishment failure of new practices during less favourable climate periods.
- Alteration of transport and market infrastructure to support altered production. (McKeon et al. 1993).

Adaptation strategies that incorporate the above considerations are more likely to be of value, as they will be more readily incorporated into existing on-farm management strategies.

Provided climate changes are not overly large nor rapid, many cropping enterprises in developed nations are likely to be largely self-adapting if adequate information and appropriate technologies are available, costs/price ratios and the policy environment are favourable and there is altered infrastructure suited to the new conditions. For example, there is emerging evidence that at least some farmers in some regions of Australia are making rational adaptations to the existing trends in climate (e.g. frost reductions) which maintain risk at historical levels but which enhance economic returns (Howden et al. 2003b). However, for farmers in less-developed nations, climate change is likely to provide a significant challenge, interacting with a range of other pressures on those systems.

6.4.4 Regional Level Adaptations

Historically, there have been substantial and quite rapid changes in land management and landuse with climate variations in many parts of the globe (e.g. Meinig 1962; Meinke and Hammer 1997). The scope and scale of potential future climate changes suggests that significantly greater landuse change may happen in the future, particularly at the margins of current industry distributions. A key reason for problems that may arise from any negative climate changes may be either inappropriate policy or unrealistic expectations that encourage people to 'hang on' in extended dry periods, leading to substantial degradation of the soil and vegetation resources (McKeon and Hall 2000). One way to avoid such problems is to integrate climate change into regional planning (e.g. Olesen and Bindi 2002), however, there are significant issues in (1) identifying climate change thresholds given the complexities of climate change interacting with the many ongoing issues (e.g. dryland salinisation, change in irrigation water allocation processes etc.) and (2) the high levels of uncertainty inherent in climate change scenarios due to large ranges in greenhouse emissions (from uncertain socio-economic, political and technological developments) and fundamental uncertainty in the science of the global climate system. There are emerging approaches to deal with this uncertainty (e.g. Howden and Jones 2004) but these have yet to be applied to a regional context.

6.4.5 National Level Adaptations

The high levels of uncertainty in future climate changes suggest that rather than try to manage for a particular climate regime, we need more resilient agricultural systems (including socio-economic and cultural/institutional structures) to cope with a broad range of possible changes. There is a substantial body of both theory and practice on resilient systems (e.g. Gunderson et al. 1995). However, enhanced resilience usually comes with various types of costs or overheads such as building in redundancy, increasing enterprise diversity and moving away from systems that maximise efficiency of production at the cost of broader sustainability goals. One approach to developing more resilient agricultural regions is to develop an adaptive management strategy where policy is structured as a series of experiments that have formal learning and review processes. However, this could provide a serious challenge to some institutions that are based on precedent (and hence only look 'backwards' not 'forward'), have a short-term focus only and which are risk averse (e.g. Abel et al. 2002). Nevertheless, there is a large range of policy activities that could be undertaken which will enhance the capacity of agricultural systems to deal with a changing climate. These include (Howden et al. 2003a):

- Policy: linkages to existing initiatives to enhance resilience
- Managing transitions: support during transitions to new systems
- Communication: industry-specific and region-specific information
- R&D and training: participatory approach to improve self-reliance and provide the knowledge base for adaptation
- Model development and application: systems modelling to integrate and extrapolate anticipated changes
- Climate data and monitoring: to link into ongoing evaluation and adaptation
- Seasonal climate forecasting: for incremental adaptation linked to other information
- Breeding and selection: support and ensure access to global gene pools
- Pests, diseases and weeds: enhanced quarantine, sentinel monitoring, forecasting and management
- Water: trading systems that allow for climate variability and climate change, distribution systems, water management tools and technologies
- Landuse change and diversification: risk assessments and support

6.5 Key Research Challenges

Many needs for future research emerge from the arguments reported in the previous sections of this chapter. Key research challenges for the forthcoming years include:

- To perform studies of the integrated impacts of climate change and CO_2 increase on cropping *systems* more than on single crops and on mixed farming systems more than on monoculture farms (e.g. Tubiello et al. 2007).
- Develop linkages between global climate models and farming systems models to undertake these studies at a range of scales (e.g. Meinke et al. 2006).
- To conduct similar studies on the effects of climate change and CO_2 increase on specialised farming systems, which have a particular reliance on product quality.
- To undertake more research on adaptation at the farm level to influence strategies for improving the sustainability of farming systems that can deal with a large array of climate changes. There is a need to assess the rates of adaptation of new management strategies relative to rates of climate change for different farming

systems and different cropping regions. This will need to include exploration of a range of technological and policy adjustments (short and long term strategies) available in agriculture, in order to evaluate their efficiency in mitigating negative impacts or exploring new options offered by climate change.
- To develop integrated assessments using climatic and non-climatic conditions (economic, social, technological, environmental, institutional) to identify necessary changes in agriculture in a changing climate.
- To develop reliable weather forecasts at seasonal and other timescales and further methods for adapting such forecasts in farm management. These climate forecasts may provide one of the most efficient ways of adapting to climate change (McKeon et al. 1993).
- To better understand how farmers perceive climatic related risks and how they respond, in both the short and long term, to variable climatic conditions, including the magnitude and frequency of extreme events. A key element here will be in reducing the uncertainty of the climate change forecasts.
- To scope how biotechnology could be better used for coping with drought, heat and other climate related problems much more rapidly than it is possible today by means of the identification of genotypes and species not previously used for agricultural purpose or others already identified that may be quickly developed into widely available varieties. Genetic traits from other species may also be introduced in domesticated species to overcome some of the anticipated problems.

Sound research on the impact of climate change on agriculture, as well in other sectors, requires extensive and good data on biophysical and socio-economic responses. Where these responses can be reliably quantified, the identification of changes in agriculture by integrated assessment using both climatic and non-climatic conditions should be performed. These analyses should be structured to provide a detailed exploration of a range of technological and policy adjustments in agriculture for reducing negative impacts, increasing positive impacts and exploring new options for agriculture. Finally, these results need to be communicated effectively so as to be useful for research and policy making, taking into account the often different cultural, institutional and capacity differences between regions.

References

Abel N., Langston A., Ive J., Tatnell B., Howden S.M. and Stol J. (2002). Institutional change for sustainable land use: a participatory approach from Australia. In: Complexity and Ecosystem Management: The Theory and Practice of Multi-Agent Systems, Janssen MA (Ed). Edward Elgar, Cheltenham, UK. pp. 286–342.
Aggarwal P.K., Joshi P.K., Ingramand J.S., Gupta R.K. (2004). Adapting food systems of the Indo-Gangetic plains to global environmental change: key information needs to improve policy formulation. Environ. Sci. Policy 7, 487–498.
Ainsworth E.A. and Long S.P. (2005). What have we learned from 15 years of free-air CO_2 enrichment (FACE)? A meta-analysis of the responses of photosynthesis, canopy properties and plant production to rising CO_2. New Phytol. 165, 351–372.

Ainsworth E.A., Rogers A., Nelson R. and Long S.P. (2004). Testing the source–sink hypothesis of down-regulation of photosynthesis in elevated CO2 in the field with single gene substitutions in Glycine max. Agric. Forest Meteorol. 122, 85–94.

Antle J.M., Capalbo S.M., Elliott E.T. and Paustian K.H. (2004). Adaptation, spatial heterogeneity, and the vulnerability of agricultural systems to climate change and CO2 fertilization: an integrated assessment approach. Climate Change 64, 289–315.

Bartolini G., Morabito M., Crisci A., Grifoni D., Torrigiani T., Petralli M., Maracchi G. and Orlandini S. (2008). Recent trends in Tuscany (Italy) summer temperature and indices of extremes. Int. J. Climatol. DOI: 10.1002/joc.1673.

Bindi M. and Fibbi L. (2000). Modelling climate change impacts at the site scale on grapevine. In: Climate Change, Climate Variability and Agriculture in Europe, Downing TE, Harrison PA, Butterfield RE and Lonsdale KG (Eds.). Research Report no. 21, Environmental Change Unit, University of Oxford, UK, pp. 117–134.

Bindi M., Ferrini F. and Miglietta F. (1992). Climatic change and the shift in the cultivated area of olive trees. J. Agric. Mediter. 22, 41–44.

Bindi M., Ferrise R., Moriondo M. and Brandani G. (2008). Climate Change as a risk to potato production. Proceedings of the Potato Science for the Poor – Challenges for the New Millennium "A Working Conference to Celebrate the International Year of the Potato", Cuzco, Peru, pp. 25–28, March 2008 (in press).

Bindi M., Fibbi L. and Miglietta F. (2001). Free air CO_2 enrichment (FACE) of grapevine (Vitis vinifera L.): II. Growth and quality of grape and wine in reponse to elevated CO_2 concentrations. Eur. J. Agron. 14, 145–155.

Bindi M., Fibbi L., Gozzini B., Orlandini S. and Miglietta F. (1996). Modeling the impact of future climate scenarios on yield and yield variability of grapevine. Climate Res. 7, 213–224.

Bindi M., Miglietta F., Vaccari F., Magliulo E., Giuntoli A. (2005) Charter 6. Growth and quality responses of potato to elevated [CO2]. In: Managed Ecosystems and CO2: Case Studies, Processes and Perspectives, Josef N (Ed.). Ecological Studies of sprinter, pp. 105–120.

Blumenthal C.S., Bekes F., Batey I.L., Wrigley C.W., Moss H.J., Mares D.J. and Barlow E.W.R. (1991). Interpretation of grain quality results from wheat variety trial with reference to high temperature stress. Aust. J. Agric. Res. 43, 325–334.

Brown R.A., Rosenberg N.J., Hays C.J., Easterling W.E. and Mearns L.O. (2000). Potential production and environmental effects of switchgrass and traditional crops under current and greenhouse-altered climate in the central United States: a simulation study. Agric. Ecosyst. Environ. 78, 31–47.

Calanca P., Kaifez Bogataj L., Halenka T., Cloppet E., Mika J. (2008). Use of climate change scenarios in agrometeorological studies: past experiences and future need. In: Survey of Agrometeorological Practices and Applications in Europe Regarding Climate Change Impacts, Nejedlik P, Orlandini S (Eds.). COST 734 – EC, Firenze, pp. 246–278.

Cannell M.G.R., Palutikof J.P., Sparks T.H. (1999). Indicators of climate change in the UK. National Environmental Research Council, DETR, Wetherby, UK.

COPA-COGECA (2003). Valutazione delle conseguenze sull'agricoltura e la silvicoltura dell'ondata di caldo e della siccità che hanno contrassegnato l'estate del 2003. www.copa-cogeca.be.

Dalla M.A., Gozzini B., Grifoni D., Orlandini S. (2003). Applications of RAMS models for the creation of agrometeorological maps at territorial level. In: Proceedings of the Sixth European Conference on Applications of Meteorology (CD-ROM). Roma, 15–19 September 2003.

Das H.P., Adamenko T.I., Anaman K.A., Gommes R.G., Johnson G. (2003). Agrometeorology related to extreme events. World Meteorological Organization Technical Note n. 201, WMO n. 943.

Döll P. (2002) Impact of Climate Change and Variability on Irrigation Requirements: A Global Perspective. Climatic Change 54, 269–293.

Easterling W.E. (1996). Adapting North American agriculture to climate change in review. Agric. Forest Meteorol. 80, 1–53.

Easterling W.E., Hurd B.H. and Smith J.B. (2004): Coping with global climate change: the role of adaptation in the United States, Pew Center on Global Climate Change, Arlington, Virginia, 52 pp. [Accessed 30.06.08: http://www.pewclimate.org/docUploads/Adaptation.pdf]

Easterling W.E., Aggarwal P.K., Batima P., Brander K.M., Erda L., Howden S.M., Kirilenko A., Morton J., Soussana J.-F., Schmidhuber J. and Tubiello F.N. (2007) Food, fibre and forest products. In: Climate Change 2007: Impacts, Adaptation and Vulnerability. Contribution of Working Group II to the Fourth Assessment Report of the Intergovernmental Panel on Climate, Change, Parry ML, Canziani OF, Palutikof JP, van der Linden PJ and Hanson CE (Eds.). Cambridge University Press, Cambridge, UK, pp. 273–313.

Easterling W.E., Aggarwal P.K., Batima P., Brander K.M., Erda L., Howden S.M., Kirilenko A., Morton J., Soussana J.-F., Schmidhuber J. and Tubiello F.N. (2007) Food, fibre and forest products. In: Climate Change 2007: Impacts, Adaptation and Vulnerability. Contribution of Working Group II to the Fourth Assessment Report of the Intergovernmental Panel on Climate, Change, Parry ML, Canziani OF, Palutikof JP, van der Linden PJ and Hanson CE (Eds.). Cambridge University Press, Cambridge, pp. 273–313.

Eitzinger J., Thaler S., Orlandini S., Nejedlik P., Kazandjiev V., Vucetic V., Sivertsen T.H., Mihailovic D.T., Lalic B., Tsiros E., Dalezios N. (2008). Agroclimatic indices and simulation models. In: Survey of Agrometeorological Practices and Applications in Europe Regarding Climate Change Impacts, Nejedlik P, Orlandini S (Eds.). COST 734 – EC, Firenze, pp. 14–113.

European Environment Agency (EEA) (2003). Mapping the impacts of recent natural disasters and technological accidents in Europe. Environmental issue report n. 35.

Ferris R., Wheeler T.R., Ellis R.H, Hadley P., Wollenweber B., Porter J.R., Karacostas T.S., Papadopoulos M. N. and Schellberg J. (1999). Effects of high temperature extremes on wheat. In: Climate Change, Climate Variability and Agriculture in Europe: An Integrated Assessment, Downing TE, Harrison PA, Butterfield RE and Lonsdale KG (Eds.). Research Report 21. Environmental Change Unit, University of Oxford.

Galan C. Garcia-Mozo H., Vazquez L., Ruiz L., de la Guardia C.D., Trigo M.M. (2005) Heat requirement for the onset of the Olea europaea L. pollen season in several sites in Andalusia and the effect of the expected future climate change. Int. J. Biometeorol. 49(3), 184–188.

Gifford R.M. (2004). The CO2 fertilising effect – does it occur in the real world? New Phytol. 163, 221–225.

Gunderson L., Holling C.S. and Light S. (Eds.) (1995). Barriers and Bridges to the Renewal of Ecosystems and Institutions. Columbia University Press. New York.

Gutierrez, A.P., Ponti L., d'Oultremont T., Ellis C.K. (2008). Climate change effects on poikilotherm tritrophic interactions. Climatic Change 87(Suppl. 1): S167–S192, Mar.

Harrison P.A., Butterfield R.E. and Orr J.L. (2000). Modelling climate change impacts on wheat, potato and grapevine in Europe. In: Climate Change, Climatic Variability and Agriculture in Europe, Downing TE, Harrison PA, Butterfield RE and Lonsdale KG (Eds.). Environmental Change Unit, University of Oxford, UK, pp. 367–390.

Howden S.M. (2002) Potential global change impacts on Australia's wheat cropping systems. In: Effects of Climate Change and Variability on Agricultural Production Systems, Doering OC, Randolph JC, Southworth J and Pfeifer RA (Eds.). Kluwer, Dordrecht, pp. 219–247.

Howden S.M. and Jones R.N. (2004): Risk assessment of climate change impacts on Australia's wheat industry. New directions for a diverse planet. In: Proceedings of the 4th International Crop Science Congress, Fischer T, Turner N, Angus J, McIntyre J, Robertson L, Borrell A and Lloyd D (Eds.). Brisbane, Australia [Accessed 30.06.08: http://www.cropscience.org.au/icsc2004/symposia/6/2/1848_howdensm.htm]

Howden S.M., Ash A.J., Barlow E.W.R., Booth Charles S., Cechet R., Crimp S., Gifford R.M., Hennessy K., Jones R.N., Kirschbaum M.U.F., McKeon G.M., Meinke H., Park S., Sutherst R., Webb L. and Whetton P.J. (2003a). An overview of the adaptive capacity of the Australian agricultural sector to climate change – options, costs and benefits. Report to the Australian Greenhouse Office, Canberra, Australia, pp.157.

Howden S.M., Meinke H., Power B. and McKeon G.M (2003b) Risk management of wheat in a non-stationary climate: frost in Central Queensland. In: Integrative Modelling of Biophysical, Social and Economic Systems for Resource Management Solutions, Post DA (Ed.). Proceedings of the International Congress on Modelling and Simulation, July 2003, Townsville, pp. 17–22.

Howden S.M., Soussana J.F., Tubiello F.N., Chhetri N., Dunlop M., Meinke H.M. (2007) Adapting agriculture to climate change. In: Proceedings of the National Academy of Sciences, 104, 19691–19696.

IPCC (2001). Climate Change 2001: Impacts, Adaptation & Vulnerability. Contribution of Working Group II to the Third Assessment Report of the Intergovernmental Panel on Climate Change (IPCC), McCarthy JJ, Canziani OF, Leary NA, Dokken DJ and White KS (Eds.). Cambridge University Press, UK.

IPCC (2007a). Climate Change 2007: The Physical Science Basis. Contribution of Working Group I to the Fourth Assessment Report of the Intergovernmental Panel on Climate Change, Solomon S, Qin D, Manning M, Chen Z, Marquis M, Averyt KB, Tignor M, Miller HL (Eds.). Cambridge University Press, Cambridge

IPCC (2007b) Climate Change 2007: Impacts, Adaptation and Vulnerability. Contribution of Working Group II to the Fourth Assessment Report of the Intergovernmental Panel on Climate Change, Parry ML, Canziani OF, Palutikof JP, van der Linden PJ, Hanson CE (Eds.). Cambridge University Press, Cambridge

Jones G.V., White M.A., Cooper O.R., Storchmann K. (2005). Climate change and global wine quality. Climatic Change 73(3), 319–343.

Kalnay, E. et al. (1996). The NCEP/NCAR 40-Year Reanalysis Project. Bull. Am. Meteorol. Soc. 77, 437–471.

Kenny G.J. and Harrison P.A. (1992). The effects of climate variability and change on grape suitability in Europe. J. Wine Res. 3, 163–183.

Kimball B.A., Kobayashi K. and Bindi M. (2002). Responses of agricultural crops to free-air CO_2 enrichment. Adv. Agron. 77, 293–368.

Komor E., Orlich G., Weig A. and Kockenberger W. (1996). Phloem loading – not metaphysical, only complex: towards a unified model of phloem loading. J. Exp. Bot. 47, 1155–1164.

Kurukulasuriya, P. and Rosenthal S. (2003). Climate change and agriculture: a review of impacts and adaptations. World Bank Climate Change Series, vol. 91, World Bank Environment Department, Washington, DC, District of Columbia, 96 pp.

Lobell D.B., Field C.B., Cahill K.N., Bonfils C. (2006). Impacts of future climate change on California perennial crop yields: Model projections with climate and crop uncertainties. Agric. Forest Meteorol. 141(2–4), 208–218.

Long S.P., Ainsworth E.A., Rogers A. and Ort D.R. (2004). Rising atmospheric carbon dioxide: plants FACE the future. Annu. Rev. Plant Biol. 55, 591–628.

Matsui T., Namuco O.S., Ziska L.H., Horie T. (1997). Effects of high temperature and CO_2 concentration on spikelet sterility in indica rice. Field Crops Res. 51, 213–219.

McKeon G. and Hall W. (2000). Learning from history: preventing land and pasture degradation under climate change. Final Report to the Australian Greenhouse Office. Queensland Department of Natural Resources and Mines, Brisbane, Australia. www.longpaddock.qld.gov.au/AboutUs/Publications/ByType/Reports/LearningFromHistory/

McKeon G.M., Howden S.M., Abel N.O.J. and King J.M. (1993). Climate change: adapting tropical and subtropical grasslands. In: Proceedings of the XVII International Grassland Congress, Palmerston NZ, 13–16 February, Vol. 2, pp. 1181–1190.

Meinig D.W. (1962). On the margins of the good earth. The south Australian wheat frontier 1869–1884. The monograph series of the Association of American Geographers, Rand McNally, Chicago, p. 231.

Meinke H., Nelson R., Kokic P., Stone R., Selvaraju R., Baethgen W. (2006). Actionable climate knowledge: from analysis to synthesis. Climate Res. 33, 101–110.

Meinke H. and Hammer G.L. (1997). Forecasting regional crop production using SOI phases: an example for the Australian peanut industry. Aust. J. Agric. Res. 48, 789–93.

Meinke H., Nelson R., Kokic P., Stone R., Selvaraju R., Baethgen W. (2006) Actionable climate knowledge: from analysis to synthesis. Climate Res. 33, 101–110.

Miglietta F., Bindi M., Vaccari F., Schapendonk A.H.C.M., Wolf J., Butterfield R. (1998). Climate change and global productivity: root and tuber crops. In: Climate Change and Global Crop Productivity. Reddy KR, Hodges HF (Eds.). CABI, UK, pp. 189–212.

Monirul M. and Mirza Q. (2002). Global warming and changes in the probability of occurrence of floods in Bangladesh and implications. Global Environ. Change 12, 127–138.

Monteith J.L. (1980). Development and extension of Penman's evaporation formula. In: Applications of Soil Physics, Hillel D (Ed.). Academic, Orlando, FL.

Moriondo M. and Bindi M. (2008). Impact of climate change on grapevine (Vitis vinifera L.) at regional scale: phenology, yield and biotic stress responses (submitted Australian Journal of Grape and Wine Research).

Moriondo M., Stefanini F.M., Bindi M (2008). Reproduction of olive tree habitat suitability for global change impact assessment. Ecological Modelling, 218(1–2), 95–109.

Nearing M.A., Pruski F.F. and O'Neal M.R. (2004). Expected climate change impacts on soil erosion rates: a review. J. Soil Water Conserv. 59, 43–50.

Olesen J.E. and Bindi M. (2002). Consequences of climate change for European agricultural productivity, land use and policy. Euro. J. Agronomy 16, 239–262.

Olesen J.E. and Grevsen K. (1993). Simulated effects of climate change on summer cauliflower production in Europe. Eur. J. Agron. 2, 313–323.

Olesen J.E., Carter T.R., Díaz-Ambrona C.H., Fronzek S., Heidmann T., Hickler T., Holt T., Mínguez, P. Morales, J. Palutikof, M. Quemada, M. Ruiz-Ramos, G. Rubæk, F. Sau, B. Smith M.I. and Sykes M. (2007). Uncertainties in projected impacts of climate change on European agriculture and terrestrial ecosystems based on scenarios from regional climate models. Climatic Change 81, S123–S143.

Olesen J.E., Jensen T, Petersen J. (2000). Sensitivity of field-scale winter wheat production in Denmark to climate variability and climate change. Climate Res. 15, 221–238.

Pearson S, Wheeler T.R., Hadley P. and Wheldon A.E. (1997). A validated model to predict the effects of environment on the growth of lettuce (*Lactuca sativa* L.): implications for climate change. J. Hortic. Sci. 72, 503–517.

Porter J.R. and Gawith M. (1999). Temperatures and the growth and development of wheat: a review. Eur. J. Agron. 10, 23–36.

Rajagopal V., Kasturi Bai K.V. and Naresh Kumar S. (2002) Drought management in plantation crops. In: Plantation Crops Research and Development in the New Millennium, Rathinam P, Khan HH, Reddy VM, Mandal PK and Suresh K (Eds.). Coconut Development Board, Kochi, Kerala State, 30–35.

Reddy K.R., Doma P.R., Mearns L.O., Boone M.Y.L., Hodges H.F., Richardson A.G. and Kakani V.G. (2002). Simulating the impacts of climate change on cotton production in the Mississippi Delta. Climate Res. 22, 271–281.

Richter G.M., Qi A., Semenov M.A. and Jaggard K.W. (2006). Modelling the variability of UK sugar beet yields under climate change and husbandry adaptations. Soil Use Manag. 22, 39–47.

Rosenzweig C., Tubiello F.N., Goldberg R.A., Mills E. and Bloomfield J. (2002). Increased crop damage in the US from excess precipitation under climate change. Global Environ. Change 12, 197–202.

Rozema J. (1993). Plant responses to atmospheric carbon dioxide enrichment: interactions with some soil and atmospheric conditions. Vegetatio 104/105, 173–190.

Sinclair T.R., Pinter P.J., Kimball B.A., Adamsen F.J., LaMorte R.L., Wall G.W., Hunsaker D.J., Adam N., Brooks T.J., Garcia R.L., Thompson T., Leavitt S. and Matthias A. (2000). Leaf nitrogen concentration of wheat subjected to elevated [CO_2] and either water or N deficits. Agric. Ecosyst. Environ. 79, 53–60.

Stephens D.J. and Lyons T.J. (1998). Variability and trends in sowing dates across the Australian wheatbelt. Aust. J. Agric. Res. 49, 1111–1118.

Stokes C.J., Ash A.J. (2006). Impacts of climate change on marginal tropical animal production systems. In: Agroecosystems in a Changing Climate, Newton PCD, Carran RA, Edwards GR, Niklaus PA (Eds.). CRC Press, London, pp. 323–328.

Struzik P., Toulios L., Stancalie G., Danson M., Mika J., Domenikiotis C. (2008). Satellite remote sensing as a tool for monitoring climate and its impact on the environment – possibilities and limitations. In: Survey of Agrometeorological Practices and Applications in Europe Regarding Climate Change Impacts, Nejedlik P, Orlandini S (Eds.). COST 734 – EC, Firenze, pp. 210–245.

Tanner C.B. and Sinclair T.R. (1983). Efficient water use in crop production: Research or re-search? In: Limitations to Efficient Water Use in Crop Production, Taylor HM, Jordan WR and Sinclair TR (Eds.). ASA, Madison, WI, pp. 1–27.

Tubiello F.N., Amthor J., Boote K., Donatelli M., Easterling W., Fischer G., Gifford R., Howden S.M., Reilly J. and Rosenzweig C. (2007). Crop response to elevated CO2 and world food supply. Euro. J. Agron. 26, 215–233.

Tubiello F.N., Donatelli M., Rosenzweig C. and Stockle C.O. (2000). Effects of climate change and elevated CO_2 on cropping systems: model predictions at two Italian locations. Eur. J. Agron. 13, 179–189.

Tubiello F.N., Soussana J-F, Howden S.M. (2007). Crop and pasture response to climate change. Proc. National Academy Sci. 104, 19686–19690.

Van Duivenbooden N., Abdoussalam S. and Ben Mohamed A. (2002). Impact of climate change on agricultural production in the Sahel. Part 2. Case study for groundnut and cowpea in Niger. Climatic Change 54, 349–368.

Van Ittersum M.K., Howden S.M. and Asseng S. (2003). Sensitivity of productivity and deep drainage of wheat cropping systems in a Mediterranean environment to changes in CO_2, temperature and precipitation. Agric. Ecosyst. Environ. 97, 255–273.

Varaprasad P.V., Boote K.J., Hartwell-Allen L. and Thomas J.M.G. (2003). Super-optimal temperatures are detrimental to peanut (Arachis hypogaea L.) reproductive processes and yield at both ambient and elevated carbon dioxide. Glob. Change Biol. 9, 1775–1787.

Webb L.B., Whetton P.H., Barlow E.W.R. (2008). Climate change and winegrape quality in Australia. Climate Res. 36(2), 99–111.

Webb L.B., Whetton P.H., Barlow E.W.R. (2007). Modelled impact of future climate change on the phenology of winegrapes in Australia. Aust. J. Grape Wine Res. 13(3), 165–175.

Wheeler T.R., Ellis R.H., Hadley P., Morison J.I.L., Batts G.R. and Daymond A.J. (1996). Assessing the effects of climate change on field crop production. Aspects Appl. Biol. 45, 49–54.

White M.A., Diffenbaugh N.S., Jones G.V., Pal J.S., Giorgi F. (2006). Extreme heat reduces and shifts United States premium wine production in the 21st century. Proc. Nat. Acad. Sci. USA 103(30), 11217–11222.

Wolf J. (2002). Comparison of two potato simulation models under climate change. II. Application, of climate change scenarios. Climate Res. 21, 187–198

Wurr D.C.E, Hand D.W., Edmondson R.N., Fellows J.R., Hannah M.A. and Cribb D.M. (1998). Climate change: a response surface study of the effects of CO_2 and temperature on the growth of beetroot, carrots and onions. J. Agric. Sci., Camb. 131, 125–133.

Chapter 7
Response of Domestic Animals to Climate Challenges

John Gaughan, Nicola Lacetera, Silvia E. Valtorta, Hesham Hussein Khalifa, LeRoy Hahn, and Terry Mader

> *The climatic niche for an animal is predicted in terms of the temperature extremes it can endure.*
>
> Folk 1974, p. 4

Abstract The livestock sector is socially, culturally and politically very significant. It accounts for 40% of the world's agriculture Gross Domestic Product (GDP). It employs 1.3 billion people, and creates livelihoods for one billion of the world's population living in poverty. Climate change is seen as a major threat to the survival of many species, ecosystems and the financial sustainability of livestock production systems in many parts of the world. The potential problems are even greater in developing countries. Economic studies suggest severe losses if current management systems are not modified to reflect the shift in climate. In short, farmers/managers need to adapt to the changes. There has been considerable interest in gaining an understanding how domestic livestock respond to climatic stressors. Studies have for the most part been undertaken in developed countries. These studies have provided a wealth of knowledge on differences between genotypes, the impact of climatic stress on production, reproduction and health. However little is known about adaptation of animals to rapid changes in climatic

---

J. Gaughan
School of Animal Studies, The University of Queensland, Gatton, Australia 4343

N. Lacetera
Dipartimento di Produzioni Animali, Università degli Studi della Tuscia, Viterbo, Italy

H.H. Khalifa
Department of Animal Production, Faculty of Agriculture, Al-Azhar University, Nasr City, Cairo, Egypt

S.E. Valtorta
Acuña de Figueroa 121 Piso 7 Depto51, C1180AAA Buenos Aires, ARGENTINA

L. Hahn
1818 Home Street, Hastings, NE 68901, USA

T. Mader
Professor, Animal Science, Haskell Ag. Lab., 57905 866 Rd., Concord, NE 68728

conditions. Furthermore, little is known about the impacts of climatic stressors on many indigenous breeds used throughout Africa, Asia and South America. The uncertainty of climate change, and how changes will impact on animal production on a global scale are largely unknown.

This chapter will discuss: what is understood about animal adaptation; the current knowledge of the impacts of climate stressors on domestic animals, in terms of production, health, and nutrition; housing and management methods which can be used to alleviate heat stress; techniques used to predict animal responses to heat; and, strategies required to ensure continued viability of livestock production.

7.1 Introduction

Climate change is not a new phenomenon. In the past, animals have been subjected to major shifts in the Earths climate. Some species did not survive, while others adapted to the changes and flourished.

- The extinction of Megafauna (mammals >100kg) around the world was probably due at least in part to environmental and ecological factors.
- The extinction was almost completed by the end of the last ice age.
- It is believed that Megafauna initially came into existence in response to glacial conditions and became extinct with the onset of warmer climates.

What are the implications for animal production today? How will climate change impact modern animal production? How will farmers and domestic animals adapt to these changes?

Climate change is seen as a major threat to the survival of many species, ecosystems (Frankham 2005; Hulme 2005; King 2004), and the viability and sustainability of livestock production systems (Hahn et al. 1990; Sombroek and Gommes 1995; Smit et al. 1996; Frank et al. 2001; Turnpenny et al. 2001). The economic impact of climate changes in relation to livestock production has been considered in several studies (Adams et al. 1990; Ray et al. 1992; Bowes and Crosson 1993; Easterling et al. 1993; Rosenweig and Parry 1994; St-Pierre et al. 2003). Most of the studies predict severe losses if current management systems are not modified to reflect the shift in climate.

7.2 Impact of Climate Change on Animal Agriculture

The livestock sector is socially, culturally and politically very significant. It accounts for 40% of the world's agriculture Gross Domestic Product (GDP). It employs 1.3 billion people, and creates livelihoods for one billion of the world's population living in poverty. Global meat production is expected to more than double from

229 to 465 million tonnes between 1999/2001 and 2050. Milk production is also expected to increase from 580 to 1,043 million tonnes over the same period. In order to achieve these increases, livestock production will intensify. Production of pigs and poultry is expected to account for much of the increase. Grazing occupies 26% of ice-free terrestrial land, and crop production for animal feed accounts for 33% of all arable land. It is estimated that livestock production accounts for 70% of all agricultural land and 30% of the total land surface (Steinfeld et al. 2006). Approximately 3.5 billion hectares are being grazed compared to 1.2–1.5 billion hectares under cropping (Howden et al. 2007).

Climate affects animal agriculture in four ways (Rötter and Van de Geijn 1999) through impacts on livestock: (1) feed-grain availability and price; (2) pastures and forage crop production and quality; (3) health, growth and reproduction; and (4) diseases and pests distributions. Adaptation of practices used by farmers to changing climatic conditions is paramount (ILRI 2006). These changes may result in a redistribution of livestock in a region; changes in the types of animals that are used (e.g., a shift from cattle to buffalo, sheep, goats or camels); genotype changes (e.g., the use of breeds that will handle adverse conditions, such as Brahman cattle); and changes in housing of animals (e.g., protective structures which have allowed the expansion of the dairy industry into areas such as southern USA, Brazil, Israel, Saudi Arabia that would not otherwise be suitable [Darwin et al. 1995]). A lack of thermally-tolerant breeds of cattle is already a major constraint on production in Africa (Voh et al. 2004). Furthermore, it is possible that conflicts over resources may become a problem (Darwin et al. 1995). However, climate change may have a positive impact on livestock production in some areas. For instance, areas that are cooler and wetter may increase forage production and, in turn, livestock production. Warming of areas such as Canada may increase in agricultural production (Arthur and Abizadeh 1988). Increased rainfalls and winter temperatures in India, Pakistan, and Bangladesh may have both positive (e.g., longer growing seasons) and negative (e.g., flooding, increased animal disease risk) effects. Changing conditions in Africa may spread trypanosomosis into previously unaffected areas. Movement of parasites into previously unaffected areas could result in large production and financial losses. Any advantages that may result from climate change could be hampered by an inability (political, social and financial) to change farming practices.

The impact of climate change (higher temperatures) on pastures and rangelands may include deterioration of pasture quality (C_3 grasses) towards lower quality tropical and subtropical C_4 grasses (Barbehenn et al. 2004) in temperate regions as a result of warmer temperatures and fewer frosts (Briske and Heitschmidt 1991; Greer et al. 2000); however, there could also exist potential increases in yield and possible expansion of C_3 grasses if climate change were favorable as a result of an increase in CO_2 (Kimball et al. 1993; McKeon et al. 1993; Idso and Idso 1994; Allen-Diaz 1996; Campbell et al. 1995; Reilly 1996), and if precipitation is favorable. An increase in CO_2 is likely to have a negative effect on C_4 grasses (Collatz et al. 1998; Christin et al. 2008) resulting in declines in pasture productivity and lower carrying capacity.

The impact of climate change on wildlife is deemed to be largely negative (Thuiller et al. 2006). However, in some instances increasing ambient temperature has had little negative impact, at least to date (Johnston and Schmitz 1997; Beaumont et al. 2006), whereas in other cases the impact is largely due to changes in vegetation (Johnston and Schmitz 1997). In some scenarios, a species may be able to extend their current range.

In the animal context, climate change needs to be viewed as more than global warming. As previously mentioned, some areas will become cooler, and this may only have minor impacts on the animals (but could alter feed availability). On the other hand, extreme events such as heat waves can have major impacts on non-adapted animals. Heat waves are recurring events in many current climates, and are projected to increase in number and intensity (Mearns et al. 1984; Gaffen and Ross 1998; IUC 2002). The risk of floods and droughts are also predicted to increase. While there may be little change *per se* in a region, extreme events may increase in both intensity and duration leading to substantial changes in animal management practices. Climatic variables which need to be assessed include: ambient temperature, relative humidity, the day to night and seasonal variations in ambient temperature, rainfall, wind speed, solar and terrestrial radiation, evaporation rates, and atmospheric CO_2 (Folk 1974; Hulme 2005). It is likely that heat and drought will be the major contributing factors to changes in animal production over the next 50 years. Some of the effects will be direct (e.g., heat stress of livestock), and others indirect (e.g., changing pasture composition). In the context of this chapter, we will concentrate on the impact of increasing heat load on livestock, and how animals adapt to increasing heat stress.

Livestock production involves a relatively small group of domesticated animals. Diamond (1999) reported that of the 148 non-carnivorous species weighing more than 45 kg as adults, only 14 have been domesticated. Even fewer bird species (0.001%) have been domesticated (Mignon-Grasteau et al. 2005). However, this does not necessarily make the task any easier.

Mammals are homeothermic endotherms and maintain a core body temperature between 35°C and 40°C depending on the species (Langlois 1994). They are able, through irradiative, conductive, convective, and evaporative exchanges, to generally maintain core body temperature within a fairly narrow range (Langlois 1994; Folk et al. 1998). In many species 5–7°C deviations from core body temperature may cause death, and at least reductions in productive performance. Mammals have a greater capacity for dealing with cold environmental conditions than they do with hot conditions (Folk et al. 1998). The lethal limit of core temperature is about 6°C above normal for healthy animals, and depression of central nervous activity, particularly in the respiratory center, occurs before that (Schmidt-Nielsen 1975). In horses, death may occur if core body temperature decreases by 10°C (27% deviation from normal) or increases by 5°C (13% deviation) (Langlois 1994). In cattle, death has occurred when rectal temperature exceeds 43.5°C (6°C above normal) (J. Gaughan, 2006, personal communication). There is a paucity of information on upper critical body temperature in livestock. Body temperature of some species is more labile, with the capacity to survive large changes in body temperature. For example, the core body temperature of camels can vary between 34.0°C and 42°C (Schmidt-Nielsen et al. 1956; Fowler 1999). Antelope ground squirrels show large fluctuations of core temperature between 37°C to 43°C. These animals will return to burrows or seek shade and rest when core body temperature approaches 43°C. Once body temperature returns to normal (approx 37°C) activity may recommence (Willmer et al. 2000). Animals that hibernate may lower body temperature to only a few degrees above freezing.

7.3 Economic Impacts of Climate Change on Animal Agriculture

Production losses in livestock enterprises are an expected outcome of climate change. Leva et al. (1997) have determined the present production losses in the major milk producing regions in Argentina, and have projected those losses if the global climate change scenario took place. Estimations based on work by Berry et al. (1964), for cows producing 15, 20 and 25 kg milk/day suggest that, under the global climate change scenario, milk production in Argentina would decline on average by 60% (Leva et al. 1997). The economic effect of heat stress on dairy cows in Australia was estimated to be AUS$11,986 per 100 cows if no heat abatement strategies were implemented (Mayer et al. 1999). A recent investigation of the economic losses due to heat stress for a number of US livestock industries (dairy cattle, beef cattle, pigs, chickens and turkeys) was undertaken by St-Pierre et al. (2003). They concluded that without heat abatement, losses across all livestock industries would average US$2.4 billion/annum. A review by Sackett et al. (2006) estimated that the annual production losses in the Australian feedlot industry due to summer heat stress at A$16.5 million. Clearly, economic losses may be significant.

7.3.1 Effect of Heat Waves on Animals

An aspect of climate change is an increase in severe weather. Significant heat events since the mid 1990s appear to be increasing and have resulted in sizable human and animal mortality. Hahn and Mader (1997), Xin and Puma (2001) and Hahn et al. (2000, 2002) described the impact on livestock from a week long heat wave in the mid-central United States during July 1995: the heat wave also resulted in a significant number of human deaths. That heat wave was estimated to have cost the US cattle industry $28 million in animal deaths and reduced livestock performance. In Iowa over 1.8 million laying hens died during this heat wave. In July 1999, a heat wave in Nebraska was responsible for 3,000 cattle deaths and over $20 million in economic loss (http://hpccsun.unl.edu/nebraska/owh-july31.html). In Australia, a heat wave in 2000 resulted in the death of 24 people and over 2,000 cattle. Poultry losses were estimated to exceed 15,000. Horses and dogs also died during this event. During the heat wave which occurred in Europe during summer 2003, over 35,000 people, thousands of pigs, poultry and rabbits died in the French regions of Brittany and Pays-de-la-Loire (http://lists.envirolink.org/pipermail/arnews/Week-of-Mon-20030804/004707.html). In 2004 during an Australian heat wave over 900 cattle died. In 2006, a major heat wave moved across the USA and Canada. This heat wave resulted in the death of over 15 pets (not defined), 225 people, 25,000 cattle, and 700,000 poultry in California alone. Heat waves in Europe in 2006 and 2007 resulted in the deaths of more than 2000 people. However the number of animal deaths could not be established. Over 800 peacocks died during a heat wave in India in 2007. It is likely that without some form of intervention, either in terms of management or genetic change (via selection for heat tolerance), significant animal deaths will occur during future heat waves. These losses could be significantly greater if the predicted increase in the intensity and duration of heat waves is realized.

7.4 Impact of Climate Change on Animal Health

The effects of climate changes and, in particular, global warming on health status of livestock have not been considered with the same attention as given to humans (http://www.who.int/globalchange/climate/en/). However, it is assumed that as in the case of humans, climate changes can affect the health of livestock and poultry, both directly and indirectly. Direct impacts include temperature-related illness and death, and the morbidity of animals during extreme weather events. Indirect impacts follow more intricate pathways and include those deriving from the influence of climate on microbial density and distribution, distribution of vector-borne diseases, host resistance to infections, food and water shortages, or food-borne diseases. Some general concepts of livestock environment and health have been presented by Simensen (1984) and these may serve as a guide to management of disease during climate change.

A series of studies carried out in dairy cows indicated a higher occurrence of mastitis during periods of hot weather (Giesecke 1985; Smith et al. 1985; Morse et al. 1988; Waage et al. 1998; Cook et al. 2002; Yeruham et al. 2003). However, the mechanisms responsible for the higher occurrence of mastitis during summer have not been elucidated. The hypothesis to explain these observations include the possibility that high temperatures can facilitate survival and multiplication of pathogens (Hogan et al. 1989) or their vectors (Chirico et al. 1997), or a negative action of heat stress on defensive mechanisms (Giesecke 1985).

During summer, ketosis is more prevalent due to increased maintenance requirements for thermoregulation and lower feed intake (Lacetera et al. 1996), and the incidence of lameness increases as a consequence of metabolic acidosis (Shearer 1999). Furthermore, analysis of metabolic parameters in the blood of dairy cows indicates that high environmental temperatures may be responsible for alteration of liver function, mineral metabolism and oxidative status (Bernabucci et al. 2002) (Table 7.1), which may also lead to animals having clinical or sub-clinical disease.

Results from an epidemiology study carried out in California (Martin et al. 1975) documented higher mortality rates of calves born during summer. Others

Table 7.1 Effects of high environmental temperatures on blood indexes of energy and mineral metabolism, liver function and oxidative balance in dairy cows (Adapted from Lacetera et al. 1996; Ronchi et al. 1999, Bernabucci et al. 2002)

Items	Changes
Energy metabolism	
Body condition score	Loss
Glycemia	Decrease
Non esterified fatty acids	Increase
Ketone bodies	Increase
Urea	Increase/decrease
Liver function	
Albumin	Decrease
Cholesterol	Decrease
Bilirubine	Increase
AST	Decrease/increase
γGT	Decrease/increase
LDH	Decrease/no changes
AlPh	Decrease
Mineral metabolism	
Ca	Decrease/no changes
P	Decrease/no changes
Mg	Decrease
Na	Decrease
K	Decrease
Cl	Increase
CAB (Na + K − Cl)	Decrease
Oxidative balance	
Pro-oxidants (TBARS, ROMs)	Increase
Antioxidants (GSH, Thiols)	Decrease

have reported that heat stress may be responsible for impairment of the protective value of colostrum both in cows (Nardone et al. 1997) and pigs (Machado-Neto et al. 1987), and also for alteration of passive immunization of calves (Donovan et al. 1986; Lacetera 1998). On the other hand, results on the negative influence of heat stress on colostral immunoglobulins may provide an explanation for the higher mortality rate of newborns observed during hot months.

Several studies have assessed the relationships between heat stress and immune responses in cattle, chickens or pigs. However, results of those studies are conflicting. In particular, some authors reported an improvement (Soper et al. 1978; Regnier and Kelley 1981; Beard and Mitchell 1987), others described an impairment (Regnier and Kelley 1981; Elvinger et al. 1991; Kamwanja et al. 1994; Morrow-Tesch et al. 1994), and others indicated no effects (Regnier et al. 1980; Kelley et al. 1982; Bonnette et al. 1990; Donker et al. 1990; Lacetera et al. 2002) of high environmental temperatures on immune function. Recently, in a field study carried out in Italy during the summer 2003 (Lacetera et al. 2005), which was characterized by the occurrence of at

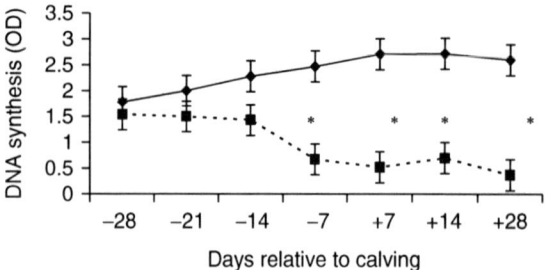

Fig. 7.1 DNA synthesis in peripheral blood mononuclear cells (PBMC) isolated from spring (*solid line*) or summer (*dotted line*) cows. The PBMC were stimulated with concanavalin A. Values are the means ± SEM of the optical density (OD). *Asterisks* indicate significant differences ($P < 0.01$) (Adapted from Lacetera et al. 2005)

least three heat waves, there was a profound impairment of cell-mediated immunity in high yielding dairy cows (Fig. 7.1). Interestingly, such results suggest that immunosuppression during hot periods may be responsible for the failure of vaccine interventions and for reduced reliability of diagnostic tests based on immune system reaction (i.e., tuberculin skin test). The large variety of experimental conditions in terms of species, severity and length of heat stress, recovery opportunities, and also of the specific immune functions taken into consideration are likely to explain the discrepancy among results of different studies. In addition other factors such as photoperiod may impact on immune function (Auchtung et al. 2004).

Global warming will affect the biology and distribution of vector-borne infections. Wittmann et al. (2001) simulated an increase of temperature values by 2°C. Under these conditions, their model indicated the possibility of an extensive spread of *Culicoides imicola*, which represents the major vector of the bluetongue virus. The distribution of ticks and flies is also likely to change.

Another mechanism through which climate changes can impair livestock health is represented by the favorable effects that high temperature and moisture have on growth of mycotoxin-producing fungi. Their growth and the associated toxin production are closely correlated to the degree of moisture to which they are exposed, which itself is dependent on weather conditions at harvest, and techniques for drying and storage (Frank 1991). With regard to alteration of animal health, mycotoxins can cause acute disease episodes when animals consume critical quantities. Specific toxins affect specific organs or tissues such as the liver, kidney, oral and gastric mucosa, brain, or reproductive tract. In acute mycotoxicoses, the signs of disease often are marked and directly referable to the affected target organs. Most frequently, however, concentrations of mycotoxin in feeds are below those that cause acute disease. At lower concentrations, mycotoxins reduce the growth rate of young animals, and some interfere with native mechanisms of resistance and impair immunologic responsiveness, making the animals more susceptible to infection. Studies have shown that some mycotoxins can alter lymphocyte functions in domestic ruminants through alteration of DNA structure and functions (Lacetera et al. 2003, 2006; Vitali et al. 2004).

7.5 Animal Adaptation

Adaptation has the potential to reduce some of the damage caused by climate change (Hulme 2005). However, little work has been undertaken to identify strategies which will allow domestic animals to adapt to climate change (King 2004).

How is adaptation in domestic animals defined? Is it simply the ability of the animals to survive, grow, and reproduce? Or, is it maintenance of productive performance at some predetermined level? Numerous terms are used to describe animal responses to adverse environments (Folk 1974; Yousef 1987).

In its broadest form, adaptation is defined as a change which reduces the physiological strain produced by a stressful component of the total environment. The change may occur within the lifetime of an organism (phenotypic) or be the result of genetic selection in a species or subspecies (genotypic) (Bligh and Johnson 1973).

- Genetic or Biological Adaptation: Adaptation is achieved through genetic change over time (generations), which involves evolutionary processes, and also through environmental stimulation and experiences during an animal's lifetime (Hafez 1968; Price 1984 cited by Mignon-Grasteau et al. 2005). This comes about via natural selection, and selection of animals by humans (Hafez 1968). Identification of heat tolerant phenotypes within existing breeds, or infusion of genes for heat tolerance may be a partial solution. The biological properties of animals are a result of interactions between stress intensity, magnitude of environmental fluctuations, and the energy available from resources (Parsons 1994).
- Phenotypic or Physiological Adaptation: Animals have the ability to respond to acute or sudden environmental change (e.g. shivering when exposed to cold) (Folk 1974; Hafez 1968; Langlois 1994), and with longer exposure to climate change (although this may be somewhat limited).

Other terms commonly used to describe an animal's response to climatic variables include acclimation, acclimatization and habituation. These were defined by Folk (1974) as follows:

- Acclimatization – the functional compensation over a period of days to weeks in response to a complex of environmental factors, as in seasonal or climatic change
- Acclimation – the functional compensation over a period of days to weeks in response to a single environmental factor only, as in controlled experiments
- Habituation – (i) Specific – specific to a particular repeated stimulus and specific to the part of the body which has been repeatedly stimulated, and (ii) General – a change in the physiological set of the organism relevant to the repeated stimulus and the conditions incidental to its application

Animals are regularly exposed to climatic stress (Parsons 1994). The extents to which they are able to adapt are limited by physiological (genetic) constraints (Devendra 1987; Parsons 1994). Selection criteria for livestock and poultry (and possibly domestic animals in general) need to be considered in the context of climate change, and whether the location (habitat) is likely to be favorable or unfavorable to the species of concern. There is a necessity to select livestock, and use livestock systems (e.g. pasture management) on the basis of expected climatic conditions.

Animal performance may be limited where there is climatic adversity. Animals that have evolved to survive in adverse conditions generally have the following characteristics: high resistance to stress, low metabolic rate, low fecundity, long lives, behavioral differences, late maturing, smaller mature size, and slow rate of development (Devendra 1987; Parsons 1994; Hansen 2004). This suggests that selection or use of animals (often indigenous breeds) that are adapted to adverse climates will have lower productivity than those selected for less stressful climates. In general this is true; however, these animals survive, grow and continue to provide food, fibre and fuel under conditions where the animals higher production potential may at the worst die or at best produce at levels at or below the indigenous breeds. In many of the developing countries located in the tropics, the poor reproductive performance of cows (both indigenous and imported), for example, results from a combination of genetics, management, and environmental factors (Agyemang et al. 1991). Depending on location, climate change may improve the local environment or have major negative impacts. In order to take advantage of positive changes or reduce the impact of negative changes, farmers will need to adapt. Improved genetics (including suitability to the environment) and improved management may go a long way to take advantage of changes or minimize the impact. The use of housing, microclimate modification (e.g. shade, sprinklers), improved nutritional management, disease control, and new reproductive technologies are usually needed if animals are to meet their genetic potential (Champak Bhakat et al. 2004; Voh et al. 2004; Magana et al. 2006). However, the cost of implementation of these processes may be too high to be economically viable, especially in developing countries.

Mechanisms of animal adaptation have been defined by Devendra (1987) as: anatomical, morphological, physiological, feeding behavior, metabolism, and performance. Physiological and behavioral adaptations are employed first in response to environmental changes. Animals employ multiple strategies in order to adapt to the environment. For example, Sudanese Desert goats tolerate thermal stress and nutritional shortage (food and water) by their capacity to lose heat via panting and cutaneous evaporation, as well as their ability to concentrate urine to levels above 3,200 mosmols/kg (Ahmed and Elkheir 2004). Additional characteristics of adaptation by goats under different climates are presented in Table 7.2. Camels also use multiple strategies to cope with thermal stress and nutritional shortages. They use sweating to control body temperature in an environment where water loss needs to be minimized. However, they have the ability to increase body temperature during the day (up to 41°C) and then dissipate the heat during the night when desert temperatures may approach or fall below 0°C. Storing heat during the day and dissipating the heat at night is a method of conserving energy and water loss. Camels have an ability to consume large amounts of water when dehydrated. Guerouali and Filali (1995) reported that the water intake of hydrated camels averaged 1.33% of body weight. The camels were then exposed to a 27 day dehydration period. When re-hydrated, water intake increased to 19.12% of body weight within a couple of minutes. Water intakes of up to one third of body weight have been reported. Large water intakes

Table 7.2 Adaptive properties of goats

Climate	Anatomical	Morphological	Physiological	Metabolism	Feeding[a]	Performance
Arid/semi-arid	Large size (30–50 kg), long legs and ears, scrotum shows two distinct sacs	White, black or brown coat color, shiny surface, white colored goats absorb less solar radiation	Panting, sweating	Increased mobilization of fat during periods of feed shortage, low water turnover rate	Browse over long distances, resistance to dehydration, dessication of feces, increased urine concentration and reduced urine volume, rumen acts as water reservoir, high digestive efficiency of coarse roughages, efficient utilization and retention of nitrogen	Meat/milk/fiber
Sub-tropical	Intermediate size (25–30 kg)	White, black or brown coat, less shiny coat	Panting, sweating	Lower water turnover rate	Intermediate	Meat/milk/fiber
Humid/sub-humid	Small size (10–25 kg), short legs, small ears	Mainly black or brown coat color, shiny coat	Reduced panting, more access to shade	Low metabolic rate	Reduced walking due to increased availability of forages and crop residues	Meat

[a] Feeding behaviour and feed utilization.

may lead to osmotic shock. However, camels are able to store large amounts of water in the stomach. The camel can also dehydrate without affecting blood viscosity and composition. This may be due in part to the shape of camel erythrocytes, which are oval rather than bi-concave as seen in most mammals (Fowler 1999).

Cattle of Indian origin (*Bos indicus*) and those from Europe and parts of Africa (*Bos taurus*) have undergone a separate evolution for several hundred thousand years (Hansen 2004). The Indian or Zebu cattle have during their genetic adaptation acquired genes for thermotolerance (Hansen 2004), and therefore have a higher degree of heat tolerance compared to *Bos taurus* cattle (Allen 1962; Finch 1986;

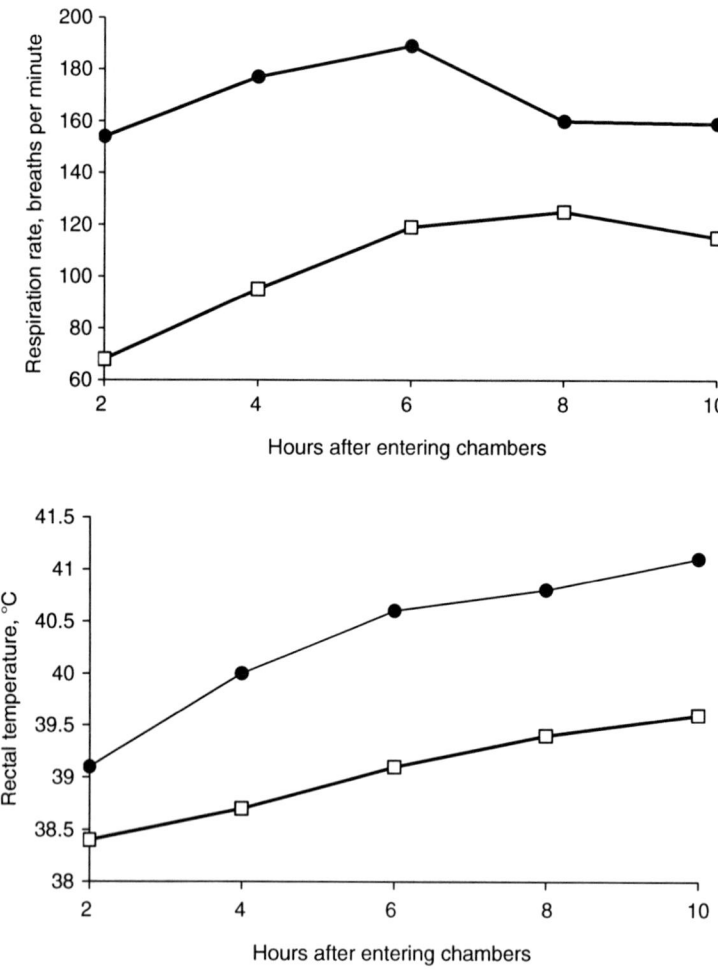

Fig. 7.2 Differences between Hereford (closed circles) and Brahman (open squares) for respiration rate and rectal temperature over 10 h in an environmental chamber when THI > 90 (Adapted from Gaughan et al. 1999)

Spiers et al. 1994; Hammond et al. 1996, 1998; Gaughan et al. 1999; Burrows and Prayaga 2004). However, some *Bos taurus* African breeds such as Tuli, and D'Nama have developed heat tolerance, and appear to be as good as *Bos indicus* cattle in this regard. Under hot climatic conditions, the genetic adaptation of *Bos indicus* cattle allows them to have a lower respiration rate and rectal temperature than *Bos taurus* cattle (Fig. 7.2). An excellent review on the adaptation of zebu cattle to thermal stress was undertaken by Hansen (2004). Adaptation to hot conditions has resulted in animal acquiring specific genes, some of which have been identified (see discussion below).

Sheep and goats are thought to be less susceptible to environmental stress than other domesticated ruminant species (Khalifa et al. 2005). They are widely distributed in regions with diverse climatic conditions and possess unique characteristics such as water conservation capability, higher sweating rate, lower basal heat metabolism, higher respiration rate, higher skin temperature, constant heart rate and constant cardiac output (Borut et al. 1979; D'miel et al. 1979; Shkolnik et al. 1980; Feistkorn et al. 1981).

Differences in physiological responses of sheep adapted to hot conditions (Omani – indigenous breed of Oman) and non-adapted to hot conditions (Merino – Australian) were reported by Srikandakumar et al. (2003). When exposed to hot conditions, the Omani sheep had lower respiration rate than the Merino (65 vs. 128 breaths/min, respectively) (Fig. 7.3). There were no differences in rectal temperature during exposure to hot conditions, but the rectal temperature of the Omani sheep was significantly lower during exposure to cool conditions. The rectal temperature

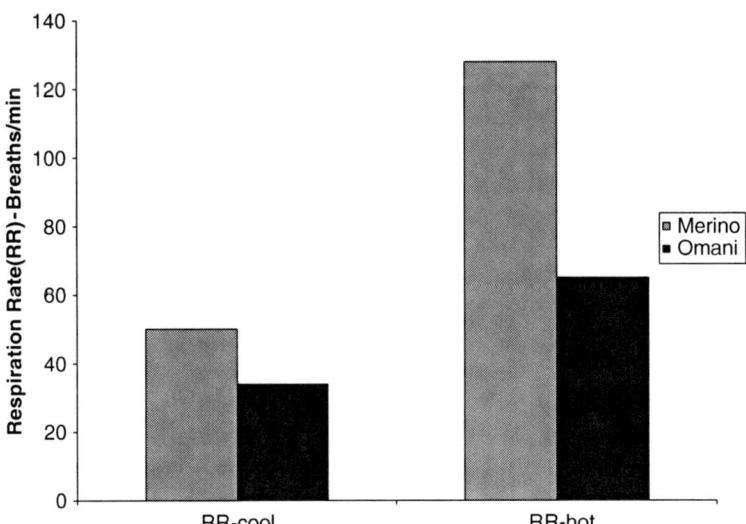

Fig. 7.3 Differences in respiration rate (RR) between Australian merino and Omani sheep when exposed to cool and hot conditions (Adapted from Srikandakumar et al. 2003)

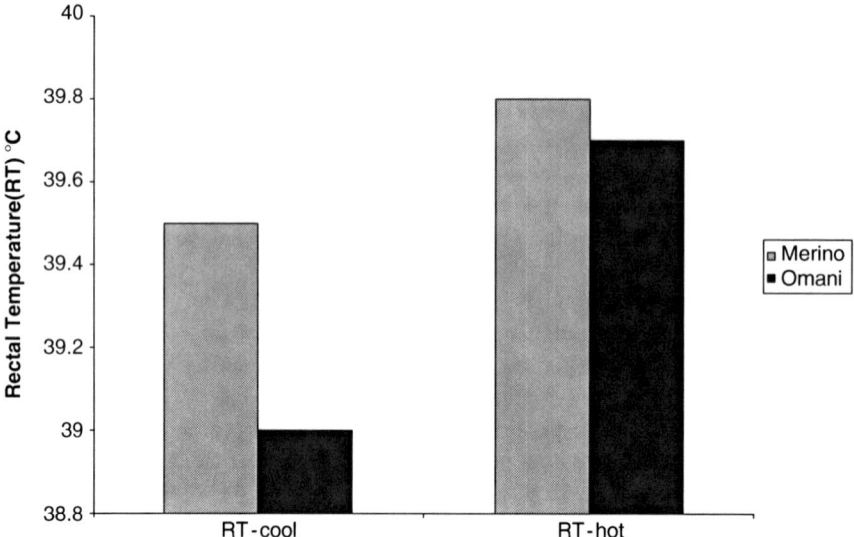

Fig. 7.4 Differences in rectal temperature (RT) between Australian merino and Omani sheep when exposed to cool and hot conditions (Adapted from Srikandakumar et al. 2003)

of the Omani sheep increased by 0.7°C during hot conditions compared to 0.3°C in the Merino sheep (Fig. 7.4).

Goats are a very good example of a domestic animal that is highly adapted to harsh conditions. Silanikove (2000) postulated that goats living in harsh environments represent a climax in the capacity of domestic ruminants to adjust to such areas. Again this ability is multifactorial. While performance in terms of growth rate is greatly reduced, low body mass and low metabolic requirements of goats can be regarded as important assets in minimising their maintenance and water requirements in areas where water sources are widely distributed and food sources are limited by their quantity and quality. An ability to reduce metabolism allows goats to survive even after prolonged periods of severely limited food availability. A skillful grazing behaviour and efficient digestive system enable goats to attain maximal food intake and maximal food utilization in a given condition. There is a positive interaction between the recycling rate of urea and a better digestive capacity of desert goats. The rumen plays an important role in the evolved adaptations by serving as a relatively large fermentation vat and water reservoir. The water stored in the rumen is utilized during dehydration, and the rumen serves as a container which accommodates the ingested water upon re-hydration. The rumen, salivary glands, and kidney coordinate functions in the regulation of water intake and water distribution following acute dehydration and rapid re-hydration.

Animals that have been exposed to non-lethal thermal stress will usually adapt to the conditions. The adaptation may be of short duration (e.g. reduction in feed intake; Mader et al. 2002) or long-term (e.g. reproductive failure). Long-term adaptation has been shown in chickens that were exposed to hot conditions at 4–7 days of age. The exposure reduced the effects (reduced heat production, lower mortality) of heat stress at a later age (May et al. 1987; Wiernusz and Teeter 1996; Yahav and Plavnik 1999; Yalchin et al. 2001).

7.5.1 General Animal Responses

Climate change deals with variations both short- and long-term. Those variations can come in many forms involving both temporal and spatial variations, ranging from the number and severity of acute, dynamic events (such as heat waves) impacting localized microclimates, to long-term chronic (decadal) global changes that may result from changes in atmospheric constituents. This broad perspective is taken here as we consider the impact of thermal environmental challenges associated with climate change on adaptive responses of animals and the management of livestock production systems. Thus, the focus is on how elements of acute and chronic climate change are linked to the ability of the animals to cope with environmental challenges, and the limits of that ability are discussed in the context of sustainable management practices.

Responses of animals vary according to the type of thermal challenge: short-term adaptive changes in behavioral, physiological, and immunological functions (survival-oriented) are the initial responses to acute events, while longer-term challenges impact performance-oriented responses. Within limits delineated by thresholds for disrupted behavior and maladapted physiology and immune functions, domestic animals can cope with many acute thermal challenges through acclimatization to minimize adverse effects and compensation for reduced performance during moderate environmental challenges. These responses to environmental challenges are illustrated in Fig. 7.5 (Hahn 1999, as adapted from Hahn and Morrow-Tesch 1993). The interrelationship between potential environmental challenges and the dynamic response of an animal is apparent. As an aside, it is important to recognize that while management strategies will likely alter some of the biological and adaptive responses, the laws of physics will still apply – heat production and heat losses must balance within the limits of heat storage capacity of the animals.

7.5.2 Animal Responses to Heat Load

Direct effects involve heat exchanges between the animal and the surrounding environment are related to radiation, temperature, humidity, and wind speed (Johnson 1987). Because the thermal environment is more than just heat, the term heat stress is somewhat misleading. The term heat load has been used to highlight the importance of the interactive effects of the fore-mentioned factors. Within breed, animal variation (phenotypes), differences among breeds (genotypes), management factors such as housing and nutrition, physiological status (stage of pregnancy, stage of

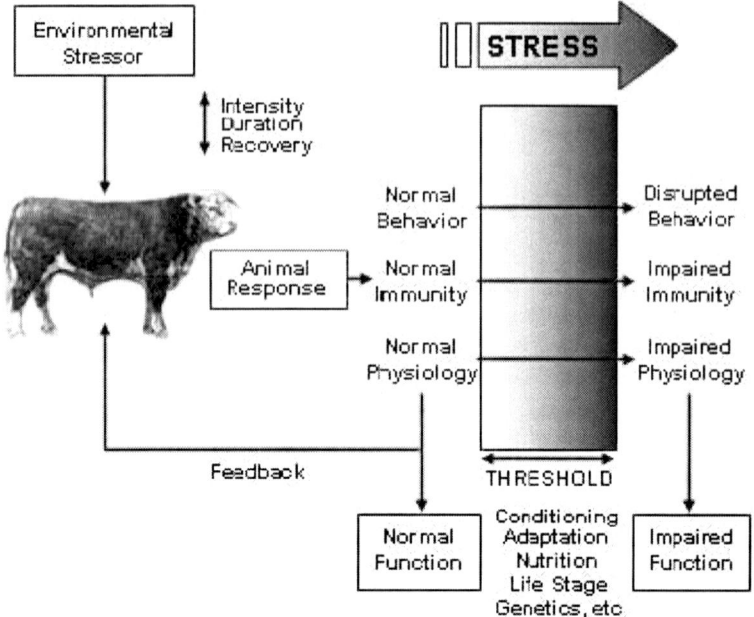

Fig. 7.5 Response model for farm animals with thermal environmental challenges (Hahn 1999)

lactation, growth rate), age and previous exposure to hot conditions may increase or decrease the impact of hot conditions.

High heat load (and environmental stress in general) has the potential for detrimental effects on susceptible animals. The negative effects on health, growth rate, feed intake, feed efficiency, tissue deposition, milk yield, health status, reproduction, and egg production are well documented (Brody 1956; El-Fouley et al. 1976; Biggers et al. 1987; Fuquay 1981; Johnson 1987; Nienaber et al. 1987a, b; Hahn et al. 1990, 1993; Liao and Veum 1994; Valtorta and Maciel 1998; Mader et al. 1999a, b; Nienaber et al. 1999, 2001; West 1999; Hansen et al. 2001; Wolfenson et al. 2001; Yalchin et al. 2001; Valtorta et al. 2002; Kerr et al. 2003; Faurie et al. 2004; Gaughan et al. 2004; Holt et al. 2004; Mader and Davis 2004; Kerr et al. 2005; Huynh et al. 2005; Wettemann and Bazer 1985). However, the actual numerical impacts are unknown. Furthermore genetic change in livestock animals especially in regards to increase productivity has resulted in animals that more likely to be susceptible to the negative impacts of heat stress.

Differences in their ability to withstand environmental stressors should allow selection of animals (within breeds and between breeds/species) better suited to particular environmental conditions (Scott and Slee 1987; Slee et al. 1991; Langlois 1994; Hammond et al. 1996, 1998; Gaughan et al. 1999; Herpin et al. 2002; Abdel Khalek and Khalifa 2004; Koga et al. 2004; Brown-Brandl et al. 2005; Hamadeh et al. 2006). However, as previously discussed, selection of such animals may result in improved welfare and ability to cope at the expense of lower productivity.

Livestock and poultry are remarkable in their ability to mobilize coping mechanisms when challenged by environmental stressors. However, not all coping capabilities are mobilized at the same time. As a general model for bovines, sheep and goats, respiration rate serves as an easily recognized early warning of increasing thermal stress (Khalifa et al. 1997; Butswat et al. 2000; Gaughan et al. 2000; Eigenberg et al. 2005), and increases markedly above a baseline as the animals try to maintain homeothermy by dissipating excess heat through respiratory evaporation. However, Starling et al. (2002) stated that the use of physiological parameters such as rectal temperature and respiration rate for selection is not enough to evaluate the level of adaptive capability. Clearly this is the case as there are many physiological factors which need to be assessed. However, a full assessment (i.e. changes in body temperature, respiration rate, heat shock proteins, hormones etc. – all of which are indicators of heat load status) of animals is difficult, especially under farming conditions. Increased respiration rate and body temperature do not necessarily indicate that an animal is not coping with the environmental conditions to which it is exposed.

7.5.3 Body Heat – Animal x Climate Interactions

There are several components to body heat load which can be divided into two broad categories *viz.* internal or metabolic heat load (ruminal fermentation and nutrient metabolism) and environmental heat load (Armsby and Kriss 1921; Duckworth and Rattray 1946; Shearer and Beede 1990). Metabolic heat load is typically a result of: (i) basal body functions (heart, lungs and liver), (ii) maintenance, (iii) activity, and (iv) performance (e.g., daily gain, milk, eggs) (McDowell 1974).

Basal body functions contribute between 35% and 70% of daily heat production (McDowell 1974), and will tend to the higher levels during non-basal periods of work (e.g., walking, high respiration rate) or high levels of production. Importantly, core body temperature is dynamic even under thermoneutral conditions (Hahn 1989, 1999), and follows a diurnal pattern which is influenced by interactions between animal and environmental factors.

There are a range of thermal conditions within which animals are able to maintain a relatively stable body temperature by behavioral and physiological means (Johnson 1987; Bucklin et al. 1992). This range is defined for a species based on upper critical and lower critical temperatures. Bligh and Johnson (1973) defined the upper critical temperature (UCT) as 'the ambient temperature above which thermal balance cannot be maintained for a long period and animals become progressively hyperthermic'. This definition was revised in 1987 as 'the ambient temperature above which the rate of evaporative heat loss of a resting thermoregulating animal must be increased (e.g., by thermal tachypnea or by thermal sweating) in order to maintain thermal balance' (IUPS Thermal Commission 1987). The lower critical temperature (LCT) is defined by the IUPS Thermal Commission (1987) as 'the ambient temperature below which the rate of metabolic heat production of a resting thermoregulating tachymetabolic animal must be increased by shivering and/or nonshivering thermogenesis in order to maintain thermal balance'. The thermoneutral zone (TNZ) is defined as the range of ambient temperature at which temperature regulation is achieved only by control of sensible heat loss, i.e. without regulatory changes in metabolic heat production or evaporative heat loss. The TNZ will therefore be different when insulation or basal metabolic rate varies (IUPS Thermal Commission 2001). The UCT, LCT and TNZ of a species are influenced by insulation, nutrition and exercise (Ames 1980; McArthur 1987; Morgan 1997).

Heat stress results from the animal's inability to dissipate sufficient heat or reduce heat influx to maintain homeothermy (Folk 1974). High ambient temperature, relative humidity and radiant energy, particularly with concurrent low air speed, compromise the ability of animals to dissipate heat. As a result, there is an increase in core body temperature, which in turn initiates compensatory and adaptive mechanisms in an attempt to re-establish homeothermy and homeostasis (El-Nouty et al. 1990; Khalifa et al. 1997; Horowitz 1998, 2002; Lin et al. 2006). These readjustments, generally referred to as adaptations, may be favorable or unfavorable to economic interests of humans, but are essential for survival of domestic animals (Stott 1981). However, it is likely that continued genetic selection for improved levels of production (e.g. growth rate, feed intake and milk production) will result in animals that are generally less heat tolerant (Joubert 1954; Young 1985; Johnson 1987; Yahav et al. 2005; Lin et al. 2006).

When animals are exposed to environmental conditions above their UCT core body temperature begins to increase as a result of the animal's inability to adequately dissipate the excess heat load. There is a concomitant decrease in feed intake as core body temperature increases, which ultimately results in reduced performance (production, reproduction), health and well-being if adverse conditions persist (Hahn et al. 1993). Thresholds are genotype/phenotype/species dependent, and are affected by many factors, as noted in Fig. 7.5. For shaded *Bos taurus* feeder cattle, Hahn (1999) reported respiration rate typically increases above a threshold of about 21°C air temperature, with a threshold for increasing core body temperature and decreasing feed intake at about 25°C. A recent study (Brown-Brandl et al. 2005) showed the influence of condition, genotype, respiratory pneumonia, and temperament on respiration rate of un-shaded *Bos taurus* heifers). Figure 7.6 illustrates the respiration rate response of different genotypes to hot environmental temperatures.

The lower and upper critical temperatures of both Arabi and Zaraiby goats were 20–25°C and 20–30°C, respectively (El-Sherbiny et al. 1983). Lu (1989) found that the upper critical temperature of goats in maintenance is 25–30°C, and heat stress occurs when they are exposed to ambient temperature above 30°C. He stated that although rectal temperature rose significantly when goats were exposed to 30°C, compared to 20°C, the limit of heat tolerance for goats is between 35°C and 40°C. Dahlanuddin and Thwaites (1993) stated that goats reached the limit of their heat tolerance at 40–45°C ambient temperature. Furthermore, D'miel et al. (1980) stated that goats had a high lower critical temperature of 26°C. Therefore, they must rely mostly on metabolic energy rather than on insulation to keep their body temperature constant during cold weather.

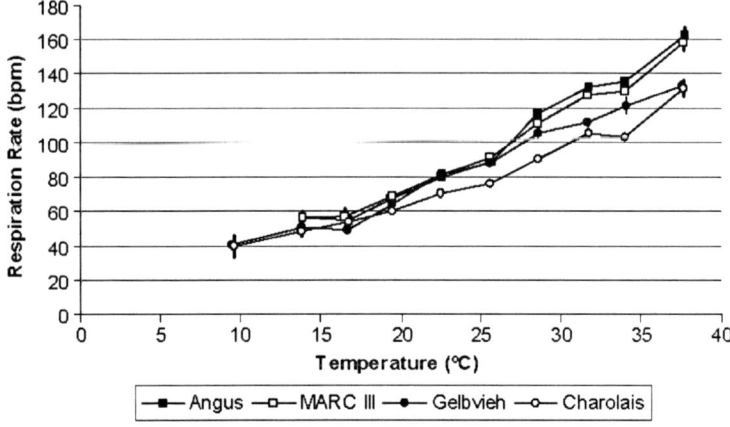

Fig. 7.6 Respiration rates as a function of ambient temperature for unshaded cattle of four genotypes (Brown-Brandl et al. 2005)

There also appears to be a time-dependency aspect of responses in some species. For example, Hahn et al. (1997) reported that for beef cattle with access to shade, respiration rate lags behind changes in dry bulb temperature, with the highest correlations obtained for a lag of 2 h between respiration rate and dry bulb temperature. For un-shaded beef cattle, respiration rate closely tracks solar radiation; increasing or decreasing with solar radiation. There is also a delay in acute body temperature responses (during the first 3–4 days of exposure) to a heat challenge, with an increasing mean and amplitude, along with a phase shift reflecting entrainment by the ambient conditions (Hahn et al. 1997; Hahn and Mader 1997; Hahn 1999). Even though feed intake reduction usually occurs on the first day of exposure to hot conditions, the endogenous metabolic heat load from existing rumen contents adds to the increased exogenous environmental heat load. Nighttime recovery also has been shown to be an essential element of survival for cattle when severe heat challenges occur (Hahn and Mader 1997). After 3 days, the animal enters the chronic response stage, with mean body temperature declining slightly and feed intake reduced in line with heat dissipation capabilities. Diurnal body temperature amplitude and phase remain altered. These typical thermoregulatory responses (discussed more fully in Hahn 1999), when left unchecked during a severe heat wave with excessive heat loads, can lead to a pathological state resulting in impaired performance or death (Hahn and Mader 1997). The intensity and duration of exposure to a given thermal stress will also determine animal responses (Hahn and Mader 1997; Gaughan and Holt 2004; Beatty et al. 2006). Further studies are required to determine species and breed responses.

Thus, an increase in air temperature, such as that expected in different scenarios of climate change, would directly affect animal performance by affecting animal heat balance. The thermal environment influences animal performance primarily through the net effects of energy exchanges between the animal and its surroundings

(Folk 1974; Hahn 1989; Yahav et al. 2005). There are four modes of energy transfer: radiation (gain or loss of heat from the animal), convection (gain or loss), conduction (gain or loss), evaporation (loss only), all of which are governed by physical laws. Several physical parameters control heat transfer by each mode. Air temperature affects energy exchanges through convective, conductive, and radiative exchanges (not evaporation) (Hahn 1976). In hot conditions, evaporation becomes the most important method of heat loss, as it is not dependent on a temperature gradient (Ingram and Mount 1975). Therefore, the combination of temperature and humidity acquire more relevance, since humidity increases the magnitude of the thermal strain especially at high ambient temperatures.

The temperature humidity index (THI; Thom 1959) is commonly used as an indicator of the intensity of climatic stress on animals, where a THI of 72 and below is considered as no heat stress, 73–77 as mild heat stress, 78–89 as moderate, and above 90 as severe (Fuquay 1981). On the other hand, the Livestock Weather Safety Index (LCI 1970) categories associated with THI are normal (THI $\leq$ 74), Alert (75–78 THI), danger (79–83 THI) and emergency (THI $\geq$ 84). Davis et al. (2003) suggest that there is no heat stress for beef cattle when average THI < 70, mild heat stress when $70 \leq$ THI < 74, moderate heat stress when $74 \leq$ THI < 77, and severe heat stress when THI $\geq$ 77. Khalifa et al. (2005) indicated that for sheep and goats, there is no heat stress when average THI < 70, mild heat stress when $70 \leq$ THI < 74 in sheep and $70 \leq$ THI < 78 in goats, moderate heat stress when $74 \leq$ THI < 88 in sheep and $78 \leq$ THI < 84 in goats and severe heat stress when THI $\geq$ 84 in goats. It is worth noting that these data were obtained on crossbred sheep and goats which are acclimatized to Egyptian conditions but not well adapted to the subtropical environment like native breeds. Dairy cattle show signs of heat stress when THI is higher than 72 (Johnson 1987; Armstrong 1994); however the actual threshold will be associated with a decline in milk production (Berry et al. 1964; Kadzere et al. 2002), and whether or not heat abatement strategies are implemented (Mayer et al. 1999). Cows with higher levels of milk production are more sensitive to heat load (Johnson 1987; Hahn 1989). Amundson et al. (2006) indicated that a THI threshold of 73 pregnancy rates of beef cattle became negatively affected. Conception rate of dairy cows was affected by just 1 day exposure to THI between 65 and 70 (Ingraham et al. 1974; and Du Preez et al. 1990). The conception rate of water buffalo was significantly lower when THI > 79 (Pagthinathan et al. 2003). Whether these beef cattle, dairy, or buffalo could adapt to a greater THI is not known. A review of heat stress in lactating dairy cows has been published by Kadzere et al. (2002).

Although THI is widely-accepted for evaluating the climatic environment, it is limited because it does not take into account the effects of thermal radiation (solar and long-wave) or wind speed. Modifications to the existing THI to account for wind speed and solar load (Mader et al. 2006) and the development of new indices (Eigenberg et al. 2000, 2005; Khalifa et al. 2005; Gaughan 2008) have been reported. A review of thermal indices used with livestock was undertaken by Hahn et al. (2003). A new heat load index (HLI) which incorporates the effects of solar radiation and wind speed on the heat load status of feedlot cattle has been established (Gaughan et al. 2008a). This index is based on the establishment of thresholds above which cattle gain heat and below which heat is dissipated. The thresholds are adjusted based on genotype, health status, nutritional management, pen management and the provision of shade.

Current indices do not account for cumulative effects of heat load, and/or natural cooling. Cattle may 'accumulate' heat during the day (body temperature rises) and dissipate the heat at night. If there is insufficient night cooling, cattle may enter the following day with an 'accumulated' heat load (Hahn and Mader 1997). The THI-hours model was developed to account for the impact of intensity x duration on thermal status (Hahn and Mader 1997). This concept was further developed for feedlot cattle as the accumulated heat load model (Gaughan et al. 2008a). This model is able to account for genotype differences, management factors and housing factors (e.g. provision of shade).

It is not only the intensity and duration of the thermal challenge, but also the amount of time animals have to recover from the challenge that determines their response (Mendel et al. 1971; Hahn et al. 2001; Gaughan et al. 2008a). In the central Santa Fe region, a major dairy area in Argentina, THI > 72 for 13 h a day is common in January (Valtorta and Leva 1998). These conditions result in poor reproductive performance and milk yield; de la Casa and Ravelo (2003) have estimated the impacts on milk production in Argentina. When considering a global climate change scenario, determined by paleoclimatological studies (Budyko et al. 1994), the hours when THI > 72 would increase to approximately 16 h by 2025 (Valtorta et al. 1996a). The implications of such a change are that the already compromised summer dairy performance measured in terms of reduced milk production (Valtorta et al. 1996b, 1997) and lower conception rates (Valtorta and Maciel 1998), could be further impaired.

7.6 Animal Management Adaptations

7.6.1 Genetic Modifications

The use of genomics may hold the key to improving heat tolerance in a number of species. How do we identify cattle with superior heat tolerance? Basically there are three broad genetic options for improving heat tolerance: (i) select phenotypes within the breed types preferred that have high heat tolerance e.g. identify heat tolerant Angus within the Angus breed, (ii) identify phenotypes within the heat tolerant breeds (e.g. Brahman) that will meet current and future market specifications, and (iii) identify breeds that currently meet market requirements and are heat tolerant. The focus of many researchers has been to identify genes, biological markers, or molecular markers that can be used to assess heat tolerance in animals (Collier et al. 2002; Mariasegaram et al. 2007; Regitono et al. 2006). An example is the "slick" gene in cattle. Cattle which have shorter hair, have hair of greater diameter and are of lighter color are more adapted to heat than those with longer hair coats and darker colors. The cattle with the shorter hair carry the "slick coat" gene. The slick coat phenotype has been observed in tropical *Bos taurus* breeds (e.g. Senepol and Carona) in the Americas. The adaptation is manifested in the shorter haired cattle by lower rectal temperatures, lower respiration rates, increased sweating rate (Olson et al. 2006) and better fertility (Bertipaglia et al. 2005) compared to long haired cattle under high heat load. The slick gene appears to be a simple dominant gene. Therefore, as long as a crossbred animal carries the dominant allele they will be heat tolerant. In a recent study, Angus x Senepol and Charolais x Senepol were shown to be as heat tolerant as Brahmans (Mariasegaram et al. 2007). However, there was no mention of carcass attributes. Selection of cattle for this gene may be a useful mechanism for improving heat tolerance provided carcass and other performance attributes are not compromised.

7.6.2 Environmental Modifications

Global climate change models predict an increase of heat stress events, as well as general warming in some areas. Therefore, methods that will alleviate the impact of these events should be considered, especially if alternative land use or species/breed use is not an option. Beede and Collier (1986) suggest three management options for reducing the effect of thermal stress in cattle which have application for all livestock and poultry. The options are: (1) physical modification of the environment; (2) genetic development of breeds with greater heat tolerance and (3) improved nutritional management during periods of high heat load.

Numerous methods of environmental modifications to ameliorate heat stress in livestock and poultry are found in the literature, ranging from provision of shade through environmental control using mechanical air conditioning (e.g., Hahn and McQuigg 1970; Hahn 1989; Bucklin et al. 1991; Bull et al. 1997; Mader et al. 1999a; Valtorta and Gallardo 1998; Mitlöhner et al. 2001; Spiers et al. 2001; Pagthinathan et al. 2003; Champak Bhakat et al. 2004; Correa-Calderon et al. 2004; Mader and Davis 2004; Lin et al. 2006). However, while all are technologically feasible, not all are economically viable or acceptable from a management perspective.

Shade: In many cases, the most economical solution to high heat load is the provision of shade. Shade is a simple method of reducing the impact of high solar radiation (Bond et al. 1967; Fuquay 1981; Curtis 1983), but has little impact on air temperature. Black globe temperatures under shade structures can be as much as 12°C lower than the black globe temperature in the sun (36°C vs. 49°C) (J. Gaughan, 2007, personal communication). Shade can be either natural or artificial. It has been suggested that shade from trees is more effective (Hahn 1985) and is preferred by cattle (Shearer et al. 1991). However, Gaughan et al. (1998) reported that dairy cows preferred the shade from a solid iron roof even when they had access to shade trees. This result may have been a result of differing shade density, leading to a greater reduction in radiation heat load under the roof. Aspects concerning design and orientation of shades have been widely published (Buffington et al. 1983; Hahn 1989; Bucklin et al. 1991; Valtorta and Gallardo 1998). Many species, even those deemed to be heat tolerant will seek shade under hot conditions if it is an option. Shades are effective in reducing heat stress and the effects of heat stress in cattle (Davison et al. 1988). Valtorta et al. (1996b) found that cattle with access to shade had lower rectal temperature and respiration rate in the afternoon, and yielded more milk and greater milk protein compared to unshaded cows. Khalifa et al. (2000) reported that exposure to solar radiation (46°C black globe temperature) and 27% RH, significantly increased rectal temperature, skin temperature, ear temperature, respiration rate and pulse rate of goats without access to shade compared to those in shade. However, exposure to solar radiation significantly decreased temperature gradients from the skin to air and from rectal to ear temperatures.

Air movement: Air movement is an important factor in the relief of heat stress, since it affects convective and evaporative heat losses (Armsby and Kriss 1921; Mader et al. 1997; Yahav et al. 2005). Air movement whether outside or in buildings is critical if cooling is to be effective. The use of natural ventilation in animal buildings should be maximized by the construction of open-sided sheds (Ferguson 1970; Bucklin et al. 1991), sheds with ridge top ventilation (Baxter 1984) and good separation distance between buildings (Ferguson 1970). Forced or mechanical ventilation, provided by fans, is an effective method for enhancing air flow, if properly designed and maintained (Baxter 1984; Xin and Puma 2001). Numerous methods of increasing air movement in animal buildings ranging from simple overhead fans, mechanically driven curtains which open or close depending on ambient temperature, tunnel ventilation systems to fully controlled computer operated systems can be found in the literature. Shelters, shade and wind breaks, if not designed correctly, can lead to micro-climate conditions that may induce severe heat stress in animals.

Using water for cooling livestock: Direct access to water such as in dams, ponds and rivers is effective in cooling animals in grazing situations. It is not unusual to see cattle standing belly deep in water. Intensively housed dairy cattle, feedlot cattle and pigs will also use water troughs for similar purposes if they can gain access. Where access is limited water splashing and dunking the head in the trough is a common practice. Pigs will use nipple drinkers to spray water over their bodies in an effort to keep cool. Where animals are housed there are several methods which are commonly used, including misting, fogging, and sprinkling systems. A large number of studies have investigated the efficacy of these systems in reducing the incidence of heat stress in domestic animals (Berman et al. 1985; Hahn 1985; Armstrong and Wiersma 1986; Schultz 1988; Turner et al. 1989; Strickland et al. 1989; Bucklin et al. 1991; Armstrong 1994; Armstrong et al. 1999; Brouk et al. 2001, 2003a; Pagthinathan et al. 2003; Gaughan et al. 2004; Marcillac et al. 2004; Barbari and Sorbetti Guerri 2005; Calegari et al. 2005; Gaughan and Tait 2005).

Evaporative coolers may be effective in reducing air temperature especially when relative humidity is low and there is adequate ventilation and air movement. Evaporative coolers are effectively used to cool the air in cattle, pig, sheep and poultry buildings in areas characterized by a hot dry environment. Misting is routinely used in the dairy industry and is especially effective in dry climates (e.g. Israel, Saudi Arabia, and Arizona) (Armstrong et al. 1993), and is also used for cooling the ventilation air entering poultry and swine buildings (Xin and Puma 2001; Brouk et al. 2003b). However, misting systems can be effective even when relative humidity is high e.g. Hawaii (dairy cattle) (Armstrong et al. 1993), Florida (dairy cattle) (Taylor et al. 1986; Beede 1993; Mearns et al. 1992), Missouri (dairy cattle, swine and poultry) (Brouk et al. 2003b) and Iowa (poultry) (Xin and Puma 2001) provided that there is sufficient air movement. Misting or fogger systems are not generally recommended in hot humid environments (Bucklin et al. 1991; Bottcher et al. 1993; Turner et al. 1993). However, misters do have the advantage of low water usage (Lin et al. 1998).

Direct water application: Direct application of water to the skin is an effective method of cooling buffalo, cattle, pigs, and poultry. Adding water to the body surface increases the latent heat loss from an animal. It is the evaporation of the water from the surface that results in the cooling of the animal. Under the right conditions, water application will reduce heat load on animals (Fig. 7.7). However, the use of water for cooling livestock may lead to an increase in relative humidity, especially where there is limited air movement, and this reduces the ability of the animal to dissipate heat via evaporation (Frazzi et al. 1997; Xin and Puma 2001; Correa-Calderon et al. 2004). Gaughan et al. (2003) demonstrated that wetting cattle exposed to high temperature and humidity had only minor short term effects

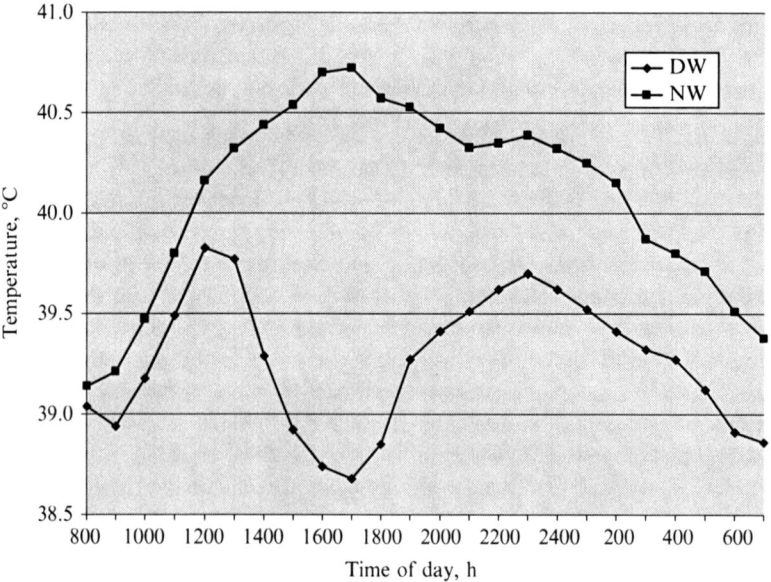

Fig. 7.7 The effect on rectal temperature when cattle have been sprinkled (DW) between 1,200 and 1,600 h or not sprinkled (NW) (Adapted from Gaughan et al. 2004)

on relative humidity, provided that ventilation was adequate. The negative impacts of high relative humidity and/or limited air movement on the animals ability to dissipate heat may be magnified when insufficient water is used e.g. foggers or misters or there is insufficient ventilation to remove the moisture laden air. In these circumstances water particles may form a cover over the hair or pelage of the animal trapping heat and thereby increasing the level of heat stress (Hahn 1985).

Continuous application of water is not required to achieve heat stress alleviation. Morrison et al. (1973) used a 30 min cycle where cattle were wetted for 30 s and then exposed to forced air ventilation for 4.5 min. The 30 min cycle was repeated nine times a day. The dairy cows which were exposed to the cooling strategy had a lower (0.5–0.9°C) rectal temperatures than those not cooled. In a later study Morrison et al. (1981) using feed conversion efficiency and rate of gain as indicators did not find any benefit from wetting cattle. However, feed intake was greater in the cooled cattle. Flamenbaum et al. (1986) wetted dairy cows for either 10, 20 or 30 s followed by forced ventilation (1.5 m/s at cow height) for either 15, 30 or 45 min. They found that the 20 and 30 s wetting followed by the 30 and 45 min forced ventilation reduced rectal temperature by 0.7°C and 1.0°C respectively. Igono et al. (1987) reported effective cooling when where sprinklers were used for were used for 20 minutes on and were then off for 10 min. Using beef cattle in Florida, Garner et al. (1989) used a 3 min on 30 min off when temperature was greater than 26.7°C. Fans were also used in this study however air speed at animal height was not mentioned. Lin et al. (1989) used 3 min on and 15 min off. Two and a half minutes on and 7 min off was used by Turner et al. (1992). Beede (1993) recommended approximately 1–2 mm per dairy cow per 15 min wetting cycle, or just enough water to wet the back. In a study by Brouk et al. (2001) used a cycle of 3 min on and 12 min off, while Gaughan et al. (2008b) used 5 min on and 15 min off.

Brouk et al. (2001) suggests that the coat of dairy cows should be allowed to dry between water applications, however Gaughan et al. (2008b) reported that beef cattle which were completely dry within 10–15 min of water application had an increase in respiration rate. It is likely that these animals were under some degree of heat stress, albeit for short periods of time, because water was evaporating from the skin but not removing sufficient body heat (Frazzi et al. 2000). This suggests that the 15 min interval between wettings was too long given the ambient conditions to which the cattle were exposed or that the duration of water application was too short or that the amount of water supplied was insufficient.

Before considering water application, a number of factors need to be considered. These include: infrastructure and running cost, water availability, how will water be removed from the site, provision of sufficient air movement, and micro-climate effects.

Livestock managers need to be especially vigilant when applying water to animals. Changes in micro-climate (e.g., an increase in relative humidity) may reduce evaporative cooling from the animal's surface. Therefore increased wetting frequency may be required. Furthermore, consistency in application is important. Once started, wetting needs to continue until high heat load has abated (Gaughan et al. 2004, 2008b).

7.6.3 Nutritional Modification

Excellent reviews of nutritional strategies for managing heat-stressed dairy cows (West 1999), and poultry (Lin et al. 2006) have been published. Dietary manipulation has been shown to be beneficial for reducing the effects of heat stress in cattle (Beede and Collier 1986; Schneider et al. 1986; West et al. 1991; Mader et al. 1999b; Granzin and Gaughan 2002). A 'cold' dairy cow diet generates a high net nutrient proportion for milk production and lower heat increment (Gallardo 1998). The author indicates that some outstanding characteristics of 'cold' diets are: (1) higher energy contents per unit volume; (2) highly fermentable fiber; (3) lower protein degradability; and (4) high by-pass nutrients contents. These recommendations are useful when feeding totally mixed rations. However, diet manipulation may be useful even under grazing systems. Gallardo et al. (2001) found that hydrogenated fish fat could be a good ingredient to sustain high yields and elevated maintenance requirements in a grazing system during hot conditions. These are in reality, however, only short-term solutions. In many areas of the world, feeding grains and other high energy ingredients will not be financially sustainable. In many areas, animal production is based on grazing, foraging or browsing. A changing climate will impact on grasses, shrubs, and trees. If these effects are negative, then there will be significant changes in livestock production in the affected areas (see previous discussion).

Water is the most critical nutrient for animals. Climate change may have a number of impacts on water availability. In addition, water requirements are higher during periods of heat load (Winchester and Morris 1956; Beede and Collier 1986; Beatty et al. 2006). High production animals have a greater need for water compared to low production animals. Classical studies have demonstrated that water losses from animals increase with increasing air temperature (Kibler and Brody 1950; McDowell and Weldy 1960). Drinking behaviour is complex and is influenced by a number of factors such as diet, live weight, health status, and physiological status. Normally, it would be expected that water intake will increase when animals are exposed to hot conditions. Beatty et al. (2006) reported that water intake of *Bos taurus* heifers (331 kg) increased from approximately 19.8 L/head/day under cool conditions (wet bulb <25°C) to 31.4 L/head/day when exposed to a wet bulb temperature between 25°C and 33°C (Fig. 7.8). Water intakes reported by Gaughan and Tait (2005) for Angus steers (550 kg) exposed to hot conditions (36°C) were 41.1 L/head/day, up from 19.8 L/head/day under thermoneutral conditions (26°C). Concurrent feed intake fell from approximately 1.5% of body weight to almost zero (Beatty et al. 2006), and from approximately 2.4% of body weight to 1.8% of body weight (Gaughan and Tait 2005). For camels, feed intake was not affected when exposed to high heat load (40°C) provided they had access to water (Guerouali and Filali 1995). However, water intake increased by 300%. Other studies have shown that water intake decreases when feed intake is reduced (Chaiyabutr et al. 1980; Kadzere et al. 2002; Mader and Davis 2004). Goats (and other species) respond to water restrictions by reducing feed intake and concentrating their urine (Ahmed and

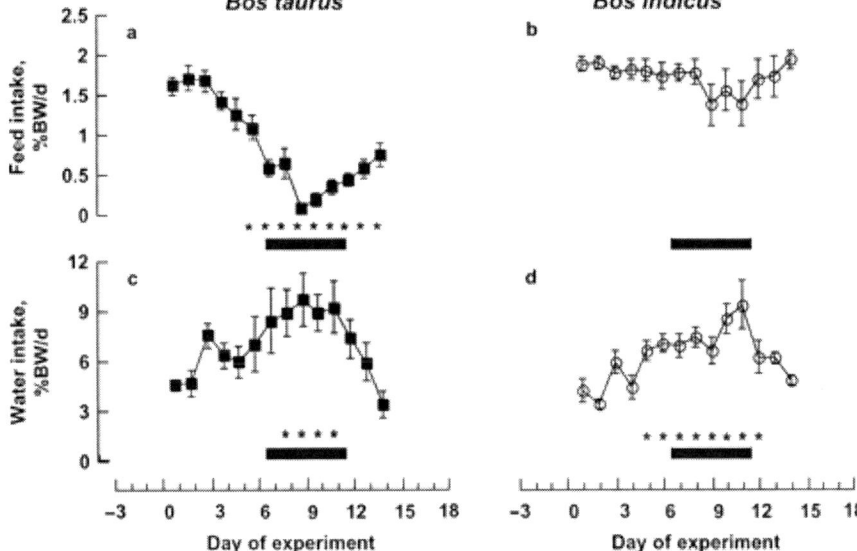

Fig. 7.8 Mean daily feed intake (**a** and **b**) and water intake (**c** and **d**) for *Bos taurus* and *Bos indicus* heifers. Points show the mean ± SEM for each of six animals. The horizontal bar under each figure indicates the hottest 5 days of the experiment. Asterisks under the data denote $P < 0.05$ for the day marked vs. the control days (days 1 and 2) (Beatty et al. 2006)

El Kheir, 2004). Offering warm rather than cool water has improved water intake of heat stressed animals in some cases (Lanham et al. 1986; Olsson and Hydbring 1996; Olsson et al. 1997). It is generally agreed that provision of good quality water is an important factor to manage nutrition during periods of high heat load.

7.7 Conclusions

The adaptive capabilities of animals and livestock production systems have been emphasized in this report. Biometeorology has a key role in rational management to meet the challenges of thermal environments for livestock production systems, whether in current or altered climates.

Understanding the responses of animals to environmental challenges is paramount to successful implementation of strategies to ameliorate negative impacts of climate change. Livestock managers have routinely dealt with intra- and interannual climate variability. The challenge will be dealing with on-going change and possible major climatic shifts. Livestock managers will need to consider both climatic conditions and resource availability (e.g. feed, water, veterinary care, financial, animals, people) when determining the best strategies to adopt. Animal managers need to be proactive, as we cannot wait for changes to occur – there is a need to act now to reduce the risk of longer-term changes to livestock industries around the world.

Summarizing, the most important element of proactive environmental management to reduce risk is preparation: (1) be informed – governments may need to fund training programs in regions where current livestock practices will no longer be viable; (2) develop a strategic plan – both short-term and long-term; (3) observe and recognize animal responses to climatic/nutritional conditions; (4) adopt farming practices to the changing conditions; and, (5) select animals (be prepared to change breeds/species) that are suited to the environmental and nutritional conditions. Livestock managers who adopt such a proactive approach will be better prepared for both current and future climates.

References

Abdel Khalek TMM, Khalifa HH (2004) Thermoregulatory mechanisms in new born kids and lambs. Egyptian J Anim Prod 41:391–402

Adams RM, Rosenweig C, Peart RM, Ritchie JT, McCarl BA, Glyer, JD, Curry RB, Jones JW, Boote KJ, Allen LH, Jr (1990) Global climate change and US agriculture. Nature 345: 219–224

Ahmed MMM, Elkheir IM (2004) Thermoregulation and water balance as affected by water and food restrictions in Sudanese Desert goats fed good-quality and poor-quality diets. Trop Animal Health Prod 36:191–204

Agyemang K, Little DA, Bah ML, Dwinger RH (1991) Effects of postpartum body weight changes on subsequent reproductive performance in N'dama cattle maintained under traditional husbandry systems. Anim Reprod Sci 26:51–59

Allen TE (1962) Responses of Zebu, Jersey and Zebu x Jersey crossbred heifers to rising temperature, with particular reference to sweating. Aust J Agric Res 13:165–179

Allen-Diaz B (1996) Rangelands in a changing climate: impacts, adaptation and mitigation. In: Watson RT, Zinyowera MC, Moss RH (eds.) Climate Change 1995: Impacts, Adaptations and Mitigation of Climate Change: Scientific-Technical Analyses. Cambridge University Press, USA

Ames D (1980) Thermal environment affects production efficiency of livestock. BioScience 30:457–460

AMS (1989) Glossary of Meteorology, 5th Edition. American Meteorological Society, Boston, MA

Amundson JL, Mader TL, Rasby RY, Hu QS (2006) Environmental effects on pregnancy rate in beef cattle. J Anim Sci 84:3415–3420

Armsby HP, Kriss M (1921) Some fundamentals of stable ventilation. J Agric Res 21:343–368

Armstrong DV (1994) Heat stress interaction with shade and cooling. J Dairy Sci 77:2044–2050

Armstrong D, Wiersma F (1986) An update on cow cooling methods in the west. ASAE Paper No. 86–4034. ASAE, St. Joseph, MI

Armstrong DV, Elchert WT, Wiersma F (1993) Environmental modification for dairy cattle housing in arid climates In: Livestock Environment IV. Proceedings of the 4th International Livestock Environment Symposium. ASAE St. Joseph, MO. pp. 1223–1231

Armstrong DV, Hillman PE, Meyer MJ, Smith JF, Stokes SR, Harner JP (1999) Heat stress management in freestall barns in the western US. In: Proceedings of the Western Dairy Management Conference, Las Vegas, NV

Auchtung TL, Salak-Johnson JL, Morin DE, Mallard CC; Dahl GE (2004) Effects of photoperiod during the dry period on cellular immune function of dairy cows. J Dairy Sci 87:3683–3689

Arthur LM, Abizadeh F (1988) Potential effects of climate change on agriculture in the prairie region of Canada. Western J Agr Econ 13:216–224

Barbari M, Sorbetti Guerri F (2005) Cooling systems for heat protection of farrowing sows. In: Livestock Environment VIII. Proceedings of the 7th International Symposium. ASAE, St. Joseph, MI, pp. 122–129

Barbehenn RV, Chen Z, Karowe DN, Spickard A (2004) C3 grasses have a higher nutritional quality than C4 grasses under ambient and elevated atmospheric CO_2 Glob Change Biol 10:1565–1575

Baxter S (1984) Intensive Pig Production: Environmental Management and Design. Granada, London

Beard CW, Mitchell BW (1987) Influence of environmental temperatures on the serologic responses of broiler chickens to inactivated and viable Newcastle disease vaccines. Avian Dis 31:321–326

Beatty DT, Barnes A, Taylor E; Pethick D, McCarthy M, Maloney SK (2006) Physiological responses of Bos taurus and Bos indicus cattle to prolonged, continuous heat and humidity. J Anim Sci 84:972–985

Beaumont LJ, McAllan IAW, Hughes L (2006) A matter of timing: changes in the first date of arrival and last date of departure of Australian migratory birds. Glob Change Biol 12:1339–1354

Beede D (1993) Management of Dairy Cattle in Warm Climates. Information Series QI93009. Queensland DPI, Brisbane

Beede DK, Collier RJ (1986) Potential nutritional strategies for intensively managed cattle during thermal stress. J Anim Sci 62:543–554

Berman A, Folman Y, Karen M, Maman M, Herz Z, Wolfenson D, Arieli A, Graber Y (1985) Upper critical temperatures and forced ventilation effects for high yielding dairy cows in a sub-tropical climate. J Dairy Sci 68:1488–1495

Bernabucci U, Ronchi B, Lacetera N, Nardone A (2002) Markers of oxidative status in plasma and erythrocytes of transition dairy cows during the hot season. J Dairy Sci 85:2173–2179

Berry IL, Shanklin MD, Johnson HD (1964) Dairy shelter design based on milk production decline as affected by temperature and humidity. Trans. ASAE 7:329

Bertipaglia ECA, Silva RG, Maia ASC (2005) Fertility and hair coat characteristics of Holstein cows in a tropical environment. Anim Reprod (Belo Horizonte) 2:187–194

Biggers BG, Geisert RP, Wetteman RP, Buchanan DS (1987) Effect heat stress on early embryonic development in the beef cow. J Anim Sci 64: 1512–1518

Bligh J, Johnson KG (1973) Glossary of terms for thermal physiology. J Appl Physiol 35:941–961

Bond TE, Kelly CF, Morrison SR, Pereira N (1967) Solar, atmosphere, and terrestrial radiation received by shaded and unshaded animals. Trans ASAE 10:622–629

Bonnette ED, Kornegay ET, Lindemann MD, Hammerberg C (1990) Humoral and cell-mediated immune response and performance of weaned pigs fed four supplemental vitamin E levels and housed at two nursery temperatures. J Anim Sci 68:1337–1345

Borut A, D'miel R, Shkolnik A (1979). Heat balance of resting and walking goats: comparison of climatic chamber and exposure in the desert. Physiol Zool 52:105–112

Bottcher RW, Singletary IB, Baughman GR (1993) Humidity effects on evaporative efficiency of misting nozzles. In: Livestock Environment IV. Proceedings of the 4th International Livestock Environment Symposium. ASAE, St. Joseph, MO, pp. 375–383

Bowes MD, Crosson P (1993) Consequences of climate change for the MINK economy: Impacts and responses. Climatic Change 24:131–158

Briske DD, Heitschmidt RK (1991) An ecological perspective. In: Heitschmidt RK, Stuth JW (eds.) Grazing Management: An Ecological Perspective. Timber Press, Portland, OR

Brody S (1956) Climate physiology of cattle. J Dairy Sci 39:715–725

Brouk MJ, Smith JF, Harner JP (2001) Efficacy of modified cooling in Midwest dairy freestall barns. In: Livestock Environment VI. Proceedings of the 4th International Livestock Environment Symposium. ASAE, St. Joseph, MO, pp. 412–418

Brouk MJ, Smith JF, Harner JP (2003a) Effect of sprinkling frequency and airflow on respiration rate, body surface temperature and body temperature of heat stressed dairy cows. In: Fifth International Dairy Housing Proceedings. Fort Worth Texas, pp. 263–268

Brouk MJ, Smith JF, Harner JP (2003b) Effect of utilizing evaporative cooling in tiestall dairy barns equipped with tunnel ventilation on respiration rate and body temperature of lactating dairy cattle. In: Fifth International Dairy Housing Proceedings, Fort Worth, TX, pp. 312–319

Brown-Brandl TA, Eigenberg RA, Nienaber JA (2005) Heat stress risk factors for feedlot heifers. In: Proceedings of the 7th International Livestock Environment Symposium. ASAE, St. Joseph, MI, pp. 600–606

Bucklin, RA, Turner LW, Beede DK, Bray DR, Hemken RW (1991) Methods to relieve heat stress for dairy cows in hot, humid climates. Appl Eng Agric 7:241–247

Bucklin RA, Hahn GL, Beede DK, Bray DR (1992) Physical facilities for warm climates. In: Van Horn HH, Wilcox CJ (eds) Large Dairy Herd Management. American Dairy Science Association, Champaign, IL 61820

Budyko MI, Borzenkova II, Menzhulin GV, Shilkomanov IA (1994) Cambios antropogénicos del clima en América del Sur. Academia Nacional de Agronomía y Veterinaria No. 19

Buffington DE, Collier RJ, Canton GH (1983) Shade management systems to reduce heat stress for dairy cows. Trans ASAE 26:1798–1802

Bull RP, Harrison PC, Riskowsi GL, Gonyou HW (1997) Preferences among cooling systems by gilts under heat stress. J Anim Sci 75:2078–2083

Burrows HM, Prayaga KC (2004) Correlated responses in production and adaptive traits and temperament following selection for growth and heat resistance in tropical cattle. Livestock Prod Sci 86:143–161

Butswat IS, Mbap ST, Ayibatonye GA (2000) Heat tolerance of sheep in Bauchi Nigeria. Trop Ag. (Trin.) 77:265–268

Calegari F, Frazzi E, Calamari L (2005) Productive response of dairy cows raised in a cooling barn located in the Po Valley (Italy). In: Livestock Environment VIII. Proceedings of the 7th International Symposium. ASAE, St. Joseph, MI, pp. 115–121.

Campbell BD, McGeon GM, Gifford RM, Clark H, Stafford Smith DM, Newton PCD, Lutze JL (1995) Impacts of atmospheric composition and climate change on temperate and tropical pastoral agriculture. In: Pearman G, Manning M (eds.) Greenhouse 94. CSIRO, Canberra, Australia.

Chaiyabutr N, Faulkner A, Peaker M (1980) Effects of starvation on the cardiovascular system, water balance and milk secretion in lactating goats. Res Vet Sci 28:291–295

Champak Bhakat, Chaturvedi D, Raghavendra S, Nagpaul PK (2004) Studies on camel management under various microenvironment of shelter systems. Indian J Dairy Sci 57:347–353

Chirico J, Jonsson P, Kjellberg S, Thomas G (1997) Summer mastitis experimentally induced by *Hydrotaea irritans* exposed to bacteria. Med Vet Entomol 11:187–192

Christin PA, Besnard G, Samaritani E, Duvall MR, Hodkinson TR, Savolainen V, Salamin N (2008) Oligocene CO_2 decline promoted C4 photosynthesis in grasses. Current Biology, 18:37–43

Collatz GJ, Berry JA, Clark JS (1998) Effects of climate and atmospheric CO_2 partial pressure on the global distribution of C4 grasses: Present, past and future. Oecologia 114:441–454

Collier R J, Kobayashi Y, Gentry P (2002) The use of genomics in genetic selection programs for environmental stress tolerance in domestic animals. In: Proceedings of the AMS 15th Biometeorology and Aerobiology Conference, Kansas, Oct 27–Nov 1, pp. 54–58

Cook NB, Bennett TB, Emery KM, Nordlund KV (2002) Monitoring nonlactating cow intramammary infection dynamics using DHI somatic cell count data. J Dairy Sci 85:1119–1126

Correa-Calderon A, Armstrong D, Ray D, DeNise S, Enns M, Howison C (2004) Thermoregulatory responses of Holstein and Brown Swiss heat stressed dairy cows to two different cooling systems. Int J Biometeorol 48:142–148

Curtis SE (1983) Environmental Management in Animal Agriculture. Iowa State University Press, Ames, IA

Dahlanuddin, Thwaites CJ (1993) Feed-water intake relations in goats at high ambient temperatures. J Anim Physiol An N 69:169–174

Darwin R, Tsigas M, Lewandrowski J, Raneses A (1995) World agriculture and climate change – economic adaptations. Agricultural Economic Report 703, ERS-NASS, Herndon, VA

Davis MS, Mader TL, Holt SM, Parkhurst AM (2003) Strategies to reduce feedlot cattle heat stress: Effects on tympanic temperature. J Anim Sci 81:649–661

Davison TM, Silver BA, Lisle AT, Orr WN (1988) The influence of shade on milk production of Holstein-Friesian cows in a tropical upland environment. Aust J Exper Agric 28:149–54

de la Casa AC, Ravelo AC (2003) Assessing temperature and humidity conditions for dairy cattle in Cordoba, Argentina. Int J Biometeorol 48:6–9

Devendra C (1987) Goats. In: Bioclimatology and the Adaptation of Livestock, Elsevier, Amsterdam, The Netherlands. Part II, Chapter 11.

Diamond J (1999) Guns, Germs and Steel: The Fate of Human Societies. Norton W.W., New York.

D'miel R, Robertshaw D, Choshniak J (1979) Sweat gland secretion in the Black Bedouin goat. Physiol Zool 52:558

D'miel R, Prevolotzky A, Ashkolnik A (1980) Is a black coat in the desert means to save metabolic energy. Nature 283:761–767

Donker RA, Nieuwland MG, van der Zijpp AJ (1990) Heat-stress influences on antibody production in chicken lines selected for high and low immune responsiveness. Poult Sci 69:599–607

Donovan GA, Badinga L, Collier RJ, Wilcox CJ, Braun RK (1986) Factors influencing passive transfer in dairy calves. J Dairy Sci 69:754–759

Duckworth J, Rattray GB (1946) Studies of diurnal variation in the body temperature of the tropical threequarter bred (Holstien-Zebu) dairy calf. Trop Agric 23:94–100

Du Preez JH, Hattingh JP, Giesecke WH, Eisenberg BE (1990) Heat stress in dairy cattle under southern African conditions. III. Monthly temperature-humidity index mean values and their significance in the performance of dairy cattle. Onderstepoort J Vet Res 57:243–248

Easterling WE, Crosson PR, Rosenberg NJ, McKenney M, Katz LA, Lemon K (1993) Agricultural impacts of and responses to climate change in the Missouri-Iowa-Nebraska-Kansas (MINK) region. Climatic Change 24:23–61

Eigenberg RA, Hahn GL, Nienaber JA, Brown-Brandl T, Spiers D (2000) Development of a new respiration rate monitor for cattle. Trans ASAE 43:723–728

Eigenberg RA, Brown-Brandl TM, Nienaber JA, Hahn GL (2005) Dynamic response indicators of heat stress in shaded and non-shaded feedlot cattle – part 2. Predictive relationships. Biosyst Eng 91:111–118

El-Fouley MA, Kotby EA, El-Sobhy HE (1976) The functional reproductive peak in Egyptian buffalo cow is related to day length and ambient temperature. Archivo Veterinaria Italiano 27:123–129

El-Nouty FD, Al-Haidary AA, Basmaeil SM (1990) Physiological responses feed intake, urine volume and serum osmolality of Aaradi Goats deprived of water during spring and summer. Asian-Aust J Anim Sci 3:331–336

El-Sherbiny AA, Yousef MK, Salem MH, Khalifa HH, Abd-El-Bary HM, Khalil MH (1983) Thermoregulatory responses of a desert and a non-desert goat breeds. Al-Azhar Agric Res Bull 89:1–11

Elvinger F, Hansen PJ, Natzke RP (1991) Modulation of function of bovine polimorphonuclear leukocytes and lymphocytes by high temperature *in vitro* and *in vivo*. Am J Vet Res 52:1692–1698

Faurie AS, Mitchell D, Laburn HP (2004) Peripartum body temperatures in free-ranging ewes (*Ovis aries*) and their lambs. J Therm Biol 29:115–122

Feistkorn G, Ritter P, Jessen C (1981) Overall cardiovascular adjustments to thermal stress in conscious goats. Pfluegers Arch 391(Suppl. 1):184

Ferguson W (1970) Poultry housing in the tropics: Applying the principles of thermal exchange. Trop Anim Health Prod 2:44–58

Finch VA (1986) Body temperature in beef cattle: its control and relevance to production in the tropics. J Anim Sci 62:513–542

Flamenbaum I, Wolfenson D, Mamen M, Berman A (1986) Cooling dairy cattle by a combination of sprinkling and forced ventilation and its implementation in the shelter system. J Dairy Sci 69:3140–3147

Folk GE (1974) Textbook of Environmental Physiology. Lea & Febiger, Philadelphia

Folk GE, Reidesel ML, Thrift DL (1998) Principles of Integrative Environmental Physiology. Austin & Winfield, San Francisco, CA

Fowler ME (1999) Medicine and Surgery of South American Camelids: Llama, Alpaca, Vicuna 2nd Edition, Blackwell Publishing Inc, Malden, USA

Frank HK (1991) Risk estimation for ochratoxin A in European countries. IARC Sci Publ 115:321–325

Frank KL, Mader TL, Harrington Jr JA, Hahn GL, Davis MS (2001) Climate change effects on livestock production in the Great Plains. In: Proceedings of the 6th International Livestock Environment Symposium. ASAE, St. Joseph, MI

Frankham R (2005) Stress and adaptation in conservation genetics. J Evol Biol 18:750–755

Frazzi E, Calamari L, Calegari F (1997) The aeration with and without misting: effects on heat stress in dairy cows. In: Livestock Environment V: Proceedings of the 5th International Livestock Symposium. ASAE, St. Joseph, MO, pp. 907–914

Frazzi E, Calamari L, Calegari F, Stefanini L (2000) Behaviour of dairy cows in response to different barn cooling systems. Trans ASAE 43:387–394

Fuquay JW (1981) Heat stress as it effects animal production. J Anim Sci 52:164–174

Gaffen DJ, Ross RJ (1998) Increased summertime heat stress in the US. Nature 396:529–530

Gallardo MR (1998) Manejo nutricional. In: Producción de leche en verano. Centro de publicaciones de la Secretaría de Extensión de la UNLitoral. Santa Fe, Argentina, pp. 47–63

Gallardo MR, Valtorta SE, Leva PE, Castro HC, Maiztegui JA (2001) Hydrogenated fish fat for grazing dairy cows in summer. Int J Biometeorol 45:111–114

Garner JC, Bucklin RA, Kunkle WE, Nordstedt RA (1989) Sprinkled water and fans to reduce heat stress of beef cattle. App Eng Agric 5:99–101

Gaughan JB, Holt MA (2004) Changes in the diurnal rhythm of rectal temperature of cattle exposed to prolonged heat stress and cooled with warm salt water. J Anim Sci 82(Suppl. 1):301

Gaughan JB, Tait LA (2005) Effectiveness of evaporative cooling of beef cattle housed in confinement. In: Livestock Environment VIII. Proceedings of the 7th International Symposium. ASAE, St. Joseph, MI, pp. 105–114

Gaughan JB, Goodwin PJ, Schoorl TA, Young BA, Imbeah M, Mader TL, Hall A (1998) Shade preferences of lactating Holstein-Friesian cows. Aust J Exp Agric 38:17–21

Gaughan JB, Mader TL, Holt SM, Josey MJ, Rowan KJ (1999) Heat tolerance of Boran and Tuli crossbred steers. J Anim Sci 77:2398–2405

Gaughan JB, Holt SM, Hahn GL, Mader TL, Eigenberg R (2000) Respiration rate – is it a good measure of heat stress in cattle. Asian-Aus. J Anim Sci 13:329–332

Gaughan JB, Lott S, Gordon G (2003) Wetting Cattle to Alleviate Heat Stress on Ships – Stage 1. Final Report LIVE.219. Meat & Livestock Australia Ltd. North Sydney, NSW

Gaughan JB, Davis MS, Mader TL (2004) Wetting and physiological responses of grain fed cattle in a heated environment. Aust J Agric Res 55:253–260

Gaughan JB, Mader TL, Holt SM, Lisle A (2008a) Development of a new heat load index for feedlot cattle. J Anim Sci 86:226–234

Gaughan JB, Mader TL, Holt SM (2008b) Cooling and feeding strategies to reduce heat load of grain-fed beef cattle in intensive housing. Livest Sci 113:226–233

Giesecke HW (1985) The effect of stress on udder health of dairy cows. Onderstepoort J Vet Res 52:175–193

Granzin BC, Gaughan JB (2002) The effect of sodium chloride supplementation on the milk production of grazing Holstein Friesian cows during summer and autumn in a humid sub-tropical environment. Anim Feed Sci Tech 96:147–160

Greer DH, Laing WA, Campbell BD, Halligan EA (2000) The effect of perturbations in temperature and photon flux density on the growth and photosynthetic responses of five pasture species. Aust J Plant Physiol 27:301–310

Guerouali A, Filali RZ (1995) Metabolic adjustments of camel during heat stress and dehydration. In: Flamant JC, Portugal AV, Costa JP, Nunes AF, Boyazoglu J (eds) Proceedings of the International Symposium on Animal Production and Rural Tourism in Mediterranean Regions. Wageningen Press, Wageningen, Netherlands pp. 57–62

Hahn GL (1976) Shelter engineering for cattle and other domestic animals. In: Johnson HD (ed.) Progress in Animal Biometeorology, Vol I, Part I. Swets and Zeitlinger, Amsterdam, pp. 496–503.

Hahn GL (1985) Management and housing of farm animals in hot environments. In: Yousef MK (ed.) Stress Physiology in Livestock, Volume II, Ungulates. CRC Press, Boca Raton, FL, pp. 151–174.
Hahn GL (1989) Bioclimatology and livestock housing: theoretical and applied aspects. In: Proceedings of the Brazilian Workshop on Animal Bioclimatology. Jaboticabal, Brazil, 15 p.
Hahn GL (1999) Dynamic responses of cattle to thermal heat loads. J. Anim Sci 77(Suppl 2): 10–20
Hahn GL, Mader TL (1997) Heat waves in relation to thermoregulation, feeding behavior and mortality of feedlot cattle. In: Bottcher RW, Hoff SJ (eds) Livestock Environment V. Proceedings of the 5th International Symposium. ASAE, St. Joseph, MI, pp. 563–571
Hahn GL, McQuigg JD (1970) Evaluation of climatological records for rational planning of livestock shelters. Agric Meteorol 7:131–141Hahn GL, Morrow-Tesch JL (1993) Improving livestock care and well-being. Agric Eng 74:14–17
Hahn GL, Eigenberg JA, Nienaber JA, Littledike ET (1990a) Measuring physiological responses of animals to environmental stressors using a micro-computer based portable datalogger. J Anim Sci 68:2658–2665
Hahn GL, Klinedinst PL, Wilhite DA (1990b) Climate change impacts on livestock production and management. American Meteorological Society Annual Meeting, Aneheim, CA
Hahn GL, Nienaber JA, Eigenberg RA (1993) Environmental influences on the dynamics of thermoregulation and feeding behaviour in cattle and swine. In: Proceedings of the 4th International Livestock Environment Symposium. ASAE, St. Joseph, MI, pp. 1106–1116
Hahn GL, Parkhurst AM, Gaughan JB (1997) Cattle respiration rate as a function of ambient temperature. ASAE Paper No. MC97–121. ASAE, St. Joseph, MI
Hahn GL, Mader TL, Gaughan JB, Hu Q, Nienaber JA (2000) Heat waves and their impacts on feedlot cattle. In: de Dear RJ, Kalma JD, Oke TR, Auliciems A (eds.) Biometeorology and Urban Climatology at the Turn of the Millennium: Selected papers from the Conference ICB-ICUC'99 (Sydney, 8–12 November 1999). WMO/TD-N° 1026. WMO, Geneva, pp. 353–357
Hahn L, Mader T, Spiers D, Gaughan J, Nienaber J, Eigenberg R, Brown-Brandl T, Hu Q, Griffin D, Hungerford L, Parkhurst A, Leonard M, Adams W, Adams L (2001) Heat waves and their impacts on feedlot cattle: considerations for improved environmental management. In: Stowell RR, Bucklin R, Bottcher RW (eds) Livestock Environment VI. Proceedings of the 6th International Livestock Environment Symposium. ASAE, St. Joseph, MO, pp. 129–139
Hahn GL, Mader TL, Harrington JA, Nienaber JA, Frank KL (2002) Living with climatic variability and potential global change: climatological analyses of impacts on livestock performance. Proceedings of the 16th International Congress on Biometeorology, Kansas, MO, pp. 45–49
Hahn GL, Mader TL, Eigenberg RA (2003) Perspective on development of thermal indices for animal studies and management. In: Interactions Between Climate and Animal Production. EAAP Technical Series No. 7. Wageningen Academic Publishers, The Netherlands, pp. 31–44.
Hamadeh SK, Rawda N, Jaber JS, Habre A, Abi Said M, Barbour EK (2006) Physiological responses to water restriction in dry and lactating Awassi ewes. Livest Sci 101(1–3):101–109
Hammond AC, Olson TA, Chase Jr. CC, Bowers EJ, Randel RD, Murphy CN, Vogt DW, Tewolde A (1996) Heat tolerance in two tropically adapted Bos taurus breeds, Senepol and Romosinuano, compared with Brahman, Angus, and Hereford cattle in Florida. J Anim Sci 74:295–303
Hammond AC, Chase Jr. CC, Bowers EJ, Olson TA, Randel RD (1998) Heat tolerance of Tuli-, Senepol-, and Brahman-sired F_1 Angus heifers in Florida. J Anim Sci 76:1568–1577
Hansen PJ (2004) Physiological and cellular adaptations of zebu cattle to thermal stress. Anim Reprod Sci 82–83:349–360
Hansen PJ, Drost M, Rivera RM, Paula-Lopes FF, al-Katanani YM, Krininger 3rd CE, Chase CC (2001) Adverse impact of heat stress on embryo production: Causes and strategies for mitigation. Theriogenology 55:91–103
Hafez ESE (1968) Principles of animal adaptation. In: Hafez ESE (ed) Adaptation of Domestic Animals. Lea & Febiger, Philadelphia
Herpin P, Damon M, Le Dividich M (2002) Development of thermoregulation and neonatal survival in pigs. Livest Prod Sci 78:25–45

Hogan JS, Smith KL, Hoblet KH, Schoenberger PS, Todhunter DA, Hueston WD, Pritchard DE, Bowman GL, Heider LE, Brockett BL (1989) Field survey of clinical mastitis in low somatic cell count herds. J Dairy Sci 72:1547–1556

Holt SM, Gaughan JB, Mader TL (2004) Feeding strategies for grain-fed cattle in a hot environment. Aust J Agric Res 55:719–725

Horowitz M (1998) Do cellular heat acclimation responses modulate central thermoregulatory activity? News Physiol Sci 13:218–225

Horowitz M (2002) From molecular and cellular to integrative heat defense during exposure to chronic heat. Comp Physiol Biochem 131:475–483

Howden SM, Soussana JF, Tubiello FN, Chhetri N, Dunlop M, Meinke H (2007) Adapting agriculture to climate change. PNAS 104:19691–19696

Hulme PH (2005) Adapting to climate change: is there scope for ecological management in the face of a global threat. J Appl Ecol 42:784–794

Huynh TTT, Aarnink AJA, Verstegen MWA, Gerrits WJJ, Heetkamp MJW, Kemp B, Canh TT (2005) Effects of increasing temperature on physiological changes in pigs at different relative humidities. J Anim Sci 83:1385–1396

Idso KE, Idso SB (1994) Plant responses to atmospheric CO2 enrichment in the face of environmental constraints: a review of the last 10 years' research. Agric For Meteorol 69:153–203

Igono MO, Johnson HD, Steevens BJ, Krause GF, Shanklin MD (1987) Physiological, productive, and economic benefits of shade, spray, and fan systems versus shade for Holstein cows during summer heat. J Dairy Sci 70:1069–1079

ILRI (International Livestock Research Institute) (2006) Climate change research by ILRI informs Stern Review on the economics of climate change. http://www.ilri.org/ILRIPubaware/ Accessed January 2007

Ingraham RH, Gillette DD, Wagner WC (1974) Relationship of temperature and humidity to conception rate in Holstein cows in a subtropical climate. J Dairy Sci 57:476–481

Ingram DL, Mount LE (1975) Heat exchange between animal and environment. In: Ingram DL, Mount LE (eds) Man and Animals in Hot Environments. Springer, Heidlberg, Germany, pp. 5–23

IUC (Information Unit for Conventions) – United Nations Environment Programme (2002) Climate disasters and extreme events. Climate change information sheet http://www.unep.ch/iuc/submenu/infokit/fact16.htm Accessed June 22nd, 2002

IUPS (The Commission for Thermal Physiology of the International Union of Physiological Sciences) (1987) Glossary of terms for thermal physiology. Pflügers Arch 410:567–587

IUPS (The Commission for Thermal Physiology of the International Union of Physiological Sciences) (2001) Glossary of Terms for Thermal Physiology, 3rd Edition. Jpn J. Physiology 51:245–280

Johnson HD (1987) Bioclimate effects on growth, reproduction and milk production. In: Bioclimatology and the adaptation of livestock. Elsevier, Amsterdam, The Netherlands. Part II, Chapter 3

Johnston KM, Schmitz OJ (1997) Wildlife and climate change: assessing the sensitivity of selected species to simulated doubling of atmospheric CO_2. Glob Change Biol 3:531–544

Joubert DM (1954) The influence of winter nutritional depression on the growth, reproduction, and production of cattle. J Agric Sci 44:5–15

Kadzere CT, Murphy MR, Silanikove N, Maltz E (2002) Heat stress in lactating dairy cows: a review. Lives Prod Sci 77:59–91

Kamwanja LA, Chase CC, Gutierrez JA, Guerriero V, Olson TA, Hammond AC, Hansen PJ (1994) Responses of bovine lymphocytes to heat shock as modified by breed and antioxidant status. J Anim Sci 72:438–444

Kelley KW, Osborne CA, Evermann JF, Parish SM, Gaskins CT (1982) Effects of chronic heat and cold stressors on plasma immunoglobulin and mitogen-induced blastogenesis in calves. J Dairy Sci 65:1514–1528

Kerr BJ, Yen JT, Nienaber JA, Easter PA (2003) Influences of dietary protein level, amino acid supplementation and environmental temperature on performance, body composition, organ weights and total heat production of growing pigs. J Anim Sci 81:1998–2007

Kerr CA, Giles LR, Jones MR, Reverter A (2005) Effects of grouping unfamiliar cohorts, high ambient temperature and stocking density on live performance of growing pigs. J Anim Sci 83:908–915

Khalifa HH, El-Sherbiny AA, Abdel-Khalek TMM (1997) Effect of seasonal variations on adaptability of goats under Egyptian environmental conditions. In: Proceedings of Earth-Atmosphere Forces for Change, Joint Assemblies of the International Association of Meteorology and Atmospheric Sciences and International Association for Physical Sciences of the Oceans, Melbourne, Australia

Khalifa HH, Ahmed AA, El-Tantawy SMT, Kicka MA and Dawoud AM (2000) Effect of heat acclimation on body fluids and plasma proteins of broilers exposed to acute heat stress in summer. Third All African Conference on Animal Agriculture and 11th Conference of the Egyptian Society of Animal Production, Alexandria, Egypt, 6–9 November, 2000

Khalifa HH, Shalaby T, Abdel-Khalek TMM (2005) An approach to develop a biometeorological thermal discomfort index for sheep and goats under Egyptian conditions. In: Proceeding of the 17th International Congress of Biometeorology, Garmisch, Germany, 5–9 September, 2005, Deutscher Wetterdienst, Kaiserleistr, 29–35, 63067 Offenbach am Main, Germany, pp. 118–122

Kibler HH, Brody S (1950) Influence of temperature, 5° to 95°F, on evaporative cooling from the respiratory tract and exterior body surfaces in Jersey and Holstein cows. Missouri Agric. Exp. Sta. Bull. No. 461. Columbia, MO

Kimball BA, Mauney JR, Nakayama FS, Idso SB (1993) Effects of elevated CO_2 and climate variables on plants. J Soil Water Cons 48:9–14

King DA (2004) Climate change science: adapt, mitigate, or ignore? Science 302:176–177

Koga A, Sugiyama M, del Barrio AN, Lapitan RM, Arenda BR, Robles AY, Cruz LC, Kanai Y (2004) Comparison of the thermoregulatory response of buffaloes and tropical cattle, using fluctuations in rectal temperature, skin temperature and haematocrit as an index. J Agric Sci 142:351–355

Lacetera N (1998) Influence of high air temperatures on colostrum composition of dairy cows and passive immunization of calves. Zoot Nutr Anim 6:239–246

Lacetera N, Bernabucci U, Ronchi B, Nardone A (1996) Body condition score, metabolic *status* and milk production of early lactating dairy cows exposed to warm environment. Riv Agr Subtrop Trop 90:43–55

Lacetera N, Bernabucci U, Ronchi B, Nardone A (2002) Moderate summer heat stress does not modify immunological parameters of Holstein dairy cows. Int J Biometeorol 46:33–37

Lacetera N, Scalia D, Bernabucci U, Ronchi B (2003) In vitro assessment of the immunotoxicity of mycotoxins in goats. Immunol Lett 87:323–324

Lacetera N, Bernabucci U, Scalia D, Ronchi B, Kuzminsky G, Nardone A (2005) Lymphocyte functions in dairy cows in hot environment. Int J Biometeorol 50:105–110

Lacetera, N, Bernabucci, U, Scalia, D, Basirico, L, Morera, P and Nardone, A (2006) Heat stress elicits different responses in peripheral blood mononuclear cells from Brown Swiss and Holstein cows. J Dairy Sci 89:4606–4612

Langlois B (1994) Inter-breed variation in the horse in regard to cold adaptation: a review. Lives Prod Sci 40:1–7

Lanham JK, Coppock CE, Milam KZ, Labore JM, Nave DH, Stermer RA, Brasington CF (1986) Effects of drinking water temperature on physiological responses on lactating Holstein cows in summer. J Dairy Sci 69:1004–1012

LCI (1970) Patterns of transit losses. Livestock Conservation, Inc. Omaha, NE

Leva PE, Valtorta SE, Fornasero LV (1997) Milk production declines during summer in Argentina: present situation and expected effects of global warming. In: Proceedings of the 14th International Congress of Biometeorology, Ljubljana, Slovenia, 1–8 September 1996. Part 2, Vol. 2, pp. 395–401

Lin JC, Moss BR, Koon JL, Flood CA, Smith III RC, Cummins KA, Coleman DA (1998) Comparison of various fan, sprinkler, and mister systems in reducing heat stress in dairy cows. App Eng Agric 14:177–182

Lin H, Jiao HC, Buyse J, Decuypere E (2006) Strategies for preventing heat stress in poultry. World's Poult Sci J 62:71–85

Liao CW, Veum TL (1994) Effects of dietary energy intake by gilts and heat stress for days 3 to 24 or 30 days after mating on embryo survival and nitrogen and energy balance. J Anim Sci 72:2369–2377

Lu CD (1989) Effects of heat stress on goat production. Small Rumin Res 2:151–162

McArthur AJ (1987) Thermal interaction between animal and microclimate: a comprehensive model. J Theor Biol 126:203–238

McDowell RE (1974) Effect of the environment on the functional efficiency of the ruminant. In: Livestock Environment. Proceedings of the 1st International Livestock Symposium. ASAE, St. Joseph, MI, pp. 220–231

McDowell RE, Weldy JR (1960) Water exchange of cattle under heat stress. In: Proceedings of the 3rd International Biometeorological Congress, London. Pergamin Press, New York pp. 414–424

McKeon GM, Hoeden SM, Abel NOJ, King JM (1993) Climate change: adapting tropical and subtropical grasslands. In: Proceedings of the XVII International Grassland Congress CSIRO, Melbourne, pp. 1181–1190

Machado-Neto R, Graves CN, Curtis SE (1987) Immunoglobulins in piglets from sows heat-stressed prepartum. J Anim Sci 65:445–455

Mader TL, Dahlquist JM, Gaughan JB (1997) Wind protection effects and airflow patterns in outside feedlots. J Anim Sci 75:26–36

Mader TL, Gaughan JB, Young BA (1999a) Feedlot diet roughage level for Herford cattle exposed to excessive heat load. The Professional Animal Scientist 15:53–62

Mader TL, Dahlquist JM, Hahn GL, Gaughan JB (1999b) Shade and wind barrier effects on summertime feedlot cattle performance. J Anim Sci 77:2065–2072

Mader TL, Holt SM, Hahn GL, Davis MS, Spiers DE (2002) Feeding strategies for managing heat load in feedlot cattle. J Anim Sci 80:2373–2382

Mader TL, Davis MS (2004) Effect of management strategies on reducing heat stress of feedlot cattle: Feed and water intake. J Anim Sci 82:3007–3087

Mader TL, Davis MS, Brown-Brandl T (2006) Environmental factors influencing heat stress in feedlot cattle. J Anim Sci 84:712–719

Magana JG, Tewolde A, Anderson S, Segura JC (2006) Productivity of different cow genetic groups in dual-purpose cattle production systems in south-eastern Mexico. Trop Anim Health Prod 38:583–591

Marcillac NM, Robinson PH, Fadel JG, Mitloehner FM (2004) Effects of shade, sprinklers, and stocking density on performance, behaviour, physiology, and environmental impact of Holstein heifers in drylot pens. J Anim Sci (Suppl1) 82:301

Mariasegaram M, Chase CC, Jr., Chaparro JX, Olson TA, Brenneman RA, Niedz, RP (2007) The slick hair coat locus maps to chromosome 20 in Senepol-derived cattle. Anim Genet 38:54–59

Martin SW, Schwabe CW, Franti CE (1975) Dairy calf mortality rate: characteristics of calf mortality rates in Tulare County, California. Am J Vet Res 36:1099–1104

May JD, Deaton JW, Branton SL (1987) Body temperature of acclimated broilers during exposure to high temperature. Poul Sci 66:378–380

Mayer DG, Davison TM, McGowan MR, Young BA, Matschoss AL, Hall AB, Goodwin PJ, Gaughan JB (1999) Extent and economic effect of heat loads on dairy cattle production in Australia. Aust Vet J 77:804–808

Mearns LO, Katz RW, Schneider SH (1984) Extreme high-temperature events; changes in their probabilities with changes in mean temperature. J Climate Appl Meteorol 23:1601–1613

Mendel VE, Morrisson SR, Bond TE, Lofgren GP (1971) Duration of heat exposure and performance of beef cattle. J Anim Sci 33:850–854

Mignon-Grasteau S, Boissy A, Bouix J, Faure J, Fisher AD, Hinch GN, Jensen P, Le Neidre P, Mormède P, Prunet P, Vandeputte M, Beaumont C (2005) Genetics of adaptation of domestic livestock. Lives Prod Sci 93:3–14

Mitlöhner FM, Morrow JL, Daily DW, Wilson SC, Galyean ML, Miller MF, McGlone JJ. (2001) Shade and water misting effects on behaviour, physiology, performance, and carcass traits of heat-stressed feedlot cattle. J Anim Sci 79:2327–2335

Morgan K (1997) Effects of short-term changes in ambient air temperature or altered insulation horses. J Therm Biol 22:187–194

Morrison SR, Givens RL, Lofgreen GP (1973) Sprinkling cattle for relief from heat stress. J Anim Sci 36:428–431

Morrison SR, Prokop M, Lofgreen GP (1981) Sprinkling cattle for heat stress relief: Activation temperature, duration of sprinkling and pen area sprinkled. Trans ASAE 24:1299–1300

Morrow-Tesch JL, McGlone JJ, Salak-Johnson JL (1994) Heat and social effects on pig immune measures. J Anim Sci 72:2599–2609

Morse D, DeLorenzo MA, Wilcox CJ, Collier RJ, Natzke RP, Bray DR (1988) Climatic effects on occurrence of clinical mastitis. J Dairy Sci 71:848–853

Nardone A, Lacetera N, Bernabucci U, Ronchi B (1997) Composition of colostrum from dairy heifers exposed to high air temperatures during late pregnancy and the early postpartum period. J Dairy Sci 80:838–844

Nienaber JA, Hahn GL, Yen JT (1987a) Thermal environment effects on growing finishing swine, Part I: growth, feed intake and heat production. Trans ASAE 30:1772–1775

Nienaber JA, Hahn GL, Yen JT (1987b) Thermal environment effects on growing finishing swine, Part II: carcass composition and organ weights. Trans ASAE 30:1776–1779

Nienaber JA, Hahn GL, Eigenberg RA (1999) Quantifying livestock responses for heat stress management: a review. Int J Biometeorol 42:183–188

Nienaber JA, Hahn GL, Eigenberg RA, Brown-Brandl TM, Gaughan JB (2001) Feed intake response of heat challenged cattle. In: Stowell RR, Bucklin R, Bottcher RW (eds.) Livestock Environment VI: Proceedings of the 6th International Symposium, Louisville, Kentucky. ASAE, St. Joseph, MI, 49085–9659, pp. 154–164

Olson TA, Chase CC, Jr. Lucena C, Codoy E, Zuniga A, Collier RJ (2006) Effect of hair characteristics on the adaptation of cattle to warm climates. In: Proceedings of the 8th World Congress on Genetics Applied to Livestock Production, Belo Horizonte. Minas Gerais, Brazil 13–18 August

Olsson K, Hydbring E (1996) The preferences for warm drinking water induces hyperhydration in heat-stressed lactating goats. Acta Physiol Scand 157:109–114

Olsson K, Cvek K, Hydbring E (1997) Preference for drinking warm water during heat stress affects milk production in food-deprived goats. Small Ruminant Res 25:69–75

Pagthinathan M, Perera ERK, Perera ANF, Kaduwela SC (2003) Relationship of environmental factors, cooling treatment, blood metabolites, and intensity of heat signs at insemination with conception rate of water buffalo (Bubalus bubalis). Trop Agric Res 15:226–234

Parsons PA (1994) Habitats, stress, and evolutionary rates. J Evolution Biol 7:387–397

Ray DE, Halbach TJ, Armstrong DV (1992) Season and lactation number effects on milk production and reproduction efficiency of dairy cattle in Arizona. J Dairy Sci 75:2976–2983

Regitono, LCA, Martinez, ML and Machado, MA (2006). Molecular aspects of bovine tropical adaptation. In: Proceedings of the 8th World Congress on Genetics Applied to Livestock Production, Belo Horizonte. Minas Gerais, Brazil 13–18 August 2006

Regnier JA, Kelley KW, Gaskins CT (1980) Acute thermal stressors and synthesis of antibodies in chickens. Poult Sci 59:985–990

Regnier JA, Kelley KW (1981) Heat- and cold-stress suppresses in vivo and in vitro cellular immune responses of chickens. Am J Vet Res 42:294–299

Reilly J (1996) Agriculture in a changing climate: impacts and adaptation. In: Watson RT, Zinyowera MC, Moss RH (eds.) Climate Change 1995: Impacts, Adaptations and Mitigation of Climate Change: Scientific-Technical Analyses. Cambridge University Press, USA, pp. 427–467

Ronchi B, Bernabucci U, Lacetera N, Verini Supplizi A, Nardone A (1999) Distinct and common effects of heat stress and restricted feeding on metabolic status of Holstein heifers. Zootec Nutr Anim 1:11–20

Rosenweig C, Parry, ML (1994) Potential impact of climate change on world food supply. Nature 367:133–138

Rötter R, Van de Geijn SC (1999) Climate change effects on plant growth, crop yield and livestock. Climatic Change 43:651–681

Sackett D, Holmes P, Abbott K, Jephcott S, Barber, B (2006) Assessing the economic cost of endemic disease on the profitability of Australian beef cattle and sheep producers–Final Report AHW.087. Meat and Livestock Australia Limited, Nth Sydney NSW 2059

Schmidt-Nielsen K (1975) Animal physiology: adaptation and environment. Cambridge University Press, New York

Schmidt-Nielsen K, Schmidt-Nielsen B, Jarnum SA, Houpt TR (1956) Body temperature of the camel and its relation to water economy. Am J Physiol 188:103–112

Schneider PL, Beede DK, Wilcox CJ (1986) Responses of lactating cows to dietary sodium source and quantity and potassium quantity during heat stress. J Dairy Sci 69:99–110

Schultz TA (1988) California dairy corral manger mister installation. ASAE Paper No. 88–4056, ASAE, St. Joseph, MI

Scott AW, Slee J (1987) The effect of litter size, sex, age, body weight, dam age and genetic selection for cold resistance on the physiological responses to cold exposure of Scottish Blackface lambs in a progressively cooled water bath. Anim Prod 45:477– 492

Shearer JK (1999) Foot health from a veterinarian's perspective. Proceedings Feed and Nutritional Management Cow College, Virginia Tech, pp. 33–43

Shearer JK, Beede DK (1990) Thermoregulation and physiological responses of dairy cattle in hot weather. Agric Practice 11:5–17

Shearer JK, Beede DK, Bucklin RA, Bray DR (1991) Environmental modifications to reduce heat stress in dairy cattle. Agric Practice 12:7–10, 13–16, 18

Shkolnik A, Maltz E, Gordon S (1980) Desert conditions and goat milk production. J Dairy Sci. 63:1749–1754

Silanikove N (2000). Goat production under harsh environmental conditions: The physiological basis and the challenge. In: Merkel RC, Abebe G, Goetsch AL (eds.) The Opportunities and Challenges of Enhancing Goat Production in East Africa. E (Kika) de la Garza Institute for Goat Research, Langston University, Langston, OK, pp. 6–28

Simensen E (1984) Livestock environment and health: general concepts and research strategies. Report 7, Department of Animal Husbandry and Genetics. Norwegian College of Veterinary Medicine, Oslo

Slee J, Alexander G, Bradley LR, Jackson N, Stevens D (1991) Genetic aspects of cold resistance and related characters in new born Merino lambs. Aust J Exp Agric 31:175 –182

Smit B, Mc Nabb D, Smihers J (1996) Agricultural adaptation to climatic variation. Climatic Change 33:7–29

Smith KL, Todhunter DA, Schoenberger PS (1985) Environmental mastitis: cause, prevalence, prevention. J Dairy Sci 68:1531–1553

Sombroek W, Gommes R (1995) The climate change – agriculture conundrum. Paper presented at the Expert Consultation on global climate change and agricultural production: direct and indirect effects of changing hydrological, soil and plant physiological processes. Rome, FAO, 7–11 December 1993, 300 p

Soper F, Muscoplat CC, Johnson DW (1978) In vitro stimulation of bovine peripheral blood lymphocytes: analysis of variation of lymphocyte blastogenic response in normal dairy cattle. Am J Vet Res 39:1039–1042

Spiers DE, Spain JN, Leonard MJ, Lucy MC (2001) Effect of cooling strategy and night temperature on dairy cow performance during heat stress. In: Livestock Environment VI. Proceedings of the 6th International Symposium. ASAE, St. Joseph, MI, pp. 45–55

Spiers DE, Vogt DW, Johnson HD, Garner GB, Murphy CN (1994) Heat stress responses of temperate and tropical breeds of Bos taurus cattle. Arch Latinoam Prod Anim 2:41–52

St-Pierre NR, Cobanov B, Schnitkey G (2003) Economic losses from heat stress by US livestock industries. J Dairy Sci 86(E Suppl):E52–E77

Starling JMC, da Silva RG, Cerón-Muñoz M, Barbosa GSSC, da Costa MJRP (2002) Analysis of some physiological variables for the evaluation of the degree of adaptation in sheep submitted to heat stress. *Revista Brasileira de Zootecnia* 31:2070–2077

Steinfeld H, Gerber P, Wassenaar T, Castel V, Rosales M, de Haan C (2006) Livestocks Long Shadow: Environmental Issues and Options. FAO, Rome, Italy

Stott GH (1981) What is animal stress and how is it measured? J Anim Sci 52:150–153

Strickland JT, Bucklin RA, Norstedt RA, Beede DK, Bray DR (1989) Sprinkler and fan cooling system for dairy cows in hot, humid climates. Appl Eng Agric 5:231–236

Srikandakumar A, Johnson EH, Mahgoub O (2003) Effect of heat stress on respiratory rate, rectal temperature and blood chemistry in Omani and Australian Merino sheep. Small Ruminant Res 49:193–198

Taylor SE, Buffington DE, Collier RJ, DeLorenzo MA (1986) Evaporative cooling for dairy cattle in Florida. ASAE Paper No. 86–4022. St. Joseph, MI

Thom EC (1959) Cooling-degree days. Air conditioning, heating and ventilation 7:65–72

Thuiller W, Broennimann O, Hughes G, Alkemades JRM, Midgley GF, Corsi F (2006) Vulnerability of African mammals to anthropogenic climate change under conservative land transformation assumptions. Glob Change Biol 12:424–440

Turner LW, Chastain JP, Hemken RW, Gates RS, Crist WL (1989) Reducing heat stress in dairy cows through sprinkler and fan cooling. ASAE Paper No 89–4025. ASAE, St. Joseph, MI

Turner LW, Warner RC, Chastain JP (1993) Reducing heat stress in dairy cows through improved facility and systems designs. In: Livestock Environment IV. Proceedings of the 4th International Livestock Environment Symposium. ASAE, St. Joseph, MI, pp. 356–364

Turnpenny JR, Parsons DJ, Armstrong AC, Clark JA, Cooper K, Mathews AM (2001) Integrated models of livestock systems for climate change studies. 2. Intensive systems. Glob Change Biol 7:163–170

Valtorta SE, Leva PE, Fornasero LV, Bardin J (1996a) Horas de estrés para el ganado de origen europeo en la República Argentina: Situación actual e impacto del cambio climático global. Anais do 1° Congresso Brasileiro de Biometeorologia pp: 275–285

Valtorta SE, Gallardo MR, Castro HC, Castelli MC (1996b) Artificial shade and supplenetation effects on grazing dairy cows in Argentina. Trans. ASAE 39:233–236

Valtorta SE, Leva PE, Gallardo MR (1997) Effect of different shades on animal well being in Argentina. Int J Biometeorol 41:65–67

Valtorta SE, Gallardo MR (1998) Modificaciones del ambiente. In: Producción de leche en verano. Centro de publicaciones de la Secretaría de Extensión de la UNLitoral. Santa Fe, Argentina, pp. 93–105

Valtorta SE, Leva PE (1998) Respuestas del animal al ambiente. In: Producción de leche en verano. Centro de publicaciones de la Secretaría de Extensión de la UNLitoral, Santa Fe, Argentina

Valtorta SE, Maciel M (1998) Respuesta reproductiva. In: "Producción de leche en verano". Centro de publicaciones de la Secretaría de Extensión de la UNLitoral, Santa Fe, Argentina, pp. 64–76

Valtorta SE, Leva PE, Gallardo MR, Scarpati OE (2002) milk production responses during heat waves events in Argentina. In: Proceedings of the 16th Congress of Biometerology. Kansas City, MI, October 27th–November 1st, pp. 98–101

Vitali A, Scalia D, Bernabucci U, Lacetera N, Ronchi B (2004) Citotoxic effect of aflatoxin B1 in goat lymphocytes. Proceedings of the 8th International Conference on Goat, p. 65

Voh Jr AA, Larbi A, Olorunju SAS, Agyemang K, Abiola BD, Williams TO (2004) Fertility of N'dama and Bunaji cattle to artificial insemination following oestrus synchronization with PRID and $PGF_{2\alpha}$ in the hot humid zone of Nigeria. Trop Anim Health Prod 36:499–511

Waage S, Sviland S, Odegaard SA (1998) Identification of risk factors for clinical mastitis in dairy heifers. J Dairy Sci 81:1275–1284

West JW (1999) Nutritional strategies for managing the heat stressed dairy cow. J. Anim Sci 77 (Suppl 2)/J Dairy Sci 82(Suppl 2):21–34

West JW, Mullinix BG, Sandifer TG (1991) Changing dietary electrolyte balance for dairy cows in cool and hot environments. J Dairy Sci 74:1662–1674

Wettemann RP, Bazer FW (1985) Influence of environmental temperature on prolificacy of pigs. J Reprod Fertil Suppl 33:199–208

Wiernusz CJ, Teeter RG (1996) Acclimation effects on fed and fasted broiler thermobalance during thermoneutral and high ambient temperature exposure. Br Poult Sci 37:677–687

Winchester CF, Morris MJ (1956) Water intake rates of cattle. J Anim Sci 15:722–739
Willmer P, Stone G, Johnston I (2000) Environmental Physiology of Animals. Blackwell, Oxford
Wittmann EJ, Mellor PS, Baylis M (2001) Using climate data to map the potential distribution of *Culicoides imicola* (Diptera: Ceratopogonidae) in Europe. Rev Sci Tech 20:731–740
Wolfenson D, Bachrach D, Maman M, Graber Y, Rozenboim I (2001) Evaporative cooling of ventral regions of the skin in heat stressed laying hens. Poul Sci 80:958–964
Xin H, Puma MC (2001) Cooling caged laying hens in high-rise house by fogging inlet air. In: Livestock Environment VI. Proceedings of the 6th International Symposium. ASAE, St. Joseph, MI, pp. 244–249
Yahav S, Plavnik I (1999) Effect of early-stage thermal conditioning and food restrictions on performance and thermotolerance of male broiler chickens. Br Poult Sci 40:120–126
Yahav S, Shinder D, Tanny J, Cohen S (2005) Sensible heat loss: the broiler's paradox. World's Poult Sci J 61:419–430
Yalchin S, Ozkan S, Turkmut L, Siegel PB (2001) Responses to heat stress in commercial and local broiler stocks. 1. Performance traits. Br J Nutr 42:149–152
Yeruham I, Elad D, Friedman S, Perl S (2003) *Corynebacterium pseudotuberculosis* infection in Israeli dairy cattle. Epidemiol Infect 131:947–955
Young BA (1985) Physiological responses and adaptations of cattle. In: Yousef MK (ed) Stress Physiology in Livestock Vol II, CRC Press, Boco Raton, FL
Yousef MK (1987) Principles of bioclimatology and adaptation. In: Bioclimatology and the Adaptation of Livestock, Elsevier, Amsterdam, The Netherlands

Chapter 8
Adaptation in the Tourism and Recreation Sector

Daniel Scott, Chris de Freitas, and Andreas Matzarakis

Abstract The tourism-recreation sector is increasingly recognized as a climate-sensitive economic sector, with both supply- (tourism operators, destination communities) and demand-side stakeholders (tourists) directly affected by climate and its indirect influence on a wide range of environmental resources that are critical to tourism. Knowledge of the affects of climate on tourism and the effectiveness of climate adaptations in this sector remains inadequate and behind that of other economic sectors that have a longer tradition of scholarly development and government involvement. This is problematic considering the continued rapid growth in tourism world-wide and its prominence in many national economies. This chapter provides an overview of the range of climate adaptations utilized in the tourism-recreation sector, comments on the state of knowledge about climate change adaptation in the sector, and discusses some important directions for future inquiry.

8.1 Introduction

Tourism is one of the largest and fastest growing global industries and is a vital contributor to national and local economies around the world. The World Travel and Tourism Council (2006) estimated that in 2005 the global travel and tourism industry, encompassing transport, accommodation, catering, recreation and services for visitors, contributed 3.6% of global GDP and 2.8% of world-wide employment. The 808 million international tourist arrivals in 2005 (United Nations World

---

D. Scott
Department of Geography, University of Waterloo, Canada

C. de Freitas
School of Geography and Environmental Science, University of Auckland, New Zealand

A. Matzarakis
Meteorological Institute, University of Freiburg, Germany

Tourism Organization [UNWTO] 2006) are projected to increase to 1.6 billion by 2020 (UNWTO 1998).

The tourism-recreation sector is highly influenced by climate (Wall 1992; de Freitas 1990, 2003; Gomez-Martin 2005; Scott 2006a, b, c). At the local scale, climate defines the length quality of multi-billion dollar outdoor recreation seasons (e.g., skiing, snowmobiling, golf, boating, beach use), while at the global scale climate is a principal resource responsible for some of the largest international tourism flows (e.g., from Northern Europe to the Mediterranean and northern North America to the Gulf of Mexico and Caribbean). Climate also affects a wide range of environmental resources that are critical to the tourism-recreation sector (e.g., glaciers, wildlife productivity and migrations, biodiversity, water levels) and affects various facets of tourism-recreation operations (e.g., snowmaking or irrigation needs, water supply).

Despite the growing global economic importance of the tourism-recreation sector and the multiple interactions between climate and the tourism-recreation sector, our understanding of how climate variability and extremes affect tourists, tourism businesses and destination communities remains highly limited. Consequently, knowledge of the potentially profound consequences of global climate change for the tourism-recreation sector remains equally limited (Intergovernmental Panel on Climate Change 2001; UNWTO 2003; Scott et al. 2005a; Gossling and Hall 2006). The title of section two of this book therefore very accurately describes the status of climate and tourism research as a 'research frontier.'

The inadequate state of biometeorological knowledge on the tourism-recreation sector may be partially explained by a combination of factors. First, robust evidence of the economic importance of the tourism-recreation sector at national and international scales has only been made available through methodological advancements over approximately the last 10–15 years. Most outside of the tourism community are still unaware of the sector's economic importance and continued growth trends. Second, with mass tourism being largely a post-World War Two phenomenon, the tourism research community is not as mature as that of other important economic sectors, such as agriculture and fisheries. In most nations, there is very limited tourism research capacity within government and what exists is largely applied, almost exclusively focusing on product/market development, marketing, and monitoring (visitation and visitor spending). Other fundamental research relating to, for example, the non-economic consequences of tourism (social, environmental), falls largely to the academic community, where the size of tourism and recreation research community remains very small relative to the economic importance of the sector. The relatively recent development of research on the tourism-recreation sector is also reflected in the history of the International Society of Biometeorology (ISB), with the Commission on Climate, Tourism and Recreation (CCTR) only being founded at the 14th Congress held at Ljubljana, Slovenia in 1996. Third, until very recently climate change had not garnered substantive attention from the tourism industry or the tourism and recreation research communities (Wall and Badke 1994; Scott et al. 2005a; Gossling and Hall 2006). Butler and Jones (2001:300), in their concluding summary of the *International Tourism and Hospitality in the 21st Century* conference, forthrightly stated, "(Climate change) could have greater effect on tomorrow's world and tourism and hospitality in particular than anything else we've discussed. "The most worrying aspect is that ... to all intents and purposes the tourism and hospitality industries ... seem intent on ignoring what could be *the* major problem of the century *(original emphasis)*." Kelly (2001:33) writing about the tourism-recreation sector from the perspective of the insurance industry, similarly concluded that, "Although a number of organizations are seeking to raise awareness, there is little evidence to suggest that the leisure industry and/or its providers is formulating a strategy or has turned its attention to the planning or close examination of the issues involved (with climate change)." Only in the last 5 years has there been recognition within the tourism industry that tourism is an important contributor to processes of global environmental change (Gössling and Hall 2006) and that climate change in particular would have important implications for the tourism industry (UNWTO 2003).

Scott et al. (2006) have compiled a comprehensive bibliography of the international climate and tourism literature, containing over 300 English, French, German and Spanish language publications (as of December 2004). Surprisingly, even though the large majority of climate change-related publications in the bibliography deal with impacts, more publications explicitly focus on climate change mitigation than adaptation. A number of publications that examine climate change impacts, particularly post-2000, implicitly mention adaptation using words like 'respond', 'cope', or 'adjust,' but without providing a thorough discussion of the full range of adaptations available to specific tourism market segments or destinations under examination or how the implementation of certain adaptations could alter the projected impacts (even qualitatively). Scott (2006b) argued that adaptation continues to be a critical research gap in the growing literature on climate change and the tourism-recreation sector. Unlike in other economic sectors, to date there has been no systematic attempt to document the range of climate adaptations currently employed in the tourism-recreation sector.

With the above in mind, this chapter represents the first effort to compile a portfolio of climate adaptations employed by tourism-recreation stakeholders (individual tourists, tourism operators, tourism destinations – communities, financial sector, and government). It remains beyond the scope of this chapter to examine the very wide range of adaptations in detail (i.e., the extent, historical evolution, and barriers to each adaptation) or attempt to evaluate the effectiveness of each adaptation. These remain important directions for future inquiry.

8.2 Climate Adaptation Portfolio for the Tourism-Recreation Sector

Climate adaptation in the tourism-recreation sector is comprised of complex mix of adaptations undertaken by diverse stakeholders at a range of spatial and temporal scales (Fig. 8.1). The climate adaptation portfolio presented in this section is not intended to be comprehensive, as this would require detailed adaptation portfolios, like that developed by Scott (2006b) for the ski industry, to be developed for several

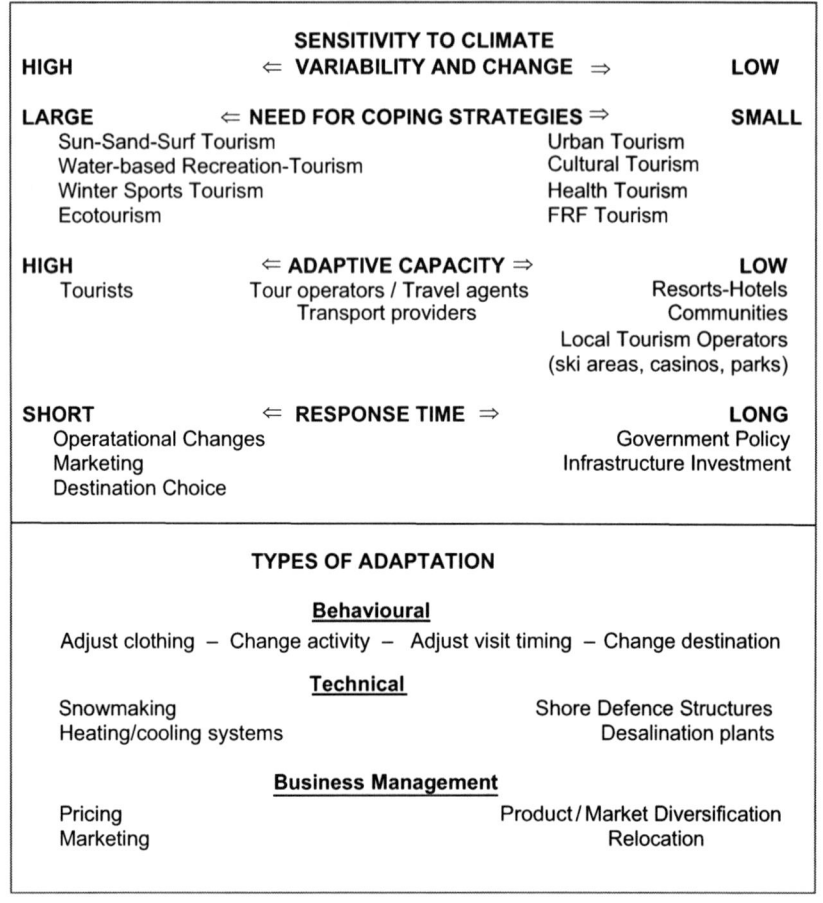

Fig. 8.1 Conceptual framework for considering adaptation to changes in tourism climate

major tourism segments in climatic zones around the world; rather it is meant to be illustrative of the diversity of climate adaptation available to this highly dynamic sector. The climate adaptation portfolio is organized by type of stakeholder (tourists, tourism operators/businesses, tourism industry associations, governments/communities, financial sector) and major types of adaptations (e.g., technical/structural, behavioural, business management, policy, research and education) utilized by each stakeholder. While climate adaptations are described individually in this chapter, rarely are adaptation options undertaken in isolation. More commonly, climate adaptation by stakeholders in the tourism-recreation sector involves multiple adaptation options (Fig. 8.1). Individual adaptation options are sometimes also undertaken by several different stakeholders in the sector (e.g., marketing by tourism businesses, communities and countries), sometimes in isolation and sometimes collaboratively.

To date, all of the studies that have specifically examined climate adaptation in some aspect of the tourism-recreation sector (Elsasser and Bürki 2002;

Scott et al. 2002, 2005b; Raksakulthai 2003; Becken 2004; Sievanen et al. 2005; Scott and Jones 2005; Scott 2006b) have concluded that there is little evidence of adaptation in anticipation of climate change. Thus, almost all of the climate adaptations identified in this portfolio represent adaptation as currently practiced. The few climate change specific adaptations in this sector are clearly identified throughout this section.

8.2.1 Tourists-Recreationists

8.2.1.1 Behavioural

Recreation and leisure tourism are by definition activities undertaken by choice during 'free time.' As such, tourists-recreationists have a great degree of freedom to choose the activities they wish engage in, as well as where and when they will do so. Spatial, temporal and activity substitution provide tourists-recreationists with tremendous adaptive capacity.

Tourists are easily able to adapt to climatic conditions or climate-related impacts at any given destination by simply going elsewhere. Giles and Perry (1998) found that the exceptionally warm and sunny summer of 1995 in the UK resulted in a drop in outbound tourism as travellers opted for domestic holidays. Extreme events regularly influence traveller decisions in regions such as the Gulf of Mexico. The four hurricanes that struck the State of Florida in 2004 caused thousands of cancellations as travellers went elsewhere and a marketing survey found that 25% were also less likely to visit Florida during hurricane season in the future (Pack 2004).

Tourists-recreationists are able to select when climatic conditions are suitable to engage in chosen activities or visit a destination. Tourists can adapt the timing of travel according to climate variability, such as unusually early or late hurricane or monsoon season, and eventually to climate change. Indeed, several climate change impact studies assume that tourists will adapt to new climatic regimes by altering the timing of visitation, particularly new opportunities during shoulder seasons (Maddison 2001; Lise and Tol 2002; Hamilton et al. 2005; Jones and Scott 2006a, b; Berrittella et al. 2006). Tourists-recreationists can also adapt the frequency of their visits in response to climate variability. Scott (2006b) found evidence of such among skiers in the US. Examining ski area 'utilization' data, which is the ratio of actual skier visits to the physical capacity of skier visits at a ski area over the ski season; it was found that utilization decreases during longer ski seasons. Greater utilization during shorter ski seasons, suggests that skiers are participating more frequently than they would in a normal year (i.e., go skiing every weekend, instead of every 2 weeks). This type of behavioural adaptation is particularly possible when the ski season starts later than usual, because skiers know they likely will have fewer opportunities that season.

Activity substitution can take place over a range of time-scales. Tourists-recreationists can also modify their activities to cope with unfavourable weather conditions. Fig. 8.2 illustrates how some tourists visiting a beach resort in Cuba

Fig. 8.2 Small scale–short term behavioural adaptation to weather conditions (Photo credit: Daniel Scott)

have adapted to strong winds and cool temperatures in ways that still allow them to engage in beach related activities (i.e., additional clothing and erecting wind screens with beach chairs). Tourists-recreationists can change from one activity to another or change the frequency of activities in response to climate variability. A 1°C warmer than average summer season has been found to increase domestic tourism expenditures in Canada by 4% (Wilton and Wirjanto 1998). Individuals can also substitute activities on a permanent basis in response to climatic changes. Scott et al. (2002) found some snowmobilers had begun to switch to All-Terrain-Vehicles (ATVs) in response to changes in snow conditions.

A limited number of studies have begun to explore the potential behavioural adaptations of tourists-recreationists to future climate change (König 1998; Braun et al. 1999; Bürki 2000; Richardson and Loomis 2004; Scott and Jones 2006a; Uyarra et al. 2005). In each study, a combination of spatial, temporal and activity substitution were found. König (1998) and Bürki (2000) utilized surveys to examine how skiers in Australia and Switzerland might respond to marginal ski conditions presented in a hypothetical climate change scenario. In Australia, 25% of respondents indicated they would continue to ski with the same frequency, nearly one-third (31%) would ski less often, but still in Australia, and the greatest portion (38%) would substitute destinations and ski overseas (mainly in New Zealand and Canada). A further 6% would not continue to ski under such conditions. In Switzerland, the majority (58%) indicated they would ski with the same frequency (30% at the same resort and 28% at a more snow reliable resort – generally at higher elevation). Almost one-third (32%) of respondents indicated they would ski less often and 4% stated they would stop skiing altogether.

In eastern North America, a climate change analogue approach has been used to understand the potential response of the ski tourism marketplace to future climate change. The winter of 2001–2002 was the record warm winter throughout much of the region and approximated the normal temperatures expected in mid-century under a mid-range warming scenario (approximately + 4.5°C). Skier visits during this record warm winter were consistently lower than in the previous climatically normal winter of 2000–2001: − 11% in the Northeast ski region of the US, − 7% in Ontario, and − 10% in Quebec (Scott 2006c). Although this finding is not surprising considering the ski season was approximately 20 days shorter in the record warm winter, what is somewhat surprising is how small the reduction in skier visits was during this climate change analogue season. It was observed that utilization levels at ski areas increased, as many skiers in the region adapted by skiing more frequently than in a normal year (i.e., skiing every weekend, instead of every 2 weeks).

Comparable studies have also been conducted on how tourists might respond to climate-induced environmental change in national parks in the Rocky Mountain region of North America. Richardson and Loomis (2004) found that between 9% and 16% of surveyed visitors to Rocky Mountain National Park (USA) would change the frequency of visitation to the park under the hypothetical environmental change scenarios (representing the 2020s). The environmental change scenarios constructed for the early and mid-decades of the twenty-first century were also found to have minimal influence on intention to visit Glacier-Waterton International Peace Park or Banff National Park, with almost all visitors still intending to visit the parks and 10% indicated they would visit more often, presumably due to improved climatic conditions (Scott et al. 2007; Jones and Scott 2006a). There is also the potential that media coverage of melting glaciers might motivate more people to visit these parks over the next 20–30 years to personally see or show children the glaciers before they disappear and in order to witness the impacts of climate change on the landscape. This 'last chance' tourism market trend is already being observed in some areas of Alaska, including Kenai Fjords National Park, where the chief ranger has described climate change as one of the new major themes for the park (Egan 2005). If such an increase is visitation is realized, it would require adaptation to accommodate larger numbers of visitors and provide new public education about the changes in natural heritage that are occurring.

In the studies that attempted to look at the potential impacts of greater environmental change (Scott et al. 2007; Jones and Scott 2006a, b), an important threshold was reached for many visitors to Glacier-Waterton International Peace Park and Banff National Park in scenarios that might occur by the end of the twenty-first century. A substantial number of tourists (19% in Glacier-Waterton and 31% in Banff) indicated they would not intend to visit the parks if the specified environmental changes occurred. The projected loss of glaciers in the region was noted as a significant heritage loss and the most important reason cited for not intending to visit the park in the future. Another 36–38% of tourists indicated they would plan to visit less often. Visitors most likely to be negatively affected by climate-induced environmental change were long-haul tourists and ecotourists, motivated by the opportunity to view pristine mountain landscapes and wildlife. As such, the impact of environmental change was more pronounced in Banff National Park, which has a much greater number of international tourists. If realized, such impacts would require these destinations to adapt to very different impacts of climate change.

Recent coral bleaching events and the imperilled future for many coral reefs under climate change are a cause for concern for diving and other related tourism. Unfortunately, there is limited information about how tourists responded to the severe coral bleaching that occurred in many reef systems around the world in 1998. A case study from El Nido, Phillippines does provide some insight into the response of different tourist market segments to coral bleaching and degraded reef environments (Cesar 2000). In El Nido and nearby islands, severe coral bleaching in 1998 led to 30–50% coral mortality and a typhoon that same year (also linked to El Niño) caused further damage to local reefs. Whether divers or not, most tourists (95%) coming to El Nido have at least some interest in the local marine environment. However, general awareness of coral bleaching among tourists was found to be low (44%). The bleaching event did not impact budget tourist arrivals, but fewer budget tourists went diving during their stay. The impact at resorts, some of which cater to the high-end dive market, was much worse. In other coastal locations, the impact of climate change was also projected to adversely affect tourist preferences for these destinations. In Bonaire and Barbados, more than 75% of tourists were unwilling to return for the same holiday price in the event that coral bleaching or reduced beach area occurred as a result of climate change (Uyarra et al. 2005). The response of tourists to recent severe bleaching events on the Great Barrier has not been systematically assessed, however a survey of tourists in Cairns (North Queensland, Australia) asked if they would visit the region if they knew that there had been a recent bleaching event – 29% were uncertain and 35% indicated they would not (Prideaux 2006).

8.2.1.2 Technical

The most common technical climate adaptations used by tourists-recreationists is the wide range of specialized equipment that allows them to engage in activities more comfortably and more safely when climate conditions are not ideal or to expand the climatic range in which activities can be undertaken. Some illustrative examples include: wetsuits for diving or windsurfing, hand and foot warmers built into snowmobiles, rain gear (clothing, equipment covers, etc.) for golf and hiking.

8.2.2 Tourism-Recreation Operators

Tourism-recreation operators is a broad category comprised of diverse stakeholders, including businesses involved in the travel planning and transportation phase (e.g., travel agents, event planners, transportation companies, international tour companies) to the wide range of businesses involved in hospitality (i.e., hotels-resorts, restaurants), attractions management (e.g., museums, golf courses, etc.), and other services (e.g., tour guides, equipment rentals, etc.) at specific destinations. While part of the same tourism-recreation sector actor group, there is an important distinction in the adaptive capacity of tourism businesses that operate in single and multiple-destinations. As examples will illustrate, tourism-recreation businesses with immobile capital assets (e.g., resort complex, marina, or casino) at individual destinations have less adaptive capacity than businesses that provide transportation or travel planning services.

8.2.2.1 Technical

A tremendous array of technical and structural climate adaptations are used in the diverse environments that the tourism-recreation sector operates in. Here the discussion will be limited to adaptations to two widespread climate-related pressures on tourism-recreation operators: snow reliability and water supply.

The ski industry uses three major types of technological adaptations to improve snow reliability: snowmaking systems, slope development and operational practices, and cloud seeding (Scott 2006b). Snowmaking is the most widespread climate adaptation used by the ski industry and has become an integral component of the ski industry in some regions (eastern North America, Australia, Japan). Over the last 30 years, hundreds of millions of dollars have been invested in snowmaking systems in order to expand operating seasons and increase the range of climate variability that ski areas could cope with. Figure 8.3 illustrates the diffusion of snowmaking technology in the in the five ski regions of the US from 1974–1975 to 2001–2002. In the mid-1970s there was much greater use of this adaptation in the Northeast and Midwest ski regions than regions with higher elevations like the Rocky Mountains and the Pacific West. Since then that difference has been gradually diminishing. A similar east-west geographic pattern exists in Canada (Scott 2006b). The implementation of snowmaking is not as extensive in Europe as in North America, but

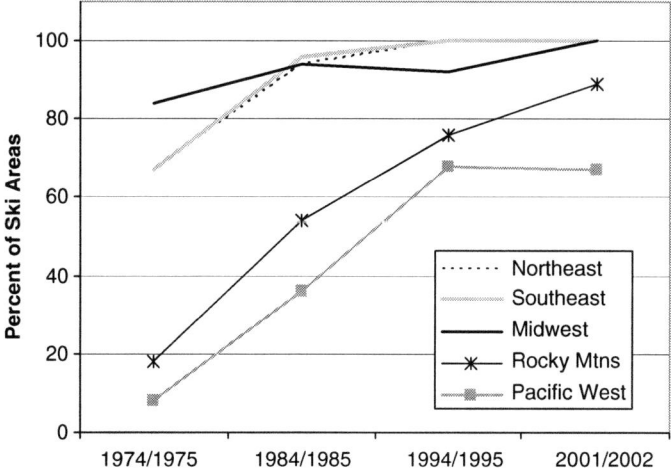

Fig. 8.3 Historical diffusion of snowmaking adaptation throughout the US (National Ski Area Association annual state of the ski industry reports)

there is no comprehensive analysis of how snowmaking differs by country. Barriers to the increased used of snowmaking in Europe include higher energy costs, challenges to securing adequate water supply, and environmental concerns (e.g., chemical additives that allow snowmaking at temperatures near 0°C are banned in Germany).

Slope development adaptations include: slope contouring, landscaping, and, interestingly, the protection of glaciers. With the increased recession of glaciers in the Alps in recent years, notably the record warm summer of 2003, ski areas in Switzerland and Austria, have needed to develop new adaptations. Initially heavy machinery was used in the autumn months to move snow to fill in gaps left by the glacier's retreat over the summer. In response to rapidly increasing fuel costs, some ski areas began to work with glaciologists on a new approach to protect critical areas of glaciers (typically cable car exist areas) from ultraviolet radiation and restrict melting during summer months with large sheets of white polyethylene (Simmons 2005; Jahn 2005). While the results have been 'fantastic' in the words of ski area operators, some environmental groups, such as Switzerland's branch of Friends of the Earth, are opposing plans to expand the use of this adaptation by ski areas (MacInnis 2006).

In addition to the modification of existing ski terrain, the development of new ski terrain in climatically advantaged locations (north facing slopes, higher elevations) is commonly cited as an adaptation to climate change. Expansion of ski areas into higher elevations appears to be the principal climate change adaptation strategy being considered in the European Alps (König and Abegg 1997; Elsasser and Bürki 2002; Breiling and Charamza 1999). High elevation mountain environments are particularly sensitive to disturbance and proposed expansion of ski areas into these environments is often met with opposition from the public and environmental groups. For example, a project to develop a world-class, four-season ski resort on Jumbo Glacier in southern British Columbia has been held up since 1991 by opposition from environmental groups and local residents (Greenwood 2004).

Cloud seeding is a weather modification technology that has been used to produce additional precipitation; although a recent US government report concluded there still is no convincing scientific evidence that this adaptation works (National Research Council 2003). Some ski areas in North America (State of Colorado) and Australia (New South Wales State) have employed this technology in an attempt to generate additional snowfall.

Inadequate water supply is a salient problem in many tourism areas around the world and is often brought about by over development (demand exceeding supply capacity) and climate variability. The following examples illustrate the three most common technical adaptations to increase water supply for tourism operations: water transfers (pipelines or tankers), reservoirs, and desalination plants. During the mid-1990s summer drought conditions on the Spanish island of Majorca threatened the operations of the tourism industry, the islands largest source of income and employment. With local aquifers suffering saline intrusion and failing to meet water supply requirements for tourism and the local population, the Spanish government implemented a yearly 10 million m^3 water transfer from the Spanish mainland via tanker ships (Wheeler 1995). The long-term adaptation strategy was the construction of two large-capacity desalination plants and additional water transfers (via pipeline) from the mountainous north side of the island (Wheeler 1995).

The Tourism Authority of Thailand (TAT), together with other national agencies, has several adaptation strategies to reduce water shortages in the heavily developed tourism island of Phuket that are often brought about by climate variability (Raksakulthai 2003). Structural elements of the multi-year water supply plan include the construction of new dams, development of abandoned mines as water sources, expanded water transmission and water recycling systems. In addition to these structural adaptations, TAT is also planning non-structural adaptations, such as a revised fee structure for water consumption and water conservation campaigns (Raksakulthai 2003).

Becken's (2004) survey of small and medium size tourism operators in Fiji recorded a number of other small-scale water supply adaptations. Over a third of the survey respondents indicated they had experienced reduced water availability during recent droughts, and as a result structural adaptations to secure on-site water supply was the second most frequent climate adaptation undertaken by respondents. Examples of small-scale structural adaptations included: retrofitting buildings with rainwater collectors, increasing storage tank capacity, converting toilets to saltwater supply, and adding diesel powered desalination capacity. Tourism operators in Fiji also utilized non-structural adaptations, including water conservation education for employees and guests, revised landscaping practices, limited use of pools.

8.2.2.2 Business Management

The business model that tourism-recreation operators chose is an important determinant in the range of climate adaptations utilized. Some tourism operators have adapted to the pronounced seasonality of tourism demand in many destinations by closing during the low season. This adaptation strategy allows operators, particularly small and medium sized family-run enterprises, to make necessary repairs, develop marketing campaigns, attend training sessions, and go on vacation themselves. Other tourism operators substantially reduce capacity and services during low season (Fig. 8.4). Still others, sometimes as part of a broader tourism industry or national tourism strategy, use a range of adaptation strategies to develop low season markets in order to operate year round.

Product and market diversification are common adaptation strategies to increase demand during low seasons. In Thailand, the low season occurs during the southwest monsoon and Raksakulthai (2003) has reported two main adaptation strategies to develop low season. TAT has concentrated on developing tourism attractions that are not climate sensitive, such as health and wellness spas, study tours on Thai culture (e.g., cooking, religion, and language classes), indoors entertainment complexes with Thai cultural performances, and shopping, and also promoted development of the MICE market (meetings, incentives, conventions, exhibitions) among business travellers. In North America, the Economist (1998) referred to the transition of major ski resorts from ski areas to winter theme parks, as the 'Disneyfication' of the winter sports industry. Many ski resorts have

Fig. 8.4 Seasonal resort shut-down as climate adaptation (Greece) (Photo credit: Daniel Scott)

made substantial investments to provide alternate activities for non-skiing visitors (e.g., snowmobiling, skating, dog sled-rides, indoor pools, health and wellness spas, fitness centres, squash and tennis, games rooms, restaurants, retail stores). A number of former 'ski resorts' have further diversified their business operations to become 'four season resorts', offering non-winter activities such as golf, boating and white-water rafting, mountain biking, paragliding, horseback riding and other business lines (spas, conference facilities).

Tourism business models can also influence vulnerability to climate variability and change. For example, an important business model to emerge in the North America ski industry over the past decade is the ski resort conglomerate. Companies like American Skiing Company, Intrawest, Booth Creek Resorts, and Boyne USA Resorts have acquired ski areas in different locations across North America. Although not intended as a climate adaptation, the conglomerate business model may prove to be one of the most effective adaptations to future climate change. The ski conglomerate business model provides greater access to capital and marketing resources, thus enhancing adaptive capacity, but also reduces the vulnerability of the conglomerate to the effects of climate variability and future climatic change, through regional diversification in business operations. The probability of poor snow conditions in one ski region of North America (e.g., New England) is much higher than for several others (e.g., New England, Quebec-Ontario, Midwest, Rocky Mountains, and California). When poor conditions occur, the financial impact can be spread out through the organization and above average economic performance in one or more regions could buffer losses in another. Companies with ski resorts in a single region or independent small-medium size ski enterprises are at greater risk to poor climatic conditions. Without substantive economic reserves or access to capital, a series of economically marginal years may be all that is required to bankrupt the business.

Marketing is another key business strategy that is used to adapt to natural seasonality, climate extremes and most recently climate change. The Caribbean Tourism Organization and individual member states have begun to actively market themselves as four-season destinations in the late 1990s with multi-million dollar advertising campaigns that target the honeymoon market and budget-conscious families (Barnes 2002). In combination with marketing messages that downplay the region's summer heat are upgraded air-conditioning, discounted room rates, and new hurricane interruption policies at many resort companies, including Sandals Resorts, Club Med, SuperClubs, TNT Vacations and Apple Vacations. The hurricane guarantees or waivers differ slightly from company to company, but basically provide a replacement stay of the same duration and equivalent value as the one originally booked (Bly 2006). The strategy has proven successful as summer occupancy rates at beach resorts are approaching or equalling winter season in many destinations (Johnson 2005).

'Last change' marketing for climate change threatened destinations is still relatively rare, but an increasing marketing trend among tourism operators in regions where climate change impacts are clearly observable. Climate change has become one of the major interpretive themes in Kenai Fjords National Park in Alaska and some Alaskan tour guides are inviting travellers who visited 25 years ago to experience how the landscape has changed (Egan 2005). The popular travel magazine *Conde Nast Traveler* (May 2004) featured an article on climate change and international tourism destinations, including at list of 'endangered wonders' that travellers were recommended to visit before they vanish. UNESCO has indicated that world heritage sites such as the Belize Barrier Reef, Waterton-Glacier International Peace Park, Australia's Great Barrier Reef and the snows of Mount Kilomanjaro are threatened by climate change (Black 2006), which will likely inspire future articles in travel magazines and climate change related marketing by opportunistic tour operators.

In contrast to tourism operators with immobile capital assets that must adapt to climate at a specific location, other tourism businesses, that provide travel planning or transportation services, operate in many locations and one of their principal adaptation strategies is to redirect the travellers they represent to alternate destinations as conditions dictate. Travel agents redirect clients away from destinations that have been recently impacted by extreme events. The Government of Mexico estimated that as a result of the late season hurricane Wilma and media coverage of damage and stranded tourists, it would lose US$800 million in tourism revenue between October and December 2004 as travel agents and individual travellers select other destinations (Williams 2005). Large-scale event planners adapt in the same manner. Convention and event businesses increased in Arizona following the four hurricanes in Florida in 2004 and were expected to remain higher during hurricane season for the next 2–5 years as event planners avoid weather risks (USA Today 2005). With the trend toward shorter-term travel planning, especially discounted 'last minute' bookings made in the week (or day) prior to departure, travel agents are increasingly able to redirect clients away from destinations that could be adversely affected by unfavourable weather (using 5–7 day forecasts) or extreme events (areas in the path of developing hurricanes).

8.2.2.3 Policy

Few corporate policies on climate change exist in the tourism-recreation sector. The first known tourism operator in the world to develop a climate change policy is the Aspen Skiing Company. In 2001, they adopted two policy statements: (1) Aspen Skiing Company acknowledges that climate change is of serious concern to the ski industry and to the environment; and (2) Aspen Skiing Company believes that a proactive approach is the most sensible method of addressing climate change. Notably, the Aspen Skiing Company joined the Chicago Climate Exchange and established a climate change action plan, which focuses on mitigation efforts and does not publicly set out the company's adaptation strategy.

8.2.2.4 Public Education

There are a number of examples of tourism-recreation operators participating in pubic education on climate change with the intention of raising awareness and influencing personal behaviours that contribute to climate change mitigation. Many individual ski areas in North America and some celebrity ski athletes participate in the annual 'Keep Winter Cool' public education campaign (described further under Tourism Industry adaptations) (National Ski Areas Association 2006). The Mountain Equipment Co-op Company has been involved with the development and sponsorship of the 'Melting Mountains' public education initiative that focuses on the impacts of climate change on mountain environments and mobilizing stakeholders of these highly important tourism-recreation resources to reduce greenhouse gas emissions at the individual and corporate level (Melting Mountains 2006). A growing number of travel companies are now educating travellers about travel related GHG emissions and promoting carbon offset programs, such as those offered through 'MyClimate' (2006).

8.2.3 *Tourism Industry Organizations*

Organizations that coordinate and act on behalf of the tourism-recreation industry exist in many forms, from local tourism cooperatives to the United Nation's World Tourism Organization. Climate adaptations by tourism industry organizations are limited and tend to focus on climate change.

8.2.3.1 Policy

One of the few tourism industry organizations that have a climate change policy is the National Ski Areas Association (NSAA) in the US. As Best (2003:57) observed, the development of this policy represented, "a remarkable turnaround for an industry that just five or 6 years ago had largely shrugged off global warming."

In support of its climate change policy, 65 ski areas lobbied government to increase political support the proposed Climate Stewardship Act in the US. Interestingly, although the membership of the NSAA was able to develop a policy on climate change mitigation, no such collaborative initiative for adaptation exists. The North America ski industry is a very competitive business environment, where the tradition of cooperation is largely limited to government lobbying, regional marketing of winter tourism, and environmental standards. As a result, Scott (2006b) argued that it is unlikely that the industry would develop a cooperative adaptation plan, that might include, for example, a national income stabilization program to spread the climate risk exposure of individual ski areas or support vulnerable, low-lying ski areas near major urban markets like Boston and New York, which play an important market development role for the industry as a whole. The more likely scenario is a continuation of the existing competitive business environment and the unplanned, market based contraction of the US ski industry that has been underway for the past 2 decades.

At the international level the United Nations World Tourism Organization has taken a lead in raising awareness about climate change within the tourism industry. UNWTO hosted the first international conference on climate change and tourism in 2003 (Djerba, Tunisia), where delegates from 45 nations developed the *Djerba Declaration on Climate Change and Tourism*, which contained several items of agreement in support of adaptation.

8.2.4 Government (International, National, Local)

8.2.4.1 Technical

Governments at the local to national level are often managers of tourism-recreation lands and other resources (e.g., national parks, reservoirs) and as such, governments utilize many of the technical climate adaptation strategies already discussed in the tourism operators section. A recent example includes government response to changes in the phenology of cherry trees in Japan. Japan's cherry blossom is a national symbol and the basis of a multi-million dollar flower viewing tourism industry. The timing of the peak bloom varies with the seasonal weather and recent warm springs have caused the peak bloom to occur too early for local festival organizers. The local government in Hirosaki have commissioned scientists to 'programme' the cherry bloom at the 'appropriate time' by experimenting with sprays and plant hormone injections as well as piling snow on the base of trees to slow the onset of blossoms (Parry 2005).

8.2.4.2 Business Management

Governments at the local to national level are also involved to varying degrees in tourism product development and marketing and have used climate adaptations that might be thought of as business strategies. For example, the State of Florida allocated US$30 million to 'hurricane recovery' marketing following the devastating sequence of four hurricanes in 2004. The State also developed a weather insurance program for convention organizers, where it pays the premiums for US$200,000 insurance coverage for rescheduling costs associated with hurricane disruption (Pain 2005).

8.2.4.3 Policy and Planning

The regulatory role of government provides it with many important climate adaptation strategies. Many of these regulatory frameworks are not specific to the tourism-recreation sector and because they are discussed in other chapters of this volume, they are only identified here: coastal management plans and set back requirements, building design standards (e.g., for heating and cooling or hurricane force winds), emergency management (e.g., tourist warning systems, evacuation plans), environmental impact assessments (e.g., influencing adaptations such as snow-making and desalination plants), wildlife management (e.g., fish and game quotes), water quality standards (e.g., establish swimming bans), and wildfire management (e.g., set open fire bans in parks and campgrounds). Other areas of government policy could have important implications for tourism-recreation sector climate change adaptation in the future, including: establishment of a 12-month school year and post-Kyoto emission reduction targets and mitigation strategies that may affect tourist mobility and demand in some destination regions.

8.2.4.4 Public Education

One of the most successful examples of the use of public education by government to adapt to climatic impacts has been the campaigns on the dangers of UV radiation. Governments around the world and international agencies like the World Health Organization (2002) have developed monitoring tools like the UV Index and combined them with adaptation recommendations for the public that have begun to have documented impacts on public perception of the desirability of a tan and the rates of some skin cancers in some countries. While these public education campaigns represent a public health adaptation and not a tourism-recreation sector specific adaptation, one of the principal target audiences for the UV Index messaging is those engaged in outdoor recreation, whether at the beach or in the backyard.

8.2.4.5 Research

Climate change adaptation research and capacity building in the tourism-recreation sector has been funded by governments at all levels over the past 5 years (e.g., international – NATO, European Science Foundation, European Union, United Nations World Tourism Organization; national governments – Australia, Canada, Finland,

New Zealand, UK; local governments – City of Aspen [USA], Town of Banff [Canada]). Adaptation in the tourism sector remains a key knowledge gap and much more government supported research is needed and expected in the decades ahead, Nowhere is this need more glaring than in developing nations where adaptive capacity is thought to be lowest and where tourism is a major component of the economy (e.g., parts of the Caribbean, Africa, and many Small Island Developing States).

8.2.4.6 Integrated Adaptation

Like tourism operators, climate adaptation by governments typically involves multiple adaptations, which are sometimes integrated as part of a broad adaptation strategy. This integration of adaptations typically occurs at the agency level. Illustrative examples of integrated adaptation can be found in very diverse climatic environments. The range of climate adaptations already in use by the National Capital Commission in Canada for recreation land management and recreation-tourism programming (Table 8.1) provides an example in a temperate nation (Scott et al. 2005b). Here unreliable snow fall, variable spring thaw and growing degree days, and summer temperature extremes each pose challenges for tourism programming.

Table 8.1 Climate adaptations used in tourism programming by the National Capital Commission of Canada (Adapted from Scott et al. 2005b)

Winterlude
- Moved attractions/programming from ice to land locations
- Used refrigerated trucks for the ice sculpture carving contest
- Was converted from a 10-day event to a 3-weekend event to increase the probability of suitable weather during the celebration
- Implemented snowmaking at Snowflake Kingdom to ensure adequate snow supply
- Removed weeds from the canal that could weaken the ice structure (e.g., strength)
- Developed collaborations with local museums to offer package deals that promote non-climate-dependent activities

Tulip festival
- Planted bulbs in shady locations
- Heavily mulched flower beds
- Erected snow fences to increase snow cover on flower beds to delay bulb maturation
- Planted bulbs with different rates of maturation
- Irrigated flower beds during warm/early springs to delay bulb maturation

Canada day
- Educated the public about heat stress
- Provided shade tents and cooling stations
- Position medical staff on stand-by at major events

Gatineau park
- Implemented snowmaking on alpine ski areas
- Developed a cross-country ski track setter for low-snow conditions
- Developed cross-country ski trails in shaded and smoothed-terrain areas that required less snow
- Implemented water quality advisory system in swimming areas

The Great Barrier Reef has experienced several mass coral bleaching events in the past decade (1998, 2002, and 2006). During the 1998 global mass bleaching event about 50% of Great Barrier reefs suffered bleaching; 87% of inshore reefs and 28% of mid-shelf and offshore reefs. Overall about 5% of reefs were severely damaged by this bleaching event (Great Barrier Reef Marine Park Authority 2007a). The Great Barrier Reef suffered the largest mass bleaching event on record in 2002, when 60% of reefs were bleached (Great Barrier Reef Marine Park Authority 2007a). The increasing threat of coral bleaching under projected climate change scenarios inspired the Great Barrier Reef Marine Park Authority to prepare a 'Coral Bleaching Response Plan' (Great Barrier Reef Marine Park Authority 2007b), with the objectives to:

- Improve ability to predict bleaching risk
- Provide early warnings of major coral bleaching events
- Measure the spatial extent of bleaching
- Assess the ecological impacts of bleaching
- Involve the community in monitoring the health of the Reef
- Communicate and raise awareness about bleaching
- Evaluate the implications of bleaching events for management policy and strategies

The Great Barrier Reef Marine Park Authority and the Australian Ministry of Tourism have also considered other technical adaptations, including spraying cooler water from deeper areas onto ocean surface at peak heat times to cool surface waters and protect the corals from being damaged or using awnings or ambrella-like structures on bouys to shade corals in high visitation tourism areas (Sulaiman 2006; Badenschier and Schmitt 2006).

8.2.5 Financial Sector

Climate adaptation by the financial sector, in particular the insurance industry, has already impacted the tourism-recreation sector and increasingly being considered in the context of adapting to future climate change.

8.2.5.1 Business Management

There is growing general consideration of climate change risk in the business community (Reuters 2004; The Wall Street Journal 2005). For example, Hypovereinsbank and Credit Suisse consider climate change in credit risk and project finance assessments (Innovest Strategic Value Advisors 2003). The investment community has also begun to adapt its lending practices to the ski industry. Swiss banks now provide very restrictive loans to ski areas at altitudes below 1,500 m above sea level (Elsasser and Bürki 2002) and banks in Canada have also discussed the issue of climate change in financial negotiations with ski operators (Scott 2006b).

Climate adaptations by the insurance industry that have affected the availability and affordability of insurance have already affected segments of the tourism-recreation sector. Insurance companies have dropped coverage of certain tourism-recreation properties in US states on the Gulf of Mexico. Following Hurricane Andrew in 1992, US insurance companies realized they had underestimated their exposure in south Florida and would not renew insurance policies for many coastal recreational properties in the region. The State of Florida then created the Florida Windstorm Underwriting Association as an insurer of last resort (Kelly 2001). A number of insurers have also indicated they would no longer insure floating casinos in the Gulf Region. Unless the State of Mississippi changes its law requiring casinos to be water-based, the multi-billion dollar gambling tourism industry that existed prior to Hurricane Katrina is unlikely to be rebuilt.

In recent years, the insurance industry has begun to provide a wider range of insurance and weather derivative products suitable for the tourism-recreation sector. The introduction of weather insurance was a potentially positive development for the ski industry. During the 1999–2000 ski season Vail Resorts in Colorado purchased snow insurance that paid the resort US$13.9 million when low snowfall affected skier visits (Bloomberg News 2004). However, insurance premiums have increased substantially in the last 5 years and large ski corporations like Intrawest and Vail Resorts no longer carry weather insurance because of the high premiums. Snow insurance costs are even more burdensome on smaller business and therefore are unlikely to be used to a great extent by the ski industry in their current form.

8.3 Conclusion

This chapter represents the first attempt to outline the scope of climate adaptation in the tourism-recreation sector and as such is offered as a starting point for future dialogue and research on adaptation in this increasingly important sector of the global economy. Nonetheless, based on this review of the available literature and the authors collective experience a number of conclusions about climate adaptation in the tourism-recreation sector can be drawn.

Climate adaptations by each of the major actor groups in the tourism-recreation sector are not undertaken in isolation as a single discrete action. As noted previously, climate adaptation in the tourism-recreation sector more commonly involves multiple adaptation options that are driven by climate and non-climate factors, sometimes over a span of several years. This is consistent with the findings of Smit et al. (2000) and Adger et al. (2005) in other economic sectors.

The tourism-recreation sector is thought to posses high adaptive capacity overall, although adaptive capacity varies substantially both between actor groups in the tourism-recreation sector (e.g., between tourists and tourism business operators) and within actor groups (e.g., between individual ski area operators). A number of authors (Wall 1992; Maddison 2001; Elsasser and Burki 2002; Scott 2006a; Gossling and Hall 2006) have theorized that a spectrum of adaptive capacity exists within the tourism-recreation sector and consistently rated the relative adaptive capacity of major actor groups, as we have illustrated conceptually in Fig. 8.1.

Adaptive capacity can also vary substantially among individuals within actor groups. For example, Scott's (2006b) analysis of climate adaptation in the ski industry indicated that ski areas with greater adaptive capacity are characterized by higher elevation terrain or advantageous micro-climates relative to local/regional competition; efficient and extensive snowmaking systems; adequate water supply (with the potential to expand); are located in a region of lower average energy costs; have well diversified resort operations (multiple winter activities and four-season operation); are part of a larger company or regionally diversified ski conglomerate that can provide financial support during poor business conditions; are located in jurisdictions with less land use restrictions (e.g., outside of national parks or in states/provinces where skiing makes a large contribution to the economy); are closer to large urban markets; and have positive relationships with host communities (which may reduce constraints to adaptation).

Within the tourism-recreation sector, adaptations undertaken by one actor group will affect other actors. In some cases, what is adaptation to one tourism-recreation sector stakeholder could be considered maladaptation by other stakeholders. The ability of tourism operators and governments in the Caribbean to use a range of climate adaptations (marketing, pricing, air conditioning, hurricane guarantees, etc.) to substantially increase summer tourism in the region represents successful adaptation by these stakeholders. However, tourist testimonials in Barnes (2002) indicate that at least some tourists believe the marketing campaigns are inaccurate and thus represent a adverse adaptation by the tourism industry.

The available studies that have examined the climate change risk appraisal of tourism operators (Scott et al. 2002; Raksakulthai 2003; Becken 2004; Scott and Jones 2005; Scott et al. 2005b; Sievanen et al. 2005) have consistently found low awareness of climate change and little evidence of long-term strategic planning in anticipation of future changes in climate. Consequently, climate change adaptation by private and public sector tourism-recreation operators is likely to remain reactive and consist mainly of incremental adjustments of existing climate adaptations. There is also some evidence that suggests tourism operators are overestimating their capacity to adapt to future climate change, especially if high emission scenarios are realized (Wolfsegger et al. 2008).

Climate adaptation research in the tourism-recreation sector is 5–7 years behind that of sectors that have been actively engaged in adaptation research (i.e., agriculture – Smit and Skinner 2002, water resources – de Loe et al. 2001, construction – Lowe 2003). In our opinion, the inadequate consideration of adaptation in climate change vulnerability studies within the tourism-recreation sector has had two important consequences. As this chapter illustrates, there exists sizable potential to adapt to climate change in the tourism-recreation sector and as a result some of the existing literature will have overestimated future damages. Future climate change vulnerability assessments in this sector need to minimally identify the range of adaptation options available and discuss how adaptation could alter projected impacts.

The limited discussion of climate change adaptation options has also posed a barrier to engagement and collaboration with the tourism-recreation community. As Grothmann and Pratt (2005:209) suggest, "If only the risks are communicated without communicating adaptation options, people will probably react by avoidant maladaptive responses like denial of risk." The tourism industry is very image sensitive and is therefore very cautious about even acknowledging concerns about climate change risks for fear of adversely affecting their reputation as a destination or sustainable business. If our understanding of the implications of climate change for the tourism-recreation sector is to advance, researchers will need to work hard to better engage the tourism-recreation community in a new generation of multi-disciplinary studies. It is the intent of the International Society for Biometeorology Commission on Climate, Tourism and Recreation to be an important facilitator in this regard.

References

Adger, W.N., Arnell, N. and Tompkins, E. (2005) Adapting to climate change: perspectives across scales, *Global Environmental Change* 15, 75–76.
Badenschier, F. and Schmitt, S. (2006) A Sun Ambrella for the Great Barrier Reef. Speigel Online, 7 November 2006. www.spieget.de/international/0,1518,drunk-446382,00.html
Barnes, B. (2002) The heat wave vacation. The *Wall Street Journal* 2 August, W1
Becken, S. (2004) Climate Change and Tourism in Fiji. Vulnerability, Adaptation and Mitigation. Final Report. Suva, Fiji, University of the South Pacific.
Berrittella, M., Bigano, A., Roson, R. and Tol, R. (2006). A general equilibrium analysis of climate change impacts on tourism. *Tourism Management* 27, 913–924.
Best, A. (2003) Is it getting hot in here? *Ski Area Management*, May, 57–76.
Black, R. (2006) Tower of London on climate list. www.news.bbc.co.uk/go/pr/fr/-/2/hi/science/nature/4810826.stm (accessed 17 March 2006).
Bloomberg News (2004) Operator betting on nature instead of snow insurance, *Financial Post* 12 January, FP5.
Bly, L. (2006) More travel firms offer hurricane 'guarantees'. USA Today, 2 June 2006. 9D.
Braun, O, Lohmann, M, Maksimovic, O, Meyer, M, Merkovic, A, Riedel, A and Turner, M. (1999) Potential impact of climate change effects on preferences for tourism destinations. A psychological pilot study. *Climate Research* 11, 247–254.
Breiling, M. and Charamza, P. (1999) The impact of global warming on winter tourism and skiing: a regionalized model for Austrian snow conditions, *Regional Environmental Change* 1(1), 4–14.
Bürki, R. (2000) Klimaaenderung und Tourismus im Alpenraum – Anpassungsprozesse von Thouristen und Tourismusverantwortlichen in der Region Ob- und Nidwalden, Ph.D. dissertation, Department of Geography, Zurich, University of Zurich.
Butler, R. and Jones, P. (2001) Conclusions – problems, challenges and solutions. In: A Lockwood and S. Medlik (Eds.). *Tourism and Hospitality in the 21st Century*. Oxford, Butterworth-Heinemann, pp. 296–309.
Cesar, H. (2000). *Impacts of the 1998 Coral Bleaching Event on Tourism in El Nido, Phillippines.* Prepared for Coastal Resources Center Coral Bleaching Initiative. Narragansett, RI, USA: University of Rhode Island.
de Freitas, C. (1990) Recreation climate assessment. *International Journal of Climatology* 10, 89–103.
de Freitas, C. (2003) Tourism climatology: evaluating environmental information for decision making and business planning in the recreation and tourism sector, *International Journal of Biometeorology* 4, 45–54.

de Loe, R., Kreutzwiser, R. and Moraru, L. (2001) Adaptation options for the near term: climate change and the Canadian water sector. *Global Environmental Change* 11, 231–245.

Egan, T. (2005) The race to Alaska before it melts. *The New York Times*, 26 June.

Elsasser, H. and Bürki, R. (2002) Climate change as a threat to tourism in the Alps, *Climate Research* 20, 253–257.

Giles, A. and Perry, A. (1998) The use of a temporal analogue to investigate the possible impact of projected global warming on the UK tourist industry. *Tourism Management* 19(1), 75–80.

Gomez-Martin, M.B. (2005) Weather, climate and tourism – a geographical perspective, *Annals of Tourism Research* 32(3), 571–591.

Gössling, S. and Hall, C.M. (2006) An introduction to tourism and global environmental change. In: S. Gössling and C.M. Hall (Eds.). *Tourism and Global Environmental Change*. London, Routledge, pp. 1–34.

Great Barrier Reef Marine Park Authority (2007a) Coral Bleaching on the Great Barrier Reef. http://www.gbrmpa.gov.au/corp_site/info_services/science/climate_change/climate_change_and_the_great_barrier_reef/coral_bleaching_on_the_great_barrier_reef

Great Barrier Reef Marine Park Authority (2007b) Great Barrier Reef Coral Bleaching Response Plan 2006–2007. http://www.gbrmpa.gov.au/__data/assets/pdf_file/0020/13169/Coral_Bleaching_Response_Plan_2006–07_Final.pdf.

Greenwood, J. (2004) Investors wait since 1991 on ski resort plan: regulatory faceoff, *Financial Post*, 12 January, FP1 and 5.

Grothmann, T. and Pratt, A. (2005) Adaptive capacity and human cognition: the process of individual adaptation to climate change. *Global Environmental Change* 15, 199–213.

Hamilton, J., Maddison, D. and Tol, R. (2005) Effects of climate change on international tourism. *Climate Research* 29, 245–254.

Innovest Strategic Value Advisors (2003) *Carbon Finance and the Global Equity Markets*. Carbon Disclosure Project, London.

Intergovernmental Panel on Climate Change (2001) Climate Change 2001: Impacts, Adaptation and Vulnerability, Third Assessment Report. Geneva, Switzerland, United Nations Intergovernmental Panel on Climate Change.

Jahn, G. (2005) Glacial Cover-Up Won't Stop Global Warming, 17 July Associated Press. http://abcnews.go.com/International/wireStory?id = 947283&CMP = OTC-RSSFeeds0312 (accessed 26 January 2006).

Johnson, A. (2005) Early start to hurricane season forces travelers, resorts to adjust. *The Wall Street Journal*, 21 July, D5.

Jones, B. and Scott, D. (2006a) Climate Change, Seasonality and Visitation to Canada's National Parks. *Journal of Parks and Recreation Administration* 24(2), 42–62.

Jones, B. and Scott, D. (2006b) Implications of Climate Change for Visitation to Ontario's Provincial Parks. *Leisure* 30(1), 233–261.

Kelly, N. (2001) Recreation and Tourism. In: *Climate Change and Insurance*. The Chartered Insurance Institute, London, pp. 27–34.

König, U. (1998) Tourism in a Warmer World: Implications of Climate Change due to Enhanced Greenhouse Effect for the Ski Industry in the Australian Alps, Wirtschaftsgeographie und Raumplanung, vol. 28, Zurich, University of Zurich.

König, U., Abegg, B. (1997) Impacts of climate change on tourism in the Swiss Alps, *Journal of Sustainable Tourism* 5(1), 46–58

Lise, W. and Tol, R. (2002) Impact of climate on tourist demand. *Climatic Change* 55, 429–449.

Lowe, R. (2003) Preparing the built environment for climate change, *Building Research and Information* 31(3–4), 195–199.

Maddison D. (2001) In search of warmer climates? The impact of climate change on flows of British tourists. *Climatic Change* 49, 193–208.

MacInnis, L (2006) Swiss ski resort tries to cover up climate change. PlanetArk World News. www.planetark.org/avantgo/dailynewsstory.cfm?newsid = 35717 (accessed 20 March 2006).

Melting Mountains (2006) http://meltingmountains.org/ (accessed 29 March 2006)

MyClimate (2006) http://www.myclimate.org/index.php?&lang = en (accessed 29 March 2006).

National Research Council (2003) Critical Issues in Weather Modification Research. Committee on the Status and Future Directions in US Weather Modification Research and Operations. Washington, DC, National Research Council.
National Ski Areas Association (2006) Available http://www.nsaa.org/nsaa/press/2005/04-05-sa-number-history.pdf (accessed 11 January 2006).
Pack, T. (2004) Florida tourism problem. Hotel Online – News for the Hospitality Executive. www.hotel-online.com/News/PR2004_4th/Oct04_FloridaNextYear.htm (accessed 28 July 2005).
Pain, J. (2005) Florida tourism flying high despite'04 hurricanes. USA Today. www.usatoday.com/weather/hurricane/2005-06-14-florida-tourism_x.htm (accessed 25 July 2005).
Parry, R. (2005) Cherry Trees Programmed to Blossom for Tourists. The Times Online. 9 June 2005. www.timesonline.co.uk/
Prideaux, B. (2006) The use of scenarios to project the imapct of global warming on future visitation to the Great Barrier Reef. ATLAS Asia Pacific 2006 Conference – Tourism After Oil. 2–5 December. Dunedin, New Zealand.
Raksakulthai, V. (2003) Climate Change Impacts and Adaptation for Tourism in Phuket, Thailand. Pathumthai, Thailand: Asian Disaster Preparedness Centre.
Reuters (2004) Climate Change Seen Climbing Agendas-Investor Group, 18 May 2004.
Richardson, R., Loomis, J. (2004) Adaptive recreation planning and climate change: a contingent visitation approach. *Ecological Economics* 50, 83–99.
Scott, D. (2006a). Climate change and sustainable tourism in the 21st century, in Cukier, J. (Ed.). *Tourism Research: Policy, Planning, and Prospects – 2003*. Department of Geography Publication Series, University of Waterloo, Waterloo, Canada.
Scott, D. (2006b) Ski Industry Adaptation to Climate Change: Hard, Soft and Policy Strategies. In: S. Gossling and M. Hall (Eds.). *Tourism and Global Environmental Change*, London, Routledge, pp. 262–285.
Scott, D. (2006c). Global environmental change and mountain tourism. In: S. Gossling and C.M. Hall (Eds.). *Tourism and Global Environmental Change*. London, Routledge, pp. 54–75.
Scott, D. and Jones, B. (2005) Climate Change and Banff: Implications for Tourism and Recreation – Executive Summary, Town of Banff, Banff, Alberta, Canada.
Scott, D. and Jones, B. (2006a) Climate Change and Nature-Based Tourism: Implications for Parks Visitation in Canada. Technical Report to the Government of Canada Climate Change Action Fund – Impacts and Adaptation Program. University of Waterloo, Faculty of Environmental Studies, Waterloo, Canada.
Scott, D. and Jones, B. (2006b) Climate Change and Outdoor Recreation in Canada. Technical Report to the Government of Canada Climate Change Action Fund – Impacts and Adaptation Program. Waterloo, Canada: University of Waterloo, Faculty of Environmental Studies.
Scott, D., Jones, B., Lemieux, C., McBoyle, G., Mills, B., Svenson, S. and Wall, G. (2002) *The Vulnerability of Winter Recreation to Climate Change in Ontario's Lakelands Tourism Region*. Waterloo, Canada, University of Waterloo, Department of Geography Publication Series, Occasional Paper 18.
Scott, D., Wall, G. and McBoyle, G.(2005a) The evolution of the climate change issue in the tourism sector. In: C.M. Hall and J. Higham (Eds.). *Tourism, Recreation and Climate Change*. Channelview Press, London, pp. 44–60.
Scott, D., Jones, B., Abi Khaled, H. (2005b) Climate Change: A Long-Term Strategic Issue for the NCC. Implications for Recreation-Tourism Business Lines. Report prepared for the National Capital Commission. University of Waterloo, Department of Geography, Waterloo, Ontario.
Scott, D., Jones, B., Konopek, J. (2007) Implications of climate and environmental change for nature-based tourism in the Canadian Rocky Mountains: A case study of Waterton Lakes National Park. *Tourism Management*, 28(2), 570–579.
Scott, D., Jones, B., McBoyle, G. (2006) Climate, Tourism and Recreation: A Bibliography 1936 to 2005. University of Waterloo, Department of Geography, Waterloo, Ontario. Available at: http://www.fes.uwaterloo.ca/geography/faculty/dscott.html
Simmons, P. (2005) Swiss plan to wrap up their glaciers, The Times, 30 March, p. 66.

Smit, B. and Skinner, M. (2002) Adaptation options in agriculture to climate change: a typology, *Mitigation and Adaptation Strategies for Global Change* 7, 85–114.

Smit, B. Burton, I., Klein, R., Wandel, J. (2000) An anatomy of adaptation to climate change and variability, *Climatic Change* 45(1): 233–51.

Sievanen, T., Tervo, K., Neuvonen, M., Pouta, E., Saarinen, J., Peltonen, A. (2005) Nature-based tourism, outdoor recreation and adaptation to climate change. FINADAPT Working Paper 11. Helsinki: Finnish Environment Institute.

Sulaiman, Y. (2006) Australia considers hosing the Great Barrier Reef. eTN Pacific, 7 November 2006.

The Economist (1998) Winter wonderlands, *The Economist*, 31 January.

The Wall Street Journal (2005) 'Investors Request Data on Company Emissions, 2 February 2005.

USA Today (2005) Florida's hurricanes boost Arizona tourism. www.usatoday.com/travel/news/2004-10-11-az-hurrican_x.htm (accessed 29 July 2005).

Uyarra M., Cote I., Gill, J., Tinch, R., Viner, D., Watkinson, A. (2005) Island-specific preferences of tourists for environmental features: Implications of climate change for tourism-dependent states. *Environmental Conservation* 32(1), 11–19.

Wall G. (1992) Tourism alternatives in an era of global climate change. In: V. Smith and W. Eadington (Eds). *Tourism Alternatives*. University of Pennsylvania, Philadelphia, pp. 194–236

Wall, G. and Badke, C. (1994) Tourism and climate change: an international perspective, *Journal of Sustainable Tourism* 2(4), 193–203.

Wheeler, D. (1995) Majorca's water shortages arouse Spanish passions. *Geography* 80, 283–286.

Williams, D. (2005) Wilma deals $800 million blow to Mexico travel industry. www.cnn.com/2005/travel/10/26/wilma.mexico.travel/index.html (accessed 27 October 2005).

Wilton D. and Wirjanto T. (1998) An analysis of the seasonal variation in the national tourism indicators. Canadian Tourism Commission, Ottawa, Canada.

Wolfsegger, C., Gossling, S., and Scott, D. (2008). Climate change risk appraisal in the Austrian ski industry. *Tourism Review International* (in press).

World Health Organization (2002) *Global Solar UV Index: A Practical Guide*. World Health Organization, Geneva, Switzerland.

World Travel and Tourism Council (2006) *Executive Summary: Travel and Tourism Climbing to New Heights*. World Travel and Tourism Council, London.

World Tourism Organization (2003) Climate Change and Tourism. Proceedings of the First International Conference on Climate Change and Tourism, Djerba, 9–11 April, Madrid, World Tourism Organization.

World Tourism Organization (2006) *UNWTO World Tourism Barometer* 4 (1).

World Tourist Organization (1998) *Tourism 2020 Vision*. WTO Publications Unit, World Tourism Organization, Madrid, Spain.

Chapter 9
Adaptation and Water Resources

Chris R. de Freitas

Abstract Adaptation to climate change in the water sector, especially changes in water management practices, will have a very significant impact on how future climate affects the water sector. The chapter starts out by describing the range of adaptive options that are available to water managers faced with changing circumstances. One classification distinguishes between "supply-side" and "demand-side" options. Another classification scheme proposed here distinguishes between 1) technological, 2) behavioural, 3) economic and 4) legal adaptive measures to manage and extend water resources. The chapter summarise these adaptive options. It assumes that the relative merits of one adaptive technique over another can be characterized in terms of the benefits and costs of the adaptation, across a spectrum of no effect ("no adaptation" or "wrong choice of adaptation") to perfectly effective ("adaptation sufficient to eliminate all effects of climate change") at an optimal level of cost effectiveness. There is also the issue of conflicting choices.

Assessment of impacts of climate change on water resources requires knowledge of future climate as well as methods capable of transforming this knowledge into likely biophysical and societal impacts. Current methods of impact assessment, however, are hampered by the unreliability of regional climate forecasts and by the incompatibility of these forecasts with the analytical tools needed to assess impacts at a scale that is useful for planning purposes. In particular, there is a mismatch between the temporal and spatial scales of available climate forecasts and the data required for impact assessment. Problems and methodologies that may cope with them are discussed. Two regions, New Zealand and the tropical Pacific, are used to illustrate application of the methods to assessment of water resources. The best adaptation strategies will depend on the impact potential of a given change in climate. This in turn will depend on the overall sensitivity of a particular water supply or demand unit to those aspects of climate that do change.

C.R. de Freitas
School of Geography, Geology and Environmental Science,
The University of Auckland, Auckland, New Zealand

9.1 Introduction

The Earth is endowed with a huge quantity of water, but less than 3% of it is fresh water and three quarters of that is locked up in the planet's ice caps and glaciers. The tiny amount of fresh water that remains is all that is available for human use, a fact that highlights the importance of managing the resource wisely. Global climate change, including changed variability, whether natural or anthropogenic, could modify the availability and spatial distribution of this scarce resource. Whatever the result we can be sure, as is already the case, the resource will unevenly distributed across continents and nations. For this reason, adapting to changes in water availability is more of a territorial than strictly global issue and most planning and investment decisions related to water resources will be made on a national or regional level. The problem is information on climate change is provided at a generalised global scale and even this is plagued with uncertainty. Thus, the important question arises: How do we plan for adaptation at the local scale to climate change at the global scale?

Over the past decade or so, and increasingly since the publication of the Second Assessment Report of the Intergovernmental Panel on Climate Change (Houghton et al. 1995), there have been many studies into climate change effects, but progress in the area of hydrology and water resources has been slow. Adaptation to climate change in the water sector, especially changes in water management practices, will have a very significant impact on how climate change affects the water sector. The best adaptation strategies will depend on the impact potential of a given change in climate. This in turn will depend on the overall sensitivity of a particular water supply or demand unit to those aspects of climate that do change. The chapter will address this and related themes.

9.2 Adaptation Options in the Water Sector

Adaptation to changing conditions in water availability and demand has always been at the centre of water management. Typically, it is assumed demand will grow and that the natural resource base is constant, except where land-use change occurs (Kundzewicz et al. 2007). Conventionally, it is also assumed that the future resource base will be similar to that of the future. Given the inevitability of climate change, this assumption is incorrect. Scenarios of future conditions vary and are not detailed. The IPCC (2007a) point to decreasing water availability in mid-latitudes semi-arid low latitudes and increased water availability in moist tropics and high latitudes. Annual average river runoff and water availability are projected to increase by 10–40% at high latitudes and in some wet tropical areas, and decrease by 10–30% over some dry regions at mid-latitudes and in the dry tropics, some of which are presently water-stressed areas (IPCC 2007b).

Climate processes powered by solar energy drive the hydrological cycle that determines the global distribution of water resources. The key question for those involved in planning and investment in the water sector is: if the climate warms in the future, will the water cycle intensify and what will be the nature of that intensification? A number of studies have addressed this question. To see how rainfall had changed with the 0.4°C global warming of the past 20 years, Wentz et al. (2007) analysed data collected by US weather satellites from 1987 to 2006. According to the results of this work, global warming will increase precipitation globally by three times the amount predicted by climate models. Wentz et al. (2007) state it impossible to predict where additional precipitation will fall; wet areas may get wetter, but drought-prone regions might also get some relief. Huntington (2006) reviewed the findings from more than 100 scientific studies that assessed trends in hydrologic variables, including precipitation, runoff, tropospheric water vapour, soil moisture, glacier mass balance, evaporation and evapotranspiration. According to Huntington, although data are not complete, and sometimes contradictory, the weight of evidence indicates an ongoing intensification of the water cycle, but the results of the work show no increase in storms or floods. Smith et al. (2006) reviewed variations in annual global precipitation for the period 1979–2004 and found that "trends have spatial variations with both positive and negative values, with a global-average near zero". This is reflected in global precipitation anomalies over the long period from 1900 to 2005 (Fig. 9.1). Based on these findings it would seem water managers have little to go on. But this might not be the case. Water management is based on minimization of risk and managers are accustomed to adapting to changing circumstances, including extreme climatic events, many of which can be regarded as analogs of future climate.

A wide range of adaptive options are available to water managers faced with changing circumstances. One widely used classification distinguishes between

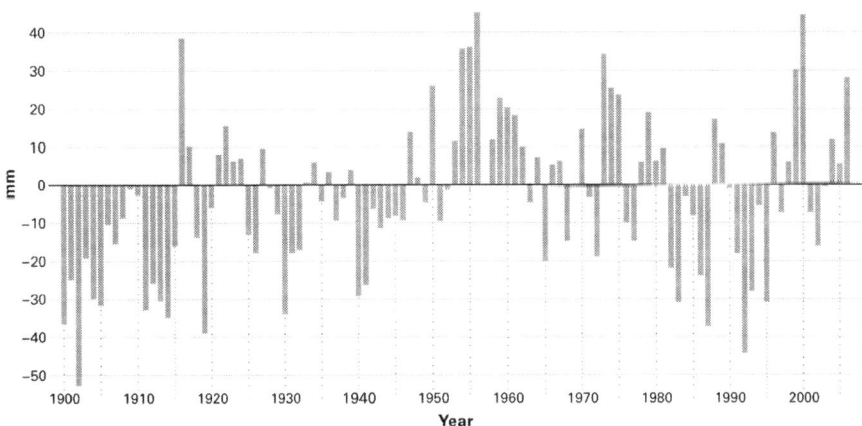

Fig. 9.1 Annual global precipitation anomalies for the period 1900–2005. (NCDC, NOAA: 'Climate of 2005 – Annual Report' 2006)

"supply-side" and "demand-side" options. Examples of supply-side options are: prospecting and extraction of groundwater; increasing storage capacity by building reservoirs and dams; expansion of rain-water storage; water transfer; desalination of sea water; and removal of invasive non-native vegetation from riparian areas. Examples of demand-side options are: improvement of water-use efficiency by recycling water; reduction in water demand for irrigation by changing the cropping calendar, crop mix, irrigation method, and area planted; reduction in water demand for irrigation by importing agricultural products, i.e., virtual water; promotion of indigenous practices for sustainable water use; expanded use of water markets to reallocate water to highly valued uses; and expanded use of economic incentives including metering and pricing to encourage water conservation (Kundzewicz et al. 2007).

Supply-side options focus on increasing capacity, while demand-side options focus on managing demand and changing institutional practices and operating rules for existing water resource systems. From the supply side, even now fresh water shortages occur virtually everywhere from time to time. Any change in local weather patterns driven by changes to global climate processes will have direct effects on water availability, but it will also have unpredictable indirect social and economic regional effects, such as impacts on agricultural productivity, availability of renewable hydroelectric power and supply of municipal water. From the demand side, agriculture and industry are the major users of water globally. Population growth, agricultural expansion, and demand for water by an expanding the industrial sector are likely to make water shortages even more prevalent in coming years.

Another classification scheme proposed here distinguishes between (1) *technological*, (2) *behavioural*, (3) *economic* and (4) *legal* adaptive measures to manage and extend water resources. The following sections summarise these adaptive options. It assumes that the relative merits of one adaptive technique over another can be characterized in terms of the benefits and costs of the adaptation, across a spectrum of no effect ("no adaptation" or "wrong choice of adaptation") to perfectly effective ("adaptation sufficient to eliminate all effects of climate change") at an optimal level of cost effectiveness. There is also the issue of conflicting choices.

9.2.1 Technological Adaptive Measures

(a) *Increased efficiency*
A major tenet in water management is that input reduction is the first choice in efforts to extend water resources. The largest gains will be made in agriculture and industry, which are the biggest water users. In agriculture, irrigation accounts for most of the water used. Micro-irrigation, or drip irrigation, is a good example of an adaptive measure. Water is conveyed to crops by pipes instead of open ditches or spray devices that encourage evaporation. This method is especially effective in water-poor, arid regions where large savings in water use can be made over more conventional and extravagant irrigation methods. Similarly, industry can easily redesign production technologies that use much less water. At the domestic level,

toilet flushing and showers are at the top the list of water use in the United States and many other developed countries. Installing dual-flush or low-flush technologies and more efficient shower heads are among the simple technical changes that can directly reduce the vast amounts of water used in what will be an increasing number of modern households.

(b) *Recycling and reclamation*
Recycling is often the next most cost effective adaptation to reduced water supplies or increased demand for water. Used water can be purified and reused in industry, on farms and domestically. "Gray water" is untreated or semi treated wastewater that can be cheaply used for such things as irrigating golf courses, lawns, parks and gardens in urban and suburban areas. It is also effective in recharging groundwater storage. Recycling the same water during production is an example of a design change from improved technology that can save large amounts of water. While such changes may temporarily increase costs, they ultimately lead to increased savings as the price of steadily decreasing water supply rises. Through improved technologies, even now many cities reuse their own wastewater directly in what is called "closed loop reclamation" through the "3Rs of return, repurify and reuse." Water recycling also reduces water pollution.

(c) *Substitution*
Desalination is the best example of substitution. Clearly, since about 97% of the Earth's water is saltwater, desalination is a means to a huge supply of freshwater. Although desalination is expensive, the costs are trending downwards as the latest and most efficient technologies become available. Clearly, this adaptive option is best suited to coastal communities.

(d) *Redistribution*
Dams and reservoirs store water from times of surplus and allow it to be used in times of deficiency. They also facilitate routine redistribution of water from water surplus regions to water deficit regions via canals and pipelines. Dams and reservoirs are the most common methods of coping with demand for water during periods of below average precipitation, but they also can have significant environmental and social impacts. They can reduce or eradicate native fish, impede fish migration routes, flood wildlife habitat and agricultural land, displace communities, diminish nutrient flow to estuarine habitats. Many of the impacts are irreversible and, ultimately, all dams have a finite life span as they eventually fill with sediment.

9.2.2 Behavioural Adaptive Measures

Conserving water by a changing human behaviour is linked to all categories of adaptive measures that require human action or a response. This category includes diverse and sundry actions, ranging from decisions by individuals to take showers rather than baths to changes in diet, say from meat to crops, which can reduce agricultural water loss, as, in general, less water is used when eating food lower on the food chain. Other simple behavioural adjustments include switching to crops with low water needs or adopting plants in landscaping designs that can save water ("xeriscaping').

9.2.3 Economic Adaptive Measures

When water resources are inexpensive, or there are no financial incentive to conserve, recycle or substitute water use, economic policy instruments can be employed to change this, usually through market forces. The largest gains will be made in agriculture and industry, which are the biggest water users and polluters. Lack of water conservations is most pronounced in agriculture, which accounts for just 70% of the water consumed globally, 85% of which is used for irrigation. Traditionally, governments worldwide heavily subsidise water supply costs for agriculture, so farmers have little incentive to conserve water, especially that used in irrigation where vast quantities are lost though evaporation. Reductions in domestic water consumption can also save significant amounts of water, especially water used for toilets, showers and laundry and garden irrigation. Taxing water use in agriculture, industry and households is a means of re-valuing it as a commodity to encourage conservation. Taxes can be levied as effluent charges, which is not only an incentive to conserve water, but also reduces pollution. Higher prices for water act as an incentive to reduce domestic water use and as an inducement for domestic water recycling. Even now, for example, Singapore and Japan use reclaimed wastewater for flush toilets.

9.2.4 Legal Adaptive Measures

Most water comes from surface water supplies, lakes, rivers and streams, but groundwater supplies are crucial in many regions globally. Even in times of normal precipitation, excessive use of ground and surface water for prolonged periods can cause streams and wetlands shrink or dry up, with profound ecological effects. Laws regulating use of these sources of water can control and extend their use. However, legal control is easiest for surface waters, which is the easiest for policing authorities to monitor. Problems of detection explain why unsustainable depletion and pollution of slow moving groundwater is so widespread. Legal protection of watersheds helps to prevent siltation in reservoirs and increases and protect groundwater recharge. Laws used for regulating surface water are categorised as "appropriation laws" or "riparian laws". The former dictates that governments can appropriate water for general use, such as water from large rivers diverted for dams or irrigation. Riparian laws apply to owners of land who have the right the withdraw water from rivers and lakes bounding their land, but on condition that the water is returned to its source in an unpolluted state. By and large, legal measures are best used in circumstance where only a small number of actors use the resource and where stricter controls are required than can be provided by market controls.

9.2.5 Other Considerations

The capacity of a country or region to adapt to climate change is determined by a range social, economic, cultural and technological factors. These include availability of (a) skilled personnel, (b) legal frameworks within which water and related agencies can work; (c) money and resources; (d) appropriate technology and technical ability; (e) hydrological and climate data; and (f) analytical land decision-making tools. Successful adaptation strategies to avert future water shortages must ensure that water supply systems are sustainable. To achieve this, the focus should be education the public on adaptive strategies, in particular conservation, recycling, reinstatement, avoidance of water pollution and overall watershed management. Fresh water conservation the most simple and cost effective short-term adaptive measure followed by measures to prevent water pollution. Government policies can greatly increase the efficiency of water use and promote water substitution strategies such as rain catchment, wastewater recycling and gray water systems.

There is a great deal of uncertainty on what future climate might be, but risks from climate change like other risks need to be taken into account in planning future water management. A useful starting point is the assumption that future climate will not create new water resource issues for a given region or area, but rather change the magnitude, frequency and areal extent of existing water resource restraints and boundaries. In planning ahead, it is important to make a distinction between (a) the various possible adaptive options available for meeting future demands and changed climate conditions and (b) resources and decisionmaking capabilities of society, of water management agencies in particular, to develop and apply decisionmaking frameworks that lead to successful adaptive outcomes. The latter is especially important because climate change challenges management techniques for assessing and selecting adaptive options. Overarching factors that affect adaptive capabilities of a country or nation include wealth, institutional competence, management strategies, planning time lines and availability of organizational arrangements of relevant professional water managers and technical personnel.

9.3 Alternative Approaches and Research Frontiers

Any assessment of adaptation to climate change effects on hydrology and water resources must first consider the impact of climate change on hydrology. This latter requires knowledge of both future climate as well as methods capable of transforming this knowledge into likely hydrological and societal effects. There are two ways of approaching this: (1) the scenario approach and (2) the sensitivity assessment approach. The scenario approach is by far the most common and is driven by our inability to adequately forecast future climate. Scenarios are effectively "what if" statements that represent plausible future climate states from a range of possibilities based on the current state of knowledge of how the global climate system works using hypothetical general circulation models (GCMs) of global climate (often referred to as global climate models) along with estimates of future rates greenhouse gas emissions and atmospheric concentrations. Scenarios project greenhouse gas emissions, which are then entered into the GCMs to project climate change. The output from are definitely not forecasts, but are frequently treated this way.

In the scenario approach, a future climate state is identified and impacts evaluated. But this method is hampered by the unreliability of GCMs, especially at the regional level. Moreover, we do not know which climate change scenario to use. Clearly, the consequences of the scenarios being 'wrong' could have serious implications for planning adaptive responses. Sometimes there is an implicit assumption is that a specific changed climate condition is predicted, reinforced by the fact that a GCM is limited to calculating an equilibrium response condition, presented in terms of a small array of climatic variables with temporal detail limited to mean monthly data at best. This is the heart of the scenario "problem" and it lies in the scale mismatch between global climate models and data required for decisionmaking at the local level. GCMs provide information generalised over months or years and at spatial resolution of several tens of thousands of square kilometres. Planners at the local level require data on at least daily scales and at a resolution of perhaps a few square kilometres. No matter how they are used, climate scenarios remain the single greatest source of uncertainty in hydrological impact assessments (Bergström et al. 2001). There are advocates for the use of regional downscaling from GCMs to assess impacts on future water resources, and although the last decade has produced a large number of publications on downscaling from climate models, very few of them consider impacts, and even fewer examine hydrological impacts (Fowler and Wilby 2007). There are some notable exceptions (Bell et al. 2007; Boé et al. 2007; Blenkinsop and Fowler 2007; Bronstert et al. 2007; Charles et al. 2007; Fowler and Blenkinsop 2007; Gachon and Dibike 2007; García-Morales and Dubus 2007; Salathé et al. 2007). However an underlying problem is that the prediction skill of the downscaling is no better than that of the GCMs.

In the alternative approach using sensitivity assessment, many of these problems are circumnavigated. First sensitivity of local-scale hydrological systems to climate is assessed, and then the question asked: What is the net effect of change on hydrology or related social exposure unit? By identifying the sensitivity to climate and evaluating it in terms of the adaptive capacity of the exposure unit, vulnerability of water resources to climate change may be determined and assessed. With this information, planning decisions would be possible without knowing precisely the magnitude of climate change that will occur. Research is needed to develop and test sensitivity assessment methodologies to cope with this, but some methods are currently in use.

To date, most investigations have focused on scenarios analysis with little regard to changes in human behaviour towards climate change including adaptation. To this end, there have been calls for real-world studies into climate impacts on water resources emphasizing that quantitative assessment of sensitivity and vulnerability of hydrological systems to climate change (Parry 2001) should be a priority research area.

9.3.1 Assessment of Sensitivity and the Need for Adaptation

The aim of climate change impact assessment is to determine how the availability of climatic resources will change and which regions will lose or gain from these changes. The nature and magnitude of impacts caused by change are the joint products of interactions between climate and society, bearing in mind that similar climatic variations may result in different impacts under different sets of social or economic conditions. However, the impact potential of a given change in climate is related to overall sensitivity of a particular water supply or demand unit to those aspects of climate that do change. Or it may be related to the particular climate type or climate regimes in which change occurs. For example, an average $1°C$ temperature rise and 10% increase in rainfall may be of little consequence in an equatorial climate region where high temperatures are commonplace and there are already extended periods of high rainfall throughout the year. On the other hand, sub-humid environment may be highly sensitive and respond dramatically to even the smallest increase in temperature or decrease in precipitation will reduce water availability and further increase water demand. There are a variety of ways of identifying sensitivity. In theory, sensitivity of a region to changes in climate does vary depending on climate type or regime. Climatic types can be characterised and assessed on the basis of this sensitivity since a given change will perturb some climatic regimes more than others (de Freitas and Fowler 1989).

Climate change impact assessment of the type described above relies on a greatly simplified picture of the role of climate, mainly because it deals with change in terms of single, secondary climatic variables which allow for only elementary statistical connections to be made with impacts. This approach is of limited use since the significance of the impact will depend on the net effect of several changed climatic variables on key environmental processes, such as those involving heat and moisture exchange which determine available soil moisture, evapotranspiration, crop yield, runoff, and so on. Impact will depend on the timing as well as the magnitude of change. For example, increases in winter precipitation may lead to flooding while increases during spring and summer may enhance plant growth, depending on the effect of changed air temperature on evapotranspiration and soil moisture. By determining the sensitivity of key variables such as actual evapotranspiration and soil water surplus to hypothetical climate changes, it may be possible to assess the effects at the regional scale of a range of climate change scenarios on the climate resource-base, evaluate impact and assess the need for adaptation. This may be achieved by way of response surface analysis or regional

9.3.2 Response Surfaces

A response surface is a two-dimensional representation of the sensitivity of a specific response variable (soil moisture in Fig. 9.2) to change in the two controlling climate features or processes. It is usually is presented as a plot of the response variable against the values of two driving climate variables, on the graph axes, for example, potential evapotranspiration (PET) and precipitation (P) in Fig. 9.2. The relationship between the response variable and climate determined from a pre-tested set of relationships, usually in the form of an empirical model, called a transfer function (de Freitas and Fowler 1989; Fowler and de Freitas 1990). Changes might be simply percentage adjustments to the each of the driving variables. The response variable is represented in the body of the graph as isolines. The three variables can be plotted using absolute values, or as values relative to the unamended baseline data representing no climate change (Fig. 9.2). The latter representation is a step removed

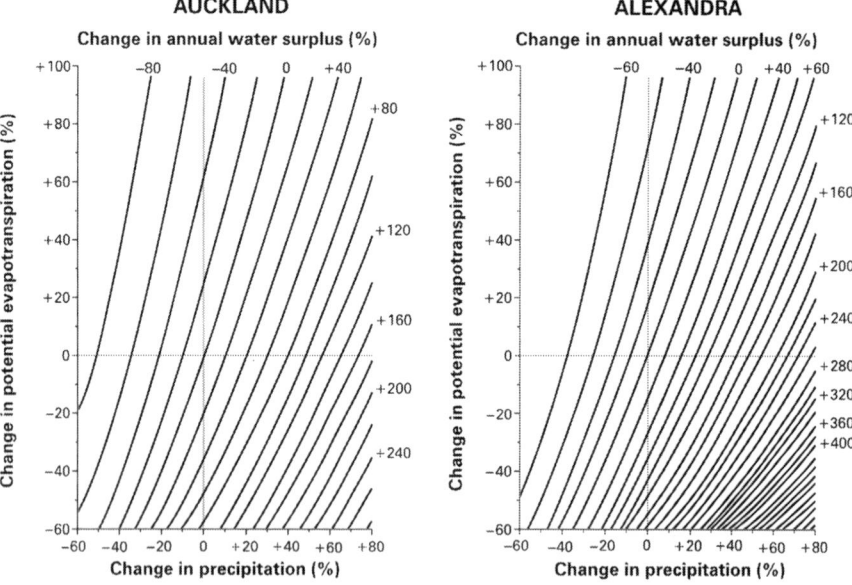

Fig. 9.2 Response surfaces showing percentage change in annual soil water surplus ('runoff' plus ground water recharge) for a range of climate change scenarios expressed as per cent change in potential evapotranspiration (PET) and precipitation (P) for a sub-temperate maritime climate (Auckland) and a semi-arid mid-latitude climate (Alexandra). The base-line reference condition, or current climate, is given by per cent change in PET = 0 and percent change in P = 0 (i.e. 0, 0 intersection marked "+"). (Fowler and de Freitas 1990)

from the input and output but does have the advantage of providing a direct measure of sensitivity. For example, a 20% response to a 10% change in a controlling climate variable is clearly an example of impact amplification. Response surface isolines are a summary of a matrix of response points associated with various combinations of changes to the two driving climate variables (Fig. 9.2). The required data are derived from repeated runs of the transfer function with the prescribed changes to the input. The slope and closeness of the isolines are an indicator of sensitivity and discontinuities an indicator of change in response (Fig. 9.2).

De Freitas and Fowler (1989), Fowler and de Freitas (1990) and Fowler (1999) used response surfaces to illustrate the impact of climate change on various aspects of water resources in the Auckland region of New Zealand. Other water resource examples of this can be found in studies by Lynch et al. (2001) and van Minnen et al. (2000). More recently, Semadeni-Davies (2004) used response surfaces for simulated waste water inflows to the Lycksele waste water treatment plant in north-central Sweden. Information such as that shown in Fig. 9.2 gives a birds-eye view of regional sensitivity in terms of key environmental responses to a range of possible changed climate conditions. The procedure provides a means by which a variety of climate change scenarios can be used to identify regions that are more or less sensitive to certain changes. This facilitates evaluation of impact potential for policy planning purposes. A 'what if' approach to climate change predictions can then be used to assess the desirability of societal adjustments in sensitive regions designed either to amplify beneficial or dampen undesirable effects of change. An advantage of the response surface method is that it is less likely to obscure inherent sensitivities to change that can occur in scenario approach. Another is its flexibility. A wide range of new or changed scenarios can be easily handled by plotting them on the response surface. This avoids the need to rerun the transfer function, thus facilitating use by non-climate specialists such as planners and policy makers wanting to reassess impacts on water resources.

Comparison of each pair of response surface diagrams illustrates the differing sensitivity of climate types to a climate change specified in terms of P and PET in each climatic region. For example, in Fig. 9.2, the closer isolines of the change in water surplus for Alexandra indicate greater sensitivity than in the Auckland region. In both cases the steepness of the isolines shows a higher sensitivity to changes in P than to PET. It should be noted that the isolines are for relative change rather than for absolute values. Absolute changes in water surplus will be greater in Auckland than in Alexandra because of a larger baseline water surplus for Auckland. Clearly, the finer points of decision making and planning will depend on resource specific considerations, the most important of which will relate to the level of climate-based resource use. In the case of availability of water resources shown in Fig. 9.2, for example, vulnerability of society to a change in climate will depend on the demand for the resource and levels of use. If there is a high level of resource use, as there is for example in the Auckland region, the society is highly vulnerable to climatic change that leads to reduced water availability. The implications for water resource planning are clearly quite serious. In contrast to this, where there is low resource use there is a larger margin of safety, it provides a buffer against both sporadic dry periods or a trend in decreasing runoff resulting from, say, gradual climatic change leading to reduced precipitation.

9.3.3 Pacific Atolls

For most atolls of the Pacific, fresh water is a seriously limited resource. Generally, water catchment areas are small and groundwater storage is in the form of a shallow fresh water lens. This essential resource is coming under increasing pressure as populations grow and rates of development increase. These things and the realisation of the possible impact of climate change and variability have highlighted the sensitivity of island communities to the availability of water. In many places there is now great concern for sustainability of groundwater reserves and the adequacy of supply of fresh water accumulated from surface catchments for drinking, irrigation and domestic, commercial or industrial use. The site specific effects of geohydrolic and soil characteristics of the land affect the availability of groundwater on particular islands. But for any atoll, the primary constraint on the fresh water resource is climate. It determines the size and temporal distribution the resource base crucial to management and planning of a wide range of activities. To assess this resource base, its sensitivity to climate change and thus the extent which adaptive measures might have to be relied on, a regional water balance model is used with average monthly climate data and uniformly applied to the Pacific Island Region. The Thornthwaite and Mather bookkeeping procedure (Thornthwaite 1948; Thornthwaite and Mather 1955) for assessing the water budget is used to calculate actual evapotranspiration (A_E) and the 'water balance' (D_S), given as:

$$D_S = P - A_E + \Delta S_m \tag{9.1}$$

where P is precipitation and ΔS_m is change in soil moisture. When $D_S \geq 0$ it is defined as water available for recharge of groundwater storage. A zero or negative value (mm) indicates a deficit. A soil moisture storage capacity (S_m) of 80 mm (Smith 1983) is used to provide a baseline for estimating A_E. To calculate A_E, the Thornthwaite bookkeeping procedure requires data for potential evapotranspiration (P_E) where

$$A_E = P_E - P + \Delta S_m. \tag{9.2}$$

The Priestly-Taylor method (Priestley and Taylor 1972) is particularly well suited to regional assessments of P_E, and has been used in the humid tropics regions (de Bruin 1983; Nullet 1987). Calculations are based on a 5° latitude-longitude grid of air temperature, rainfall, cloud amount, solar radiation and net allwave radiation.

To assess regional sensitivity to climate change, the per cent change in the current mean rainfall required to bring about a zero water balance, given as $(D_S/P)100$, is calculated. It provides a measure of the sensitivity of the region to change in rainfall that might be brought about by an enhanced greenhouse effect. The results show that areas sensitive to a 20% change in mean annual precipitation occur in a long, narrow zone at about 12°N latitude and a broad belt extending from Fiji through Tonga, the Southern Cook Islands to Pitcairn Island (Fig. 9.3). This zone identifies those atolls that will have to rely heavily on adaptive measures should climate change result in declining rainfall. More importantly, this is the zone most vulnerable to drought should rainfall decrease in the future. Most other parts of the Pacific Islands Region will require increases or decreases in rainfall well in excess of 20% of the mean annual figure for the change to have any significant effect on the water resource base.

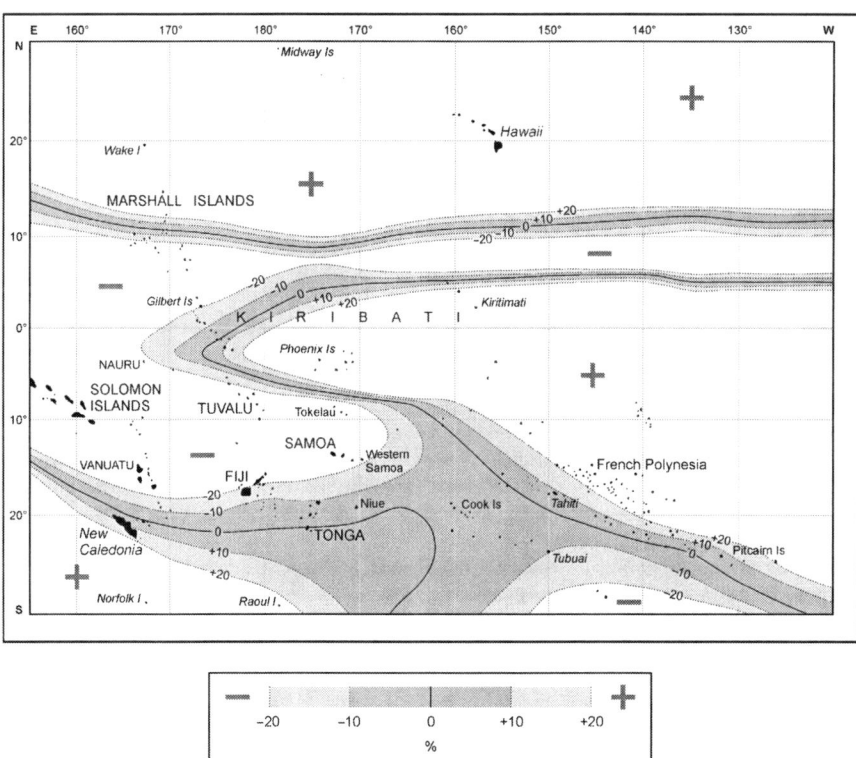

Fig. 9.3 Percentage change (±20%) in rainfall required to bring about a zero water surplus or deficit based on the annual mean. The shaded area bordered by +20% and the −20% isolines is the zone of high sensitivity to change in rainfall. In the future, should rainfall decline, atolls in this zone would suffer the most from soil moisture deficits and, under worse case conditions, they would experience an increase in the frequency of droughts. On the other hand, should rainfall increase, atolls in this zone would stand to benefit the most

9.4 Conclusion

Over the past decade or so there have been many studies into climate change effects, but progress in the area of hydrology and water resources has been slow. Adaptation to climate change in the water sector, especially changes in water management practices, will have a very significant impact on how climate change affects the water sector. But adaptive strategies adopted will depend on climate change impact potential in the absence of such strategies. Concepts of sensitivity and vulnerability of climate regions to change underlay the approaches examined here. They are put forward as integrating concepts that may be useful for identifying areas that may be affected by climate change. This approach is likely to be useful in regional assessments of vulnerability or resiliency.

It is often assumed that detailed information about the exact nature of climatic change is needed before detailed impact assessment is worthwhile. However, better use of existing climatic information along with improved conceptual models of climate-society interaction will provide useful insights that can form a basis for later in-depth analyses of specific impacts, or case by case assessments of sensitivity or vulnerability. Clearly, the finer points of decision making and planning will depend on resource specific considerations, the most important of which will relate to the level of climate-based resource use.

References

Bell V.A., Kay A.L., Jones R.G. and Moore R.J., 2007. Use of a grid-based hydrological model and regional climate model outputs to assess changing flood risk. *International Journal of Climatology*, 27 (12), 1657–1671.

Bergström S., Carlsson B., Gardelin M., Lindström G., Pettersson A. and Rummukainen M., 2001. Climate change impacts on runoff in Sweden – assessments by global climate models, dynamical downscaling and hydrological modelling, *Climate Research*, 16 (2), 101–112.

Blenkinsop S. and Fowler H.J., 2007. Changes in European drought characteristics projected by the PRUDENCE regional climate models. *International Journal of Climatology*, 27 (12), 1595–1610.

Boé J., Terray L., Habets F., Martin E., 2007. Statistical and dynamical downscaling of the Seine basin climate for hydro-meteorological studies. *International Journal of Climatology*, 27 (12), 1643–1655.

Bronstert A., Bárdossy A., Bismuth C., Buiteveld H., Disse M., Enge H., Fritsch U., Hundecha Y., Lammersen, R., Niehoff D., Ritter N., 2007. Multi-scale modelling of land-use change and river training effects on floods in the Rhine basin. *River Research and Applications*, 23 (10), 1102–1125.

Charles S.P., Bari M.S., Kitsios A. and Bates B.C., 2007. Effect of GCM bias on downscaled precipitation and runoff projections for the Serpentine Catchment, Western Australia. *International Journal of Climatology*, 27 (12), 1673–1690.

de Bruin H.A.R., 1983. Evapotranspiration in humid tropical regions. *Proceedings of IAHS Hamburg Symposium*, IAHS Pub. No. 140, 299–301.

de Freitas C.R. and Fowler A., 1989. Identifying sensitivity to climate change at the regional scale: the New Zealand example, in Welch, R. (ed.) *Proceedings of the 15th New Zealand Geography Conference*, NZ Geographical Society Conference Series, pp. 254–261.

Fowler A., 1999. Potential climate change impacts on water resources in the Auckland Region (New Zealand), *Climate Research*, 11, 221–245.

Fowler A.M. and de Freitas C.R., 1990. Climate impact studies from scenarios: help or hindrance? *Weather and Climate*, 10, 3–10.

Fowler H.J. and Blenkinsop S., 2007. Linking climate change modelling to impacts studies: recent advances in downscaling techniques for hydrological modelling. *International Journal of Climatology*, 27 (12), 1547–1578.

Fowler H.J. and Wilby R.L., 2007. Beyond the downscaling comparison study. *International Journal of Climatology*, 27 (12), 1543–1545.

Gachon P. and Dibike Y., 2007. Temperature change signals in northern Canada: Convergence of statistical downscaling results using two driving GCMs. *International Journal of Climatology*, 27 (12), 1623–1641.

García-Morales M.B. and Dubus L., 2007. Forecasting precipitation for hydroelectric power management: how to exploit GCM's seasonal ensemble forecasts? *International Journal of Climatology*, 27 (12), 1691–1705.

Houghton J.T., Meira Filho L.G., Callander B.A., Harris N., Katenberg A. and Maskell K., 1995. (eds.) *Climate Change 1995: The Science of Climate Change*, Contribution to Working Group 1 to the Second Assessment Report of the Intergovernmental Panel on Climate Change. Cambridge University Press, New York, 572 p.

Huntington T.G., 2006. Evidence for intensification of the global water cycle: Review and synthesis. *Journal of Hydrology*, 319 (1–4), 83–95.

IPCC, 2007a. Summary for Policymakers. In: *Climate Change 2007: Impacts, Adaptation and Vulnerability. Contribution of Working Group II to the Fourth Assessment Report of the Intergovernmental Panel on Climate Change*, M.L. Parry, O.F. Canziani, J.P. Palutikof, P.J. van der Linden and C.E. Hanson (eds.), Cambridge University Press, Cambridge, pp. 7–22.

IPCC, 2007b. Summary for Policymakers. In: *Climate Change 2007: Impacts, Adaptation and Vulnerability. Contribution of Working Group II to the Fourth Assessment Report of the Intergovernmental Panel on Climate Change*, M.L. Parry, O.F. Canziani, J.P. Palutikof, P.J. van der Linden and C.E. Hanson (eds.), Cambridge University Press, Cambridge, pp. 7–22.

Kundzewicz Z.W., L.J. Mata N.W. Arnell P. Döll P. Kabat B. Jiménez K.A. Miller T. Oki, Z. Sen and I.A. Shiklomanov, 2007. Freshwater resources and their management. In: *Climate Change 2007: Impacts, Adaptation and Vulnerability. Contribution of Working Group II to the Fourth Assessment Report of the Intergovernmental Panel on Climate Change*, M.L. Parry, O.F. Canziani, J.P. Palutikof, P.J. van der Linden and C.E. Hanson (eds.), Cambridge University Press, Cambridge, pp. 173–210.

Lynch A.H., Mellwaine S., Beringer J. and Bonan G.B., 2001. 'An investigation of the sensitivity of a land surface model to climate change using a reduced form model'. *Climate Dynamics*, 17, 643–652.

NCDC, 2006. Climate of 2005 – Annual Report. National Climate Data Center, NOAA, United States Department of Commerce. http://www.ncdc.noaa.gov/oa/climate/research/2005/ann/global.html

Nullet D. 1987. Water balance of Pacific atolls. *Water Resources Bulletin*, 23(6), 1125–1132.

Parry M., 2001. Climate change: where should our research priorities be? *Global Environmental Change*, 11(4), 257–260.

Priestley C.B. and Taylor R.J., 1972. On the assessment of surface heat flux and evaporation using large scale parameters. *Monthly Weather Review*, 100(2), 81–92.

Salathe E.P., Mote P.W., Wiley M.W., 2007. Review of scenario selection and downscaling methods for the assessment of climate change impacts on hydrology in the United States Pacific northwest. *International Journal of Climatology*, 27(12), 1611–1621.

Semadeni-Davies A., 2004. Urban water management vs. climate change: impacts on cold region waste water inflows. *Climate Change*, 64, 103–126.

Smith C.W., 1983. Soil Survey of the Islands of Palau, Republic of Palau. US Dept. of Agriculture Conservation Service.

Smith, Thomas M., Xungang Yin, and Arnold Gruber, 2006. Variations in annual global precipitation (1979–2004), based on the Global Precipitation Climatology Project 2.5° analysis. *Geophysical Research Letters*, 33, L06705.

Thornthwaite C.W., 1948. An approach toward a rational classification of climate. *Geographical Review*, 38(1), 55–94.

Thornthwaite C.W. and Mather J.R., 1955. The water balance. *Publications in Climatology*, 8 (1), 1–104.

van Minnen J.G., Alcamo J. and Haupt W., 2000. Deriving and applying response surface diagrams for evaluating climate change impacts on crop production. *Climatic Change*, 46, 317–338.

Wentz F. J., Ricciardulli L., Hilburn K. and Mears C., 2007. How much more rain will global warming bring? 2007. *Science*, 317, 233–235.

Chapter 10
Psychological Perspectives on Adaptation to Weather and Climate

Alan E. Stewart

Abstract The author examines the psychological dimensions of adaptation to weather and the climate events. Psychological approaches to adaptation are discussed and the distinction between primary, secondary, and tertiary adaptation are made. The author next presents a theoretical framework that lends itself well to organizing the different psychological variables that can affect adaptation, some of which have been explored in previous studies. This organizing framework is known as Protection Motivation Theory (Rogers 1975; Rogers and Prentice-Dunn 1997). The model components are discussed along with the cognitive and motivational biases that can affect the adaptive course that people might pursue. The chapter concludes with recommendations for furthering the psychological study of human adaptation to climate variability and change.

In the early morning hours of August 29, 2005 Hurricane Katrina struck the United States Gulf Coast as a strong category 3 storm. Nearly a day before, an urgent bulletin from the National Weather Service office in New Orleans communicated in ominous and unequivocal terms the dangers posed by Katrina. The message began (Guiney et al. 2006, p. 27):

> URGENT – WEATHER MESSAGE
> NATIONAL WEATHER SERVICE NEW ORLEANS LA
> 1011 A.M. CDT SUN AUG 28 2005
> ...DEVASTATING DAMAGE EXPECTED...
> HURRICANE KATRINA...A MOST POWERFUL HURRICANE WITH UNPRECEDENTED STRENGTH...RIVALING THE INTENSITY OF HURRICANE CAMILLE OF 1969.

A.E. Stewart
Department of Counseling & Human Development, University of Georgia,
402 Aderhold Hall, Athens, Georgia 30602, USA
e-mail: aeswx@uga.edu

MOST OF THE AREA WILL BE UNINHABITABLE FOR WEEKS...PERHAPS LONGER. AT LEAST ONE HALF OF WELL CONSTRUCTED HOMES WILL HAVE ROOF AND WALL FAILURE. ALL GABLED ROOFS WILL FAIL...LEAVING THOSE HOMES SEVERELY DAMAGED OR DESTROYED... AIRBORNE DEBRIS WILL BE WIDESPREAD...AND MAY INCLUDE HEAVY ITEMS SUCH AS HOUSEHOLD APPLIANCES AND EVEN LIGHT VEHICLES... PERSONS...PETS...AND LIVESTOCK EXPOSED TO THE WINDS WILL FACE CERTAIN DEATH IF STRUCK.

With this warning and prior forecasts from the National Hurricane Center, Gulf Coast residents and people living in high risk areas protected by levees had sufficient lead time to evacuate. Approximately 80% of the population of New Orleans evacuated and 83% of Jefferson Parish were evacuated (Guiney and Services Assessment Team 2006). Forecasters predicted the behaviour of Katrina with a high degree of accuracy and, as expected, the area was devastated on August 29. At least 1,833 people died, thousands of homes and businesses were destroyed, and infrastructure is still being rebuilt as of this writing (Knabb et al. 2005). Still, some residents who had the resources to evacuate chose not to leave the city as illustrated by dramatic video of survivors awaiting a rooftop rescue (Cable News Network 2005). What led this significant proportion of people to not heed the recommendations of both National Weather Service forecasters and city officials?

Despite the experiences of Hurricane Katrina, a national poll of residents who lived in states bordering the Atlantic or Gulf Coasts in May, 2007 found: 53% believed that they were not vulnerable to a hurricane or its effects, 52% had no personal disaster plan, and 61% did not have a hurricane survival kit (Mason-Dixon Polling and Research 2007). Perhaps the most noteworthy finding is that while 84% indicated that they would follow mandatory evacuation orders, similar to what was observed for the city of New Orleans, 10% indicated that would not evacuate their homes even if ordered to do so by emergency officials. Yet 73% of respondents believed it was an individual's responsibility, and not the government's, to assure their own hurricane preparedness. In this regard the situation following Hurricane Katrina is much as Mileti (1999, pp. 136–137) observed before the storm: "people typically are unaware of the hazards they face, underestimate those of which they are aware, overestimate their ability to cope when disaster strikes, often blame others for their losses, underutilize preimpact hazard strategies, and rely heavily on emergency relief when the need arises." What might explain this seeming lack of hurricane preparedness for a significant proportion of residents in the coastal states?

Increased losses of property and life from floods, hurricanes, convective storms, and winter storms are due, in part, to increases in *societal vulnerability* (Call 2005; Pielke 1998; Pielke et al. 2007; Pielke and Sarewitz 2005). Increases in population, changes in lifestyles, and demographic shifts have resulted in greater exposure of both lives and property to extreme events (Changnon et al. 2000; Kunkel et al. 1999). Further, as a majority of the world's population will dwell in urban centres in the coming years (Cohen 2003), the likelihood increases that a single extreme event can affect a large number of people at one time, as in the case of New Orleans. In many respects, disasters may be created by societal and governmental practices than by nature itself (Glantz 2005). The realization of increased vulnerability to weather and climate events has given rise to broad efforts to help society mitigate the sources of climate changes where possible and to adapt to the changes that are anticipated (Smit et al. 2001; Smith 1997; Tol et al. 1998). These efforts have thus far largely focused upon institutions, agencies, and systems of people that are assumed to be homogeneous in their compositions and adaptive behaviours.

Beyond societal vulnerability and adaptation, there is a growing realization that a fuller understanding of adaptation to climate variability and change occurs at the level of the individual person (Dash and Gladwin 2005; Dow and Cutter 2000, 1998; Grothmann and Reusswig 2006; Niemeyer et al. 2005; Stewart 2007). Society is vulnerable to a significant extent because of *individual vulnerability* and processes that occur on the human scale (Vedwan and Rhoades 2001). It makes sense to examine individual adaptation to weather and climate because organizations and institutions within society are comprised of individuals, each with their own unique histories, experiences, motivational, emotional, and cognitive systems. Such psychological variables can help us to understand and account for variability in individual behaviour that contributes to the behaviour of larger-scale societal systems.

This chapter examines the psychological dimensions of adaptation to weather and climate.[1] Psychological approaches to adaptation are discussed first followed by the presentation of a model that lends itself well to organizing the different psychological variables that can affect adaptation, some of which have been explored in previous studies. The model components are discussed along with the cognitive and motivational biases that can affect the adaptive course that people might pursue.

10.1 Psychological Conceptions of Adaptation

The adaptation to weather and climate encompasses several related psychological constructs. Overall, *psychological adaptation* refers both to the *processes* (i.e., *adjustment* and *coping*) and to the *outcomes* of attempting to respond effectively to variations in the environment, which here concerns weather and climate. *Adaptation* generally refers to the adjustment of psychological processes (cognition, affect, and behaviours) so that they reflect environmental demands and are responsive to them (American Psychological Association, APA 2007). This definition implies that people assimilate their environments into existing psychological structures and processes as well as accommodate themselves (i.e., actively change) to novel environmental requirements.

Adjustment pertains to the processes by which cognitions, behaviours, and emotions are used to achieve changes that are either imposed by the environment

---

[1] In speaking of weather and climate events the author adopts Bryson's (1997, p. 451) definition of climate as "the thermodynamic/hydrodynamic status of global boundary conditions that determine the concurrent array of weather patterns." In this way, weather derives from climate and, correspondingly, climate variability and change can alter the distribution of weather events that are significant for adaptation.

from without or that come through the person's realization of the need to create a new way of functioning (APA 2007; Wohlwill 1974). A variety of psychological variables and processes may be used in the service of the adjustment process such as receptive and expressive speech to receive and give information about weather and climate. Sensation and perception processes are necessary to gather and recognize information directly under various weather conditions. Memory (both episodic and semantic) is necessary for storing information for later use. Decision-making processes may use current sensed or communicated information along with information retrieved from memory to make choices about how to best respond behaviourally to weather or climate conditions.

Coping refers to single psychological processes or ensembles of them used together, as *coping strategies*, to meet environmental demands for adaptation when such demands tax or exceed (in time, magnitude, or nature) the person's ability to adjust (Holahan et al. 1996; APA 2007). The environmental stimuli that invoke coping processes may be experienced negatively as hassles and frustrations such as when bad weather delays one's commute home (Kanner et al. 1981). In addition, the environment may become a source of significant stressors when, for example, an extended power outage occurs following a thunderstorm (Campbell 1983; Evans and Cohen 1987). Extreme weather or climate events may even become traumatic stressors as, for example, cataclysmic conditions of a hurricane or a flood that result in the loss of life, property, and infrastructure (Norris, Murphy et al. 2004; Riad and Norris 1996).

Adaptation will involve adjustment or coping processes as a function of the magnitude of the weather or climate event and of the timeframe in which adaptation is required (i.e., before, during, or after the event, Lazarus and Cohen 1977). These variables (nature of the event and relationship to time) give rise to three types of psychological adaptation to weather and climate. First, adaptation can occur as proaction and preparation ahead of a wide variety of weather events (Easterling et al. 2004; Grothmann and Patt 2005). The author refers to this as *primary adaptation*, borrowing from a classification used in public health (Caplan 1964). Viewed in this manner, primary adaptation could involve such long-term efforts as educating children about weather, climate, and other natural disasters through a program such as the American Red Cross' *Masters of Disaster* school curriculum (American National Red Cross 2000). Other more immediate and personal, primary adaptations could involve checking the condition of one's vehicle and the road-weather conditions ahead of an anticipated drive during inclement weather (National Research Council 2004).

Second, adaptation can occur in the form of acclimation and adjustment to current weather events or to anticipated changes in climate. This is *secondary adaptation* and is probably the most common conception of adaptation insofar as people do not respond to the event until it is occurring. Because opportunities for proaction may be limited, peoples' focus may be upon preventing further effects of the weather or climate event or attempting to reduce the negative impacts of what already has occurred. Cognitive and behavioural processes that underlie adjustment may be especially active with secondary adaptation. As an example, secondary adaptation may involve mountain climbers, sailors, or golfers seeking shelter and safety when a thunderstorm brings dangerous winds and lightning.

Third, adaptation may consist of coping and recovery in the aftermath of extreme events; this is *tertiary adaptation*. If such events overwhelmed the person's attempts at primary or secondary adaptation, or if such adaptations were not undertaken, the person may be faced with recovering from myriad acute and long-term effects of the weather event. The most salient and recent example of tertiary adaptation was Hurricane Katrina in New Orleans, Louisiana, USA. To varying degrees both secondary and tertiary adaptation represents a *muddling through* response to the extent that the effects of weather upon people or property were not or prevented or at least minimised (Easterling et al. 2004).

Several considerations are noteworthy when applying this typology, the first of which concerns the relationships among the types of adaptation attempted. Although it is possible that primary adaptation may obviate secondary and tertiary adaptation, this is not guaranteed. The amount and kinds of primary adaptation along with the weather or climate events being anticipated may jointly necessitate that people undertake the adjustment and coping associated with secondary and tertiary adaptation, respectively. Another consideration is that although one is free to pursue or not pursue primary adaptation activities ahead of climate change or a particular weather event, it is increasingly difficult not to pursue secondary and tertiary adaptation as the anticipated events unfold. Certainly, a person can choose to do nothing by way of adaptation, but this is a choice that has consequences.

Finally, and perhaps most importantly, there are a wide variety of inter-individual trajectories that people follow in formulating their responses to environmental events. People may differ in how they evaluate the threat of climate change and extreme weather events, they may construe different attitudes, plans, and behaviours as sufficiently adaptive for the same event, and they may experience different objective and subjective outcomes than other people after performing very similar adaptive behaviours (Dash and Gladwin 2005; Dow and Cutter 1998, 2000). These possibilities give rise to important questions such as: What psychological variables are related to pursing different courses of adaptation? What gives rise to the diversity of observed outcomes?

10.2 Protection Motivation Theory: An Organizing Framework

Protection Motivation Theory (PMT, see Fig. 10.1) provides a valuable perspective for addressing such questions. PMT was created as a model of disease prevention and health promotion (Floyd et al. 2000; Rogers 1975; Rogers and Prentice-Dunn 1997). The author selected PMT from among other possible models (i.e., the Health Belief Model, Rosenstock 1974; Theory of Reasoned/Planned Behaviour, Ajzen 1991) because its variables and hypothesized relationships possess the greatest potential to extend our understanding of adaptation to weather and climate. Many of the

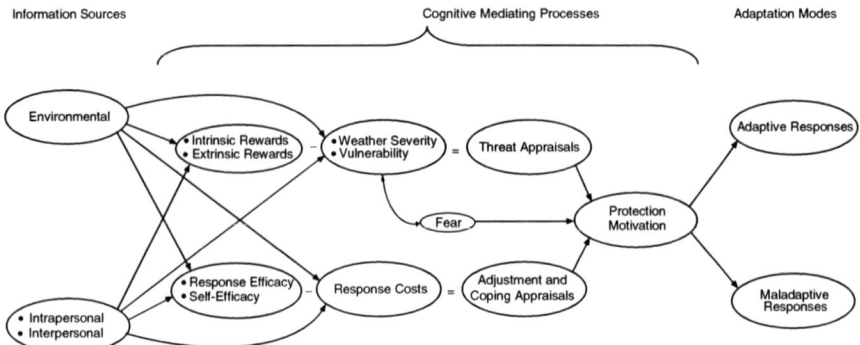

Fig. 10.1 Protection motivation theory applied to adaptation to weather and climate changes (Adapted from: Floyd, Prentice-Dunn and Rogers, 2000. Used with Permission)

variables pertaining to sources of information, cognition, and decision-making lend themselves well to organizing the research in the natural hazards and societal impacts literature. In addition to health-related behaviours, PMT is also well-suited to model adaptive behaviour for weather and climate events (Grothmann and Patt 2005; Grothmann and Reusswig 2006; Prentice-Dunn, personal communication). Two meta-analyses of studies that employed PMT have been largely supportive of the model (Floyd et al. 2000; Milne et al. 2000). In addition, PMT has been evaluated favorably alongside other health behaviour models (Weinstein 1993).

The PMT model comprises three main components: 1. information sources, 2. cognitive mediating processes, and 3. adaptation modes (see Fig. 10.1). Information sources reside within the person's environment and include directly-sensed inputs along with communications from various forms of electronic and broadcast media. Information may also exist within the person as memories of prior-experienced events (intrapersonal), gathered through verbal interactions and relationships with others, or through nonverbal communications (e.g., observing others' behaviour).

These sources inform the cognitive mediating processes, which in PMT, involve parallel paths that eventuate in threat appraisals (top of Fig. 10.1) and adjustment and coping appraisals. Intrinsic rewards (e.g., thrill of watching severe weather) and extrinsic rewards (e.g., encouragement by friends) increase the likelihood that risky and maladaptive responses will occur (e.g. not having sufficient protection from lightning). The severity of the weather or climate changes along with one's perceived vulnerabilities to these conditions are balanced against the rewards; the difference represents the appraisal of the threat posed by the behavioural response. Regarding the lower track of Fig. 10.1, the likelihood of an adaptive response is increased to the extent that the person knows about appropriate adaptive behaviours to perform for a particular weather or climate event (response efficacy) and believes that he or she personally can perform the behaviours (self-efficacy). The perceived cost(s) in performing the behaviour (response costs) diminishes the likelihood of an adaptive response. Response and personal efficacy are balanced against the response cost to create an appraisal for adaptive, adjustment, or coping responses.

The appraisals of threat and of adjustment or coping along with the emotion of fear which attend the evaluation of the weather severity and one's perceived vulnerabilities give rise to protection motivation. Protection motivation, in turn, leads to making responses that may be adaptive or maladaptive. These responses may be one-time events or may be repeated as frequently as needed.

The following sections examine a variety of psychological variables and biases that can affect adaptive behaviour. The PMT model provides a valuable framework for understanding these variables and for reviewing the different biases that can affect adaptation (Nicholls 1999). The first section below discusses the sources of information that people use. Next, the discussion of the cognitive mediating processes will examine how maladaptive and adaptive response alternatives are appraised and also will emphasize the variety of psychological biases people may exhibit in evaluating response alternatives, some of which have been identified previously in the atmospheric science literature (e.g., Nicholls 1999). Finally, the content and form of responses will be discussed briefly.

10.2.1 Information Sources

10.2.1.1 Information from the Environment

People use a wide variety of information sources and differ from each other in how much they sense and observe weather information directly (Dow and Cutter 2000; Stewart 2007; Whyte 1985). Several characteristics of weather and climate affect how people may notice and gain information about atmospheric events (Evans and Cohen 1987). First, people will notice atmospheric events if their nature and/or magnitude makes them *perceptually salient* (Stokols 1979). Second, the *valence* of the event, as either primarily positive or negative, provides basic information regarding the course of adaptive behaviour that should be pursued (Campbell 1983). Third, the *degree of predictability* of the event relates to both the nature of the event and the amount of time that may be available for primary adaptation activities to occur. Fourth, the *duration and periodicity* of the weather event can be used to inform the length and frequency of adaptations that may be necessary. Fifth, the *degree of controllability* associated with an event is quite significant for adaptation to the extent that relatively minor weather events whose effects can be controlled or easily accommodated are not as likely to promote the consumption and use of information that is associated with uncontrollable events. Sixth, different climate events can possess unique and specific kinds of information that can be used to cue the *types of adaptation* that might be required (Evans and Cohen 1987).

Weather or climate events will become psychologically significant for people to the extent that these characteristics singly, or in combination, require cognitive or emotional resources (Dow and Cutter 1998). In this regard, climate events are individually and socially construed and this affects the information that people extract from them and their uses of this information (Call 2005; Kelly 1955; Stehr and von Storch 1995). For instance, the same tropical storm that possesses a negative valence for weary coastal residents may bring welcomed relief from drought in inland farming regions. In addition, an existing climate regime such as a drought may make a tropical storm event perceptually salient for people in need of rain. The degree of controllability and types of adaptation needed may also differ among people, groups, and organizations as a function of their experiences in adapting to similar events in the past and the degree to which their experiences lead them to trust in the natural variability of the climate (Stehr 1997).

The duration and periodicity of various weather and climate events can affect the extent to which people remember the occurrence, impacts, and adaptations to prior events. Climatological events that occur gradually might span several generations such that the changed climate regime escapes notice (Glanz 2003). Farhar-Pilgrim (1985) reported results from the Metropolitan Meteorological Experiment in which a 30% increase in summer precipitation over a 30-year period went unnoticed by the population. Conversely, Moser and Dilling (2007) suggested that extreme weather events that are circumscribed in time such as tornadoes, floods, severe storms, and in the case of hurricanes, named, were more likely to be perceptually salient and to gain the notice of a larger segment of the population than events occurring over longer timeframes such as drought.

10.2.1.2 Information from Intra- and Interpersonal Sources

Just as events may possess perceptual salience, people exhibit varying degrees of *motivational salience* for weather and climate events (Stewart 2007). *Weather salience* refers to the degree of psychological importance that people attach to weather. People possessing a greater degree of weather salience reported more frequent sensing and observing of the atmosphere directly, more frequent reliance upon various forms of electronic media for weather information, and greater effects of weather upon their mood (Stewart 2007). In this way, weather-salient individuals are a source of their own (intra-personal) weather and climate information. Given the weather and climate information that is currently available, Dow and Cutter (1998) have observed that people increasingly obtain their own information rather than rely upon authorities (i.e., emergency managers). People also use their memories of past weather events that they have experienced as a guide for what to expect and how to cope with forecasted weather events and climate regimes (Strauss and Orlove 2003).

Interpersonal communications encompass a variety of sources and modalities through which people may acquire knowledge about weather and climate (Moser and Dilling 2007). Formal sources for current weather and climate information include governmental meteorological agencies along with companies that provide specialised products for public and private interests. This information may be conveyed through broadcast and other electronic modes. These formal sources also may provide programming to persuade consumers to prepare for and adapt to various conditions.

There are converging results from different researchers that people are more likely to gather and to trust information about weather events from friends and family members compared to government representatives (i.e., emergency managers, police, elected officials; Dow and Cutter 1998; Lindell et al. 2005; Mason-Dixon Polling and Research 2007; Zhang et al. 2007). That is, people are more likely to base evacuation decisions upon persuasive communications from family members than from others.

Although the preponderance of evidence supports that human activities have had an influence upon the climate system and this is resulting in globally warming temperatures (Houghton et al. 2001), contrary opinions exist and may be taken up by consumers as a justification not to adapt to weather events or climate change (McCright 2007; Michaels 2004). Popular media may confuse the occurrence of weather extremes for climate change and thus muddle the evidence of global warming for the public (Bostrom and Lashof 2007).

Interpersonal communications can be affected by a phenomenon known as the *social amplification of risk* (Kasperson et al. 1988), which involves the attenuation or accentuation of a risk communication based upon interactions of communication media, socio-cultural factors, organizations, and individuals. Media sources may frame an event in such a way as to arouse risk perceptions among people. Subsequent interpersonal risk communications and discussions among people who receive the information may amplify risk perceptions beyond their actual levels. Such amplified perceptions may lead to exaggerated behavioural responses which further affect the level of risk that is communicated to others.

Interpersonal communication can also occur nonverbally through the observation of other people (Bandura 1986). Observation can enable people to learn new behaviours (e.g., how to prepare a building for an approaching hurricane or cyclone) as well as facilitate the performance of behaviours already learned (e.g., seeing others making preparations or adapting may cue similar behaviours in the observer). Further, people socially compare themselves with others that they observe on a wide range of variables such as knowledge, attitudes, values, skills and abilities (Festinger 1954; Suls et al. 2002). People compare upwardly to those who exceed one's competencies in some domain, downwardly to people whose competencies the observer exceeds, and to peers who are on-par with the observer. Observation of others and social comparisons with them provides people with important information that they can use to guide their behaviours in adapting to climate variability and change.

10.2.2 Cognitive Mediating Processes

Once initiated by the information about a weather or climate event, the threat and adjustment/coping appraisal processes will begin. These parallel processes and the psychological biases that can affect them largely occur automatically, outside of awareness, and without much contribution of rational, and deliberative processing (Kihlstrom 1987). A variety of cognitive, social, and decision-making biases can influence the cognitive mediating processes of the PMT model (and similar components in *any* health behaviour model, for that matter, Nicholls 1999). Although the operation of such characteristics and biases precludes a true representation of the objective variables, the threat and adaptation appraisals incorporate these biases and will reflect them in people's intended or actual behaviours (Floyd et al. 2000). That is, people act on the basis of their biases and the PMT model will reflect this.

10.2.2.1 Threat Appraisals

The presence of intrinsic and extrinsic rewards for failing to prepare, muddling through, or performing risky behaviours increases the likelihood of a maladaptive outcome. People who are curious about the nature of rare or extreme weather events may experience those events as *intrinsically rewarding* (Burt 2004). Unusual weather also may provide opportunities for testing oneself regarding the use of skills or abilities in professional or recreational roles (Floyd et al. 2000). *Extrinsic rewards* may take the form of increased compensation, professional advancement, and/or greater respect for an employee who reports for work during dangerous weather.

Evaluation of weather as dangerous and perceiving one's *vulnerabilities* to the event decreases the probability of a maladaptive response. Prior negative experiences with weather events that resulted in injury, death, or property losses and a corresponding negative emotionality could lead to greater appraisals of the dangers posed by extreme weather; it also could increase people's sense of vulnerabilities (Weber 2006). In a broad review of the literature, Weinstein (1989) observed that personal experiences with hazards lead people to view hazards as more frequent and themselves as more vulnerable, to think more frequently of the hazards and with greater clarity, and to take precautions that are appropriate given the hazards people faced in the past. In this regard, Stewart (2007) observed that people whose families experienced weather-related property damages reported significantly greater overall weather salience and greater effects of weather upon their day-to-day moods. More specifically, Weinstein et al. (2000) observed that survivors from three communities struck by tornadoes took more precautions 14 months afterward to the extent people were preoccupied with tornadoes. Mulilis et al. (2003) reported similar results following a tornado outbreak in western Pennsylvania. People who are psychologically traumatized by extreme weather also may maintain a heightened state of arousal, emotional numbing, and re-experiencing symptoms that may combine to overestimate the severity of subsequent weather and their vulnerabilities to it (Norris et al. 1999).

Perceptions of rewards and vulnerabilities are seldom veridical representations of the actual weather or climate scenarios. Human cognition and decision-making processes are subject to a variety of social, cognitive, and motivational biases that are reflected in the choices people make. Such biases affect the adaptation decisions people make in the face of weather and climate events. A discussion of some of the biases that have appeared in the weather and climate literature follows below. This listing is by no means exhaustive of all the psychological biases that exist, however it provides an indication of the kinds of biases that can skew threat, adjustment, and coping appraisals.

Optimistic bias involves discounting or not attending to negative characteristics of a situation and instead placing greater emphasis upon positive, but unlikely outcomes. This bias also leads one to believe that he or she is less likely to experience negative events than are other people (Weinstein 1980; Weinstein and Klein 1996). Such a bias would serve to decrease the appraisal of impending conditions as severe or dangerous. In describing optimism about natural hazards that borders on denial, Eric Holdeman, director of emergency management for Seattle's King County, outlined four stages: "One is, it won't happen. Two is, if it does happen, it won't happen to me. Three: if it does happen to me, it won't be that bad. And four: if it happens to me and it's bad, there's nothing I can do to stop it anyway" (Ripley 2006, p. 56). Other researchers observed that people would not evacuate ahead of a hurricane because they believed that their homes were built strong enough to withstand the storm (Blendon et al. 2006) or that they could simply "ride it out" (Smith and McCarty 2007).

Second, and by way of gathering information from others or from the environment directly, people may engage in *confirmatory bias* whereby they only attend to or value information that is in accord with a belief that they want to support. For example, someone who believes that global warming does not exist may attempt to confirm this position by citing a colder than normal winter as evidence. For shorter-fused events, people may play one weather information source against another to confirm the response that they would like to make.

Third, *anchoring effects* of one's prior experiences with various weather events may serve as an erroneous standard for comparison of weather or climate conditions that are forecasted to occur (Tversky and Kahneman 1974). For example, a person previously may have experienced the outer bands of wind and rain from a hurricane and erroneously concluded that hurricanes generally *are not that bad*. Such anchoring beliefs may lead the people to discount the severity of subsequent storms so that they do not make preparations or evacuate (Dash and Gladwin 2005; Glantz 2005). A wide variety of weather events, to the extent that they are remembered, may function as evaluative anchors.

A forth phenomena, known as *hindsight bias*, also pertains to past weather events. This bias essentially involves the erroneous belief that previously-experienced events were more predictable than they were in actuality (Blank et al. 2007). Like the anchoring effect, if the hindsight bias is played forward, people may assume that weather events can be predicted with greater precision and specificity than is the case. This bias also may feed an *all's well that ends well* mentality. Here, people may take no adaptive efforts ahead of an event and experience, fortuitously, no negative consequences. This could further reinforce making no preparations or taking adaptive measures ahead of subsequent events (Glantz 2005).

A fifth bias concerns what is done with information that is obtained by observing and comparing with other people. The *actor-observer bias,* a special case of the *fundamental attribution error*, involves overweighting the role of dispositional variables in evaluating the behaviour of others (i.e., actors), while overweighting the role of situational variables in making attributions about oneself (i.e., the observer's behaviour). For example, an observer may witness person who has become stranded while attempting to drive across a flooded road. According to this bias, the observer may attribute that circumstance to the driver *being* unskilled, unknowledgeable, fool-hearty, and so forth, without considering the driver's situation that led to the behaviour. The danger of the actor-observer bias is that the observer may think him or herself more skilled or knowledgeable than the driver and perhaps more likely to attempt risky maneuvers, only to attribute later failures or losses to the constraints of the situation.

Sixth, social comparison can also induce biases in assessing the nature of weather or climate risks. Although objective and reliable information about how to prepare or evacuate ahead of a weather event may have been given, an individual may gauge his or her own levels of threat and action by engaging in downward and upward comparisons with other people. Although this may be a useful heuristic to a point, if others are not themselves responding or taking the appropriate precautions then one ultimately may be under-prepared.

10.2.2.2 Adjustment and Coping Appraisals

According to PMT, higher levels of response efficacy and self-efficacy increase the likelihood of an adaptive response whereas higher response costs may preclude performing an adaptive behaviour. Response efficacy in Fig. 10.1 pertains to a more generalized knowledge of what to do or not do that may be adaptive in responding to a weather or climate event (Kroemker and Mosler 2002). For example, response efficacy would be evident in knowing that the safest place to be during a tornado is in either a basement or storm cellar (i.e., the lowest, most protected area in a building). Similarly, knowing that one should wear light-colored clothing and adequately hydrate while also avoiding exertion in the hottest portion of the day would evidence response efficacy. Response efficacy is concerned with *knowing that* or *knowing what*.

Self-efficacy pertains to the individual's appraisal that he or she could actually perform the adaptive response; it is concerned with *knowing how*. Self-efficacy could emerge from prior experience in taking adaptive measures or from knowledge of how to assemble components of learned responses into a novel adaptive response. Others' adaptive responses may cue similar responses in an observer. This person may then evaluate his or her responses, socially compare with others, and then alter or self-correct the response (Bandura 1986). Self-efficacy perceptions may increase with successive repetitions and refinements of the response. Thus, self-efficacy for weather and climate adaptation depends upon a knowledge base that informs repeated practice.

People may learn about the appropriate adaptive responses to various weather and climate scenarios (response efficacy) during their formal education as children (i.e., Kindergarten through 12th Grade in the United States), among other sources.

Because systematic educational coverage of adaptive behaviours for natural disasters has been lacking, the American National Red Cross developed the *Masters of Disaster* (MoD) curriculum for grades K-8 (American National Red Cross 2000). English and Spanish language versions of the curricula exist in DVD form to facilitate adoption. In addition to general preparedness and prevention, the curriculum also includes specific education on how to prepare and adapt to: earthquakes, wildfires, tornadoes, hurricanes, floods, fire, and lightning. MoD is valuable in providing knowledge that underlies response efficacy and can provide practice to enhance self-efficacy.

Weather services and other governmental agencies periodically conduct educational outreach campaigns locally that are designed to educate and remind residents in affected areas about extreme weather events that may require their attention. Within the United States, the National Weather Service had developed a range of online resource materials that are available for the general population, for educators, and children. To the extent that these materials are used to build response efficacy and self-efficacy, people will be in a better position to adapt.

10.2.2.3 Response Costs

The likelihood that an adaptive response is performed will be decreased to the extent that a person perceives increasing costs, which are broadly conceived here in terms of time, money, emotional effort, convenience, and/or interpersonal resources, among other variables. With respect to secondary adaptation, a driver may believe that the time required to take an alternate route (adaptive response) is too great compared to the estimated risk of driving through a flooded roadway (risky response). People living near coastlines or rivers may not have the financial resources to purchase flood insurance or, in the case of hurricanes, to purchase storm shutters by way of primary adaptation. Response costs may also take the form of financial opportunities that are lost when one evacuates, concerns about crowded roadways, and concerns about possessions being damaged or stolen (Blendon et al. 2006; Smith and McCarty 2007). Regarding attachment costs, younger family members may not evacuate ahead of extreme weather because they want to remain and provide care for elderly or infirm family who cannot evacuate or because they have concerns about the care and safety of their pets (Blendon et al. 2006; Dow and Cutter 2000; Smith and McCarty 2007).

The general adaptational load that a person is experiencing at the time that a weather or climate-related adaptation is required also relates to response cost. That is, the range of events in a person's life that require adaptation, adjustment, and coping constitute an emotional overhead that may limit the resources available to make an adaptive response. Evidence for the existence of a *finite pool of worry* and other emotional concerns comes from clinical and personality psychology where the accumulation of stressors can lead to problems in making satisfactory emotional adaptations (Kanner et al. 1981; Lazarus 1993; Linville and Fischer 1991). Similarly, Hansen et al. (2004) observed that a finite pool of worry for events such as climate, politics and so forth existed among farmers. This means that people may only have a limited amount of adaptive capacity for the events that they face. Further, the level of adaptive capacity that people possess at any given time may help to explain, in part, the variability that exists in their adaptive behaviour for various atmospheric events (Kroemker and Mosler 2002).

Like the other decision-making components discussed thus far, perceptions of response costs are susceptible to biases and distortions. Some people may exaggerate costs or *awfulize* about the time or monetary opportunities lost by pursuing adaptive responses. Biasing response costs upward while also optimistically biasing weather severity and one's vulnerability downward may produce a risky response. Similarly, people may remember prior near-misses of storms for which they evacuated and then inflate the costs of responding adaptively to subsequent storms (Blendon et al. 2006; Dow and Cutter 1998; Glantz 2005).

10.2.2.4 Emotion and Protection Motivation

Threat appraisals, adjustment and coping appraisals, and the experienced level of fear come together in the protection motivation node of the PMT model (see Fig. 10.1). The model predicts that an adaptive response will follow to the extent that fear of one's vulnerability to severe events and the appraisal for making an adaptive response outweigh the threat appraisals associated with maladaptive responses (Rogers and Prentice-Dunn 1997). Fear, and presumably other similar affective states (e.g., worry), provides the motivation for taking protective action. In behavioural terms, fear functions as an aversive state; responses that lead to its decrease or disappearance are negatively reinforced by the abatement of fear. This reduction in fear following an adaptive behaviour makes performing that behaviour more likely in the future.

The creators of PMT were prescient in incorporating both cognitive and emotional processes in their model (Rogers 1975; Rogers and Prentice-Dunn 1997). In some respects the PMT model anticipated more recent scholarship in the field of decision-making and risk-taking under conditions of uncertainty (Loewenstein et al. 2001; Slovic et al. 2004; Slovic and Peters 2006). This research has challenged the existing consequentialist decision-making models (e.g. Kahneman and Tversky 2000) that viewed emotional processes as deriving *from* cognitive evaluations of anticipated outcomes and subjective probabilities or from behavioural outcomes following a decision. The *risk-as-feelings* perspective (see Fig. 10.2) recognizes that feelings stemming from the benefit (or harm) from taking the risk previously exists alongside cognitively-based evaluations of factual, descriptive, or abstract information about the risk (Kaplan 1991; Loewenstein et al. 2001; Slovic et al. 2004; Weber 2006). In other words, both rational, deliberate cognitive evaluations *and* intuitive, rapid, gut-level emotions contribute to decisions. Research using the risk-as-feelings perspective finds, contrary to the predictions of rational choice theories, that people tend to underweight the likelihood of

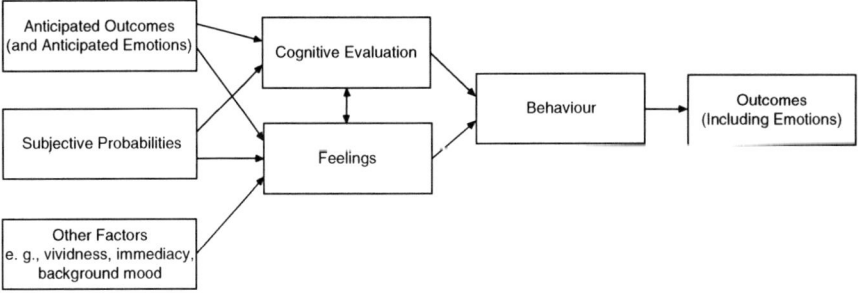

Fig. 10.2 Risk-as-feelings perspective (Adapted from Loewenstein, Weber et al. 2001. Used with Permission)

rare events occurring and overweight the likelihood of more common events (Hertwig et al. 2004; Weber et al. 2004). The reason for this stems from taking a small number of risks and not encountering the rare and negative outcome (e.g., experiences of remaining outside during a thunderstorm and not being injured by lightning) such that emotionally-based experiential estimates of the risk are lower (Hertwig et al. 2004).

The PMT model explicitly specifies fear as the emotion that stems from perceptions of weather severity and vulnerability and that serves to increase motivation for an adaptive, protective response. The more general risk-as-feelings model suggests that other emotions in addition to fear could influence decision-making. For instance, using a five-category emotional classification, it is possible that curiosities and interests about weather and climate may create the emotions of *happiness* or *desire* that stem from the intrinsic rewards node of the PMT (Fig. 10.1). Similarly, emotions of *anger* or *disgust* concerning response costs (e.g., anger at having to evacuate or disgust with conditions of the shelter) may contribute to risk-taking or other responses that have maladaptive outcomes. These possibilities suggest that the PMT model could be expanded productively to include the more general risk-as-feelings perspective. Thus, cognitive mediating processes in the model may more appropriately become *cognitive-emotional mediating processes*.

Response preferences (for either risky/maladaptive or adaptive responses) are likely constructed dynamically and reflect particular situational values of the variables that comprise the PMT (Slovic 1995). Risky or adaptive choices are also domain specific (Blais and Weber 2006; Weber et al. 2002). That is, people who report benefits in taking social risks may not experience benefits in pursuing ethical or financial risks. Weather-related risk-taking also comprises its own specific domain that is related to health and recreational risk-taking but not to other forms of risk such as social, financial or ethical (Stewart and Weber 2007). By contrast, response preferences are likely not to be as affected by longstanding personality traits (e.g., a risky personality type, a generally self-efficacious personality, Stern 1992). Although personality variables may have some slight effects on a PMT node such intrinsic or extrinsic rewards, their contributions are likely less influential than those pertaining directly to the situation in which the risk decision is being made.

10.2.3 Adaptation Modes

The final component of the PMT model concerns the nature of the responses that emerge from the cognitive-emotional mediating processes. There are two ways in which responses may be adaptive or maladaptive, the first of which concerns the direction of the response *ahead* of the event. That is, according to environmental and social information inputs (i.e., the weather or climate scenario as it exists), one response may be more clearly adaptive than another. The second way pertains to the outcome given the response and the weather or climate events that occurred. In this regard, an adaptive response for an event that occurred as expected would represent a *hit* whereas lack of adaptation responses (primary, secondary or tertiary) in the face of the event would constitute a *miss* along with its associated consequences. Those who proactively adapt for forecasted events that do not materialize experience a *false alarm*. Experiences of false alarms may bias people against responding adaptively for subsequent events (Dow and Cutter 1998). A *lucky hit* occurs for people who do not make adaptations for forecasted events that do not occur. People also may experience reinforcement for their lack of adaptation and may behave similarly during forecasts of subsequent events.

The emotional experiences that people have as a result of their level of adaptation efforts and how these helped them to manage the weather or climate events can affect subsequent adaptations. Emotions not only inform adaptive decision-making, they also attend the results of adaptive responses (see Fig. 10.2). This emotional feedback is significant in that an event may require a series of decisions and responses, not just a single one. It is conceivable that negative outcomes stemming from initial decisions could emotionally bias (e.g., excessive negativity) subsequent efforts to adapt. Thus the psychological forms of adaptation *after the event* become important.

Psychologists distinguish two primary ways of coping: *problem-focused* and *emotion-focused*. Problem focused coping involves active efforts to change a troubled person-environment relationship by working actively either upon the environmental conditions or upon oneself (Lazarus 1993). Coping behaviours that involve seeking social support, planful problem solving, positive reappraisal, and so forth exemplify a problem-focused approach. Emotion-focused coping is characterized by more passive approaches at adapting by changing the way the environment is attended to or by attempting to alter the meaning of what has occurred in the environment. Emotion-focused coping involves distancing, self-controlling, and escape-avoidance. Although it is tempting to value problem-focused over emotional-focused coping, the ultimate success in adapting may lie in knowing which type of coping efforts to use given the situation (Lazarus 1993).

10.3 Conclusions

The emphasis throughout this chapter has been upon increasing the possibilities for understanding the adaptation to climate variability and change through the perspectives afforded by an individual psychological approach. The PMT model provides a valuable heuristic for organizing the existing and emerging scholarship on adaptation from a psychological perspective. Indeed, psychological approaches to adaptation will likely become more important in the coming years because researchers are now realizing that further useful variability in human adaptive behaviour exists beyond some of the broader indicators such as gender, race, education, and socioeconomic status, among others, which have been employed up to this point (Dow and Cutter 2000; Grothmann and Reusswig 2006). By way of closing, several recommendations are provided to further the psychological study of human adaptation to climate variability and change.

10.3.1 Focus upon the Role of Emotional Processes

PMT and work by Loewenstein et al. (2001) emphasise the important contribution of emotional processes in decisions that involve risk and uncertainty. As a personal and experiential source of information that exists alongside descriptive or abstract information, emotions in decision-making promise to increase our understanding of the choices people make. In this regard, Slovic et al. (2004) have advocated for broadening the definition of rational behaviour to include the risk-as-feelings perspective. Although the rational choice family of theories have enjoyed much use over the years, increasingly, its core assumptions that people have full information and make choices to maximize their gains seem untenable in the present context. Deciding how to best adapt, adjust, or cope with a climate event given the myriad considerations and options for responding that exist are handled more precisely by models such as PMT compared to the rational choice theory. As this chapter has suggested, people at best have partial information about weather and climate events and may not know what additional data that they need or are lacking to make a decision. Further, people construe gains in many ways beyond those implied in economic conceptualizations so that across a group of people gains at best are an admixture of different end states, some of which may have questionable statuses as gains.

10.3.2 Examine Sources of Uncertainty in Conveying Weather and Climate Information

Contemporary research has been examining the best methods of conveying forecast uncertainty and the processes of decision-making under conditions of uncertainty (Committee on Estimating and Communicating Uncertainty in Weather and Climate

Forecasts 2006; Friday 2003). This work largely has focused upon uncertainties in forecast products and not how these are represented within the minds of users and decision-makers. In looking forward, a subsidiary line of research should examine the uncertainties that people possess or exhibit about probabilistic forecasts. As shown at the beginning of this chapter Hurricane Katrina was depicted with certainty to cause a great amount of destruction to New Orleans. Yet, the behaviour of a significant proportion of residents did not reflect this certainty. Some other evaluative weighting process could have discounted the certainties conveyed in forecast information (e.g., people have 60% confidence in the forecast of a certain strike). In this regard, the Mason-Dixon Poll (2007) indicated that respondents trusted friends and family more than government officials regarding evacuation orders. This finding suggests that the different sources that are pertinent to adapting to weather and climate may receive different weights according to their trustworthiness or veracity. Explicating this weighting process in addition to examining uncertainties conveyed in products and decision-making may move the field further along.

10.3.3 Use Research Methodologies That Can Assess and Model the Relationships of Psychological Variables

As conveyed in this chapter and in emerging scholarship, the PMT model has demonstrated promise in providing a more refined account of adaptive behaviour (Grothmann and Reusswig 2006). Scholarship in this area can be furthered by efforts to assess the PMT variables with psychological measures. The strength of psychological measures lies in their ability to assess individual perspectives, attitudes, and emotions with in ways that are not possible with demographic or broader sociological variables. Understanding adaptive behaviour more fully requires *getting inside the heads* of the people who are faced with adaptive or risky choices; psychological measures uniquely afford this perspective. Finally, analytical approaches such as structural equation modeling or growth curve analysis that can represent the simultaneous contributions of variables and the relationships among them, as shown in the PMT and the risk-as-feelings approach, holds promise in moving the field ahead in further understanding adaptive behaviour.

References

Ajzen, I., 1991: The theory of planned behavior. *Organizational Behavior and Human Decision Processes*, 50, 179–211.
American National Red Cross, 2000: *Masters of Disaster*. Author. http://www.redcross.org/disaster/masters/
American Psychological Association, 2007: *APA dictionary of psychology*. APA Press, Washington, DC, 1008 pp.

Bandura, A., 1986: *Social foundations of thought and action: a social cognitive theory.* Prentice-Hall, Englewood Cliffs, NJ, 617 pp.

Blais, R., and Weber, E.U., 2006: A domain-specific risk-taking (DOSPERT) scale for adult populations. *Judgment and Decision Making*, 1, 33–47.

Blank, H., Musch, J., and Pohl, R.F., 2007: Hindsight bias: On being wise after the event. *Social Cognition*, 25, 1–9.

Blendon, R.J., Benson, J.M., Buhr, T., Weldon, K.J., and Herrmann, M.J., 2006: High-risk area hurricane survey. Harvard School of Public Health Project on the Public and Biological Security. http:// www.hsph.harvard.edu/hurricane/topline.doc (Accessed August 2006).

Bostrom, A. and Lashof, D., 2007: Weather or climate change? *Creating a climate for change: communicating climate change and facilitating social change.* S.C. Moser and L. Dilling, Eds., Cambridge: New York, NY, pp. 31–43.

Bryson, R.A., 1997: The paradigm of climatology: An essay. *Bulletin of the American Meteorological Society*, 78: 449–455.

Burt, C.C., 2004: *Extreme weather: a guide and record book*, Norton, New York, NY, 304 pp.

Cable News Network, 2005: *Hurricane Katrina rooftop rescue (Sept. 4).* http://www.cnn.com/interactive/us/0509/gallery.rescue/frameset.exclude.html (Accessed June 18, 2007).

Call, D.A., 2005: Rethinking snowstorms as snow events: A regional case study from upstate New York. *Bulletin of the American Meteorological Society*, 86, 1783–1793.

Campbell, J.M., 1983: Ambient stressors. *Environment and Behavior*, 15, 355–380.

Caplan, G., 1964: *Principles of preventive psychiatry.* Basic Books, New York, 304 pp.

Changnon, S.A., Pielke, R.A., Changnon, D. Sylves, R.T. and Pulwarty, R., 2000: Human factors explain the increased losses from weather and climate extremes, *Bulletin of the American Meteorological Society*, 81, 437–442.

Cohen, J.E., 2003: Human population: The next half century. *Science*, 302, 1172–1175.

Committee on Estimating and Communicating Uncertainty in Weather and Climate Forecasts, 2006: *Completing the forecast: Characterizing and communicating uncertainty for better decisions using weather and climate forecasts.* National Research Council, 124 pp.

Dash, N., and Gladwin, H., 2005: *Evacuation decision making and behavioral responses: Individual and household.* White paper prepared for the Hurricane Forecast Socioeconomic Workshop, 23 pp.

Dow, K., and Cutter, S.L., 1998: Crying wolf: Repeat responses to hurricane evacuation orders. *Coastal Management*, 26, 237–252.

Dow, K., and Cutter, S.L., 2000: Public orders and personal opinions: household strategies for hurricane risk assessment. *Environmental Hazards*, 2, 143–155.

Easterling, W.E., Hurd, B.H, and Smith, J.B., 2004: *Coping with global climate change: The role of adaptation in the United States.* Pew Center on Global Climate Change, 40 pp.

Evans, G.W. and Cohen, S., 1987: Environmental stress. *Handbook of environmental psychology.* D. Stokols and I. Altman, Eds., Wiley, New York, pp. 571–610.

Farhar-Pilgrim, B., 1985: Social analysis. *Climate impact assessment: studies of the interaction of climate and society.* R.W. Kates, J.H. Ausubel, and M. Berberian, Eds., Wiley, New York, pp. 323–350.

Festinger, L., 1954: A theory of social comparison processes. *Human Relations*, 7, 117–140.

Floyd, D.L, Prentice-Dunn, S., and Rogers, R.W., 2000: A meta-analysis of research on Protection Motivation Theory. *Journal of Applied Social Psychology*, 30, 407–429.

Friday, E.W., 2003: *Communicating uncertainties in weather and climate information: A workshop summary.* National Research Council 68 pp.

Glanz, M.H., 2003: *Climate affairs: A primer.* Island Press, Washington, DC, 291 pp.

Glantz, M.H., 2005: Hurricane Katrina rekindles thoughts about fallacies of a so-called "natural" disaster. *Sustainability: Science, Practice, & Policy*, 1, 1–4.

Grothmann, T. and Patt, A., 2005: Adaptive capacity and human cognition: The process of individual adaptation to climate change. *Global Environmental Change*, 15, 199–213.

Grothmann, T. and Reusswig, F., 2006: People at risk for flooding: Why some residents take precautionary action while others do not. *Natural Hazards*, 38, 101–120.

Guiney, J.L. and the Service Assessment Team, 2006: *Service assessment: Hurricane Katrina, August 23–31, 2005.* United States Department of Commerce, 50 pp. http://www.weather.gov/os/assessments/pdfs/Katrina.pdf.

Hansen, J., Marx, S., and Weber, E.U., 2004: *The role of climate perceptions, expectations, and forecasts in farmer decision making: The Argentine Pampas and South Florida.* International Research Institute for Climate Prediction (IRI), Palisades, NY: Technical Report 04–01.

Hertwig, R., Barron, G., Weber, E.U., and Erev, I., 2004: Decisions from experience and the effect of rare events in risky choice. *Psychological Science, 15,* 534–539.

Holahan, C.J., Moos, R.H., and Schaeffer, J.A., 1996: Coping, stress resistance and growth: Conceptualizing adaptive functioning. *Handbook of coping: theory, research, applications.* M. Zeidner and N.S. Endler, Eds., Wiley, New York, pp. 24–43.

Houghton, J.T., Ding, Y., Griggs, D.J., Noguer, M., van der Linden, P.J., Dai, K., Maskell, K., and Johnson, C.A., Eds., 2001: *Climate change 2001: The scientific basis.* Cambridge University Press, Cambridge, 892 pp.

Kahneman, D. and Tversky, A., Eds., 2000: *Choices, values, and frames.* Cambridge University Press, New York, 840 pp.

Kanner, A.D., Coyne, J.C., Schaefer, C., and Lazarus, R.S., 1981: Comparison of two modes of stress measurement: Daily hassles and uplifts versus major life events. *Journal of Behavioural Medicine, 4,* 1–39.

Kaplan, S., 1991: Beyond rationality: Clarity-based decision making. *Environment, cognition, and action: An integrated approach.* Gärling, T. and Evans, G.W., Eds., Oxford University Press, New York, pp. 171–190.

Kasperson, R.E., Renn, O., Slovic, P., Brown, H.S., Emel, J., Goble, R., Kasperson, J.X., and Ratick, S., 1988: The social amplification of risk: A conceptual framework. *Risk Analysis, 8* 177–187.

Kelly, G.A., 1955: *The psychology of personal constructs.* Norton, New York, 556 pp.

Kihlstrom, J.F., 1987: The cognitive unconscious. *Science, 237,* 1445–1452.

Knabb, R.D., Rhome, J.R., and Brown, D.P., 2005: *Tropical cyclone report: Hurricane Katrina, 23–30 August 2005.* National Hurricane Center, 43 pp. (http://www.nhc.noaa.gov/ms-word/TCR-AL122005_Katrina.doc).

Kroemker, D. and Mosler, H., 2002: Human vulnerability-factors influencing the implementation of prevention and protection measures: An agent-based approach. *Global environmental change in alpine regions.* K.W. Steininger and H. Weck-Hannemann, Eds., Edward Elgar, Cheltenham, UK, pp. 93–112.

Kunkel, K.E., Pielke, R.A., Jr., and Changnon, S.A., 1999: Temporal fluctuations in weather and climate extremes that cause economic and human health impacts: A review. *Bulletin of the American Meteorological Society, 80,* 1077–1098.

Lazarus, R.S., 1993: Coping theory and research: Past, present and future. *Psychosomatic medicine, 55,* 234–247.

Lazarus, R.S. and Cohen, J.B., 1977: Environmental stress. *Human behavior and environment: advances in theory and research, Vol. 2.* I. Altman and J.F. Wohlwill, Eds., Plenum, New York, pp. 89–127.

Lindell, M.K., Lu, J., and Prater, C.S., 2005: Household decision making and evacuation in response to hurricane Lili. *Natural Hazards Review, 6,* 171–179.

Linville, P. and Fischer, G.W., 1991: Preferences for separating or combining events. *Journal of Personality and Social Psychology, 60,* 5–23.

Loewenstein, G.F., Weber, E.U., Hsee, C.K., and Welch, E., 2001: Risk as feelings. *Psychological Bulletin, 127,* 267–286.

Mason-Dixon Polling and Research, 2007: *May 2007 hurricane awareness poll results.* http://www.hurricanesafety.org/home2.cfm.

McCright, A.M., 2007: Dealing with climate contrarians. *Creating a climate for change: communicating climate change and facilitating social change.* S.C. Moser and L. Dilling, Eds., Cambridge: New York, NY, pp. 200–212.

Michaels, P.J., 2004: Meltdown: The predictable distortion of global warming by scientists, politicians, and the media. *Cato*, Washington, DC, 304 pp.
Mileti, D.S., 1999: *Disasters by design: A reassessment of natural hazards in the United States.* Joseph Henry Press, Washington, DC, 351 pp.
Milne, S., Sheeran, P., and Orbell, S., 2000: Prediction and intervention in health-related behaviour: A meta-analytic review of Protection Motivation Theory. *Journal of Applied Social Psychology*, 30, 1–6–143.
Moser, S.C., and Dilling, L., Eds., 2007: *Creating a climate for change: Communicating climate change and facilitating social change.* Cambridge University Press, Cambridge, 549 pp.
Mulilis, J., Duval, T.S., and Rogers, R., 2003: The effect of a swarm of local tornados on Tornado preparedness: A quasi-comparable cohort investigation. *Journal of Applied Social Psychology*, 33, 1716–1725.
National Research Council, 2004: *Where the weather meets the road: A research agenda for improving road weather services.* National Academy Press, Washington, DC, 188 pp.
Niemeyer, S., Petts, J., and Hobson, K., 2005: Rapid climate change and society: Assessing responses and thresholds. *Risk Analysis*, 25, 1443–1456.
Nicholls, N. 1999: Cognitive illusions, heuristics, and climate prediction. *Bulletin of the American Meteorological Society*, 80, 1385–1397.
Norris, F.H., Smith, T., and Kaniasty, K., 1999: Revisiting the experience-behavior hypothesis: The effects of Hurricane Hugo on hazard preparedness and other self-protective acts. *Basic and Applied Social Psychology*, 21, 37–47.
Norris, F.H., Murphy, A.D., Baker, C.K. and Perilla, J.L., 2004: Postdisaster PTSD over four waves of a panel study of Mexico's 1999 flood. *Journal of Traumatic Stress*, 17, 283–292.
Pielke, R.A., Jr., 1998: Rethinking the role of adaptation in climate policy. *Global Environmental Change*, 8, 159–170.
Pielke, R.A., Jr., and Sarewitz, D., 2005: Bringing society back into the climate debate. *Population and Environment*, 26, 255–268.
Pielke, R., Prins, G., Rayner, S., and Sarewitz, D., 2007: Lifting the taboo on adaptation. *Nature*, 445, 597–598.
Riad, J.K. and Norris, F.H., 1996: The influence of relocation on the environmental, social, and psychological stress experienced by disaster victims. *Environment and Behavior*, 28, 163–182.
Ripley, A., 2006: Floods, tornadoes, hurricanes, wildfires, earthquakes…why we don't prepare. *Time*, 168 (9), 54–58.
Rogers, R.W., 1975: A protection motivation theory of fear appeals and attitude change. *The Journal of Psychology*, 91, 93–114.
Rogers, R.W. and Prentice-Dunn, S., 1997: Protection motivation theory. *Handbook of health behavior research: vol. 1. Determinants of health behavior: personal and social.* D. Gochman, Ed., Plenum, New York, pp. 113–132.
Rosenstock, I.M., 1974: Historical origins of the Health Belief Model. *Health Education Monographs*, 2, 328–335.
Slovic, P., 1995: The construction of preference. *American Psychologist*, 50, 364–371.
Slovic, P. and Peters, E., 2006: Risk perception and affect. *Current Directions in Psychological Science*, 15, 322–325.
Slovic, P., Finucane, M.L., Peters, E., and MacGregor, D.G., 2004: Risk as analysis and risk as feelings: Some thoughts about affect, reason, risk, and rationality. *Risk Analysis*, 24, 311–322.
Smit, B., O. Pilifosova, I. Burton, B. Challenger, S. Huq, R.J.T. Klein and G. Yohe, 2001: Adaptation to climate change in the context of sustainable development and equity. *Climate change 2001 – impacts, adaptation and vulnerability.* J.J. McCarthy, O.F. Canziano and N. Leary, Eds., *Contribution of Working Group II to the Third Assessment Report of the Intergovernmental Panel on Climate Change*, Cambridge University Press, Cambridge, pp. 877–912.

Smith, J.B., 1997: Setting priorities for adapting to climate change. *Global Environmental Change*, 7, 251–264.
Smith, S.K. and McCarty, C. 2007: Fleeing the storm(s): Evacuations during Florida's 2004 hurricane season. Paper presented at the Population Association of America, New York http://paa2007.princeton.edu/download.aspx?submissionId = 7165 (accessed June 18, 2007).
Stehr, N., 1997: Trust and climate. *Climate Research*, 8, 163–169.
Stehr, N. and von Storch, H., 1995: The social construct of climate and climate change. *Climate Research*, 5, 99–105.
Stern, P.C., 1992: Psychological dimensions of global environmental change. *Annual Review of Psychology*, 43, 269–302.
Stewart, A.E., 2007: Minding the weather: The measurement of weather salience. Manuscript submitted for publication.
Stewart, A.E. and Weber, E.U., 2007: The weather risk-taking scale. Poster presented at the 2007 meeting of the American Psychological Association, San Francisco, CA.
Stokols, D., 1979: A congruence analysis of human stress. *Stress and anxiety, vol. 6*. I.G. Sarason and C.D. Spielberger, Eds., John Wiley, New York, pp. 27–53.
Strauss, S. and Orlove, B.S., 2003: *Weather, climate, culture*. Berg, Oxford, 307 pp.
Suls, J., Martin, R., and Wheeler, L., 2002: Social Comparison: Why, with whom and with what effect? *Current Directions in Psychological Science*, 11, 159–163.
Tol, R.S.J., Fankhauser, S., and Smith. J.B., 1998: The scope of adaptation to climate change: What can we learn from the impact literature? *Global Environmental Change*, 8, 109–123.
Tversky, A., and Kahneman, D., 1974: Judgment under uncertainty: Heuristics and biases. *Science*, 185, 1124–1131.
Vedwan, N., and Rhoades, R.E., 2001: Climate change in the Western Himalayas of India: a study of local perception and response. *Climate Research*, 19, 109–117.
Weber, E.U., 2006: Experience-based and description-based perceptions of long-term risk: Why global warming does not scare us (yet). *Climatic Change*, 77, 103–120.
Weber, E.U., Blais, R., and Betz, N.E., 2002: A domain-specific risk-attitude scale: Measuring risk perceptions and risk behaviors. *Journal of Behavioral Decision Making*, 15, 263–290.
Weber, E.U., Shafir, S., and Blais, A. (2004). Predicting risk sensitivity in humans and lower animals: Risk as variance or coefficient of variation. *Psychological Review*, 111, 430–445.
Weinstein, N.D., 1980: Unrealistic optimism about future life events. *Journal of Personality and Social Psychology*, 39, 806–820.
Weinstein, N.D., 1989: Effects of personal experience on self-protective behavior. *Psychological Bulletin*, 105, 31–50.
Weinstein, N.D., 1993: Testing four competing theories of health-protective behavior. *Health Psychology*, 12, 324–333.
Weinstein, N.D. and Klein, W.M., 1996: Unrealistic optimism: Present and future. *Journal of Social and Clinical Psychology*, 15, 1–8.
Weinstein, N.D., Lyon, J.E., Rothman, A.J., and Cuite, C.L., 2000: Preoccupation and affect as predictors of protective action following natural disaster. *British Journal of Health Psychology*, 5, 351–363.
Whyte, A.V.T., 1985: Perception. *Climate impact assessment: studies of the interaction of climate and society*. R.W. Kates, J.H. Ausubel, and M. Berberian, Eds., Wiley, New York, pp. 403–436.
Wohlwill, J.F., 1974: Human adaptation to levels of environmental stimulation. *Human Ecology*, 2: 127–147.
Zhang, F., Morss, R.E., Sippel, J.A., Beckman, T.K., Clements, N.C., Hampshire, N.L., Harvey, J.N., Hernandez, J.M., Morgan, Z.C., Mosier, R.M., Wang, S., and Winkley, S.D., 2007: An In-person survey investigating public perceptions of and response to hurricane Rita forecasts along the Texas coast. *Weather and Forecasting*.

Section II
Perspectives

Chapter 11
Human Adaptation within a Paradigm of Climatic Determinism and Change

Andris Auliciems

> "Among all the factors which influence people's modes of life the ... most dominant are climate and stage of culture already obtained." Ellsworth Huntington, Mainsprings of Civilization 1945:281
>
> "One must seek to understand climate in the light of what mankind can perceive, understand and adapt to." Ken Hare, SCOPE 27 1985:2.1
>
> "...we simply cannot say how the patterns of natural and human induced change might cancel or reinforce each other at the regional and local level over coming decades and centuries." John Zillman, at a book launch 2004
>
> "The philosophers have only interpreted the world,... the point is to change it." Karl Marx, Theses on Feuerbach 1845: XI

Abstract Oversimplified interpretations of climate impacts, by Ellsworth Huntington and his contemporaries, lead to academic alienation and at times misguided social policy. The purpose of this chapter is to examine some perceptions about climatic determinism and scientific attitudes, and offer a different perspective on 'adaptation' within a changing climatic environment. Man-atmosphere interrelationships and adaptations are embedded within complex homeostatic and dynamic systems, which operate at several levels of human organization. An adaptive model, based on human thermoregulatory response to climate variability and hazards is presented to allow integration between several orders of impacts and essential control processes. However, given the uncertainty of both climate change and human responses, it is emphasized that adaptability of society and individuals is preferable to attaining adaptation to particular environmental conditions.

A. Auliciems
Biometeorologist & Geographer, *Kūgures* Saldus Latvia

11.1 Constant Climate Determinism

Climate change has been of scholarly interest since Charles Darwin's (1859) writings on the causes of evolution of species. To the natural scientist, however, the issue over the following century was largely one of evolutionary and geologic time frames: the atmospheric environment at a particular location could be considered to be a given constant. In the writings of Hippocrates (circa 400 BC) *On Air, Waters and Places*, hydrometeor impacts or "meteorotropisms" could be observed, but resultant human responses were predetermined aggregates of long term exposure to particular constant environments. Subsequently the concept that culture itself was a product of long term adjustments to climate was afforded a scholarly basis by natural and human scientific exploration, and description by Alexander von Humboldt and Carl Ritter (Riebsame 1985). By the end of the nineteenth century, all embracing climate determinism had become the prevailing paradigm of climate-human interaction, leaving little room for notions of either environmental change at the sub-evolutionary scale, or free will.

Unfortunately, the tenets of climatic determinism became associated with social policies and bigotry. Following the Spanish-American War of 1898, the United States had acquired substantial tropical possessions in the Caribbean area and the Pacific Ocean, and to thousands of colonial administrators, teachers, engineers, soldiers, and missionaries, concern about the climatic conditions in the new colonies was widespread. Amongst the "core" ideas held then about the interconnection of climate, race, and empire (Schumacher 2002), the most virulent case of universal determinism was made in a widely read and influential book Effects *of Tropical Light on White Men* by US Army Medical Corps Colonel Charles Woodruff (1905) "it is quite likely that everyone who lives in the tropics over 1 year is more or less neurasthenic" He also noted the unavoidable effects of tropical exposure to include mental depression, hypochondriasis, amnesia, migraine, melancholia, skin diseases, tuberculosis, cardiac problems, anemia, menstrual irregularities and general irritability. Woodruff was an adherent to Hippocratic notions that at the larger scale, racial characteristics were determined by the climate *milieu*. He was convinced that the superior white race would never find it possible to acclimatize to tropical conditions, where they could not expect to thrive, and therefore should remain within their own cooler latitudes. They should rule from afar. Here we might be tempted also to jest that Charles Woodruff had inadvertently made a most astute observation: the adaptation of American colonists of the tropical climate zone is their inability to adapt! Or alternatively, avoidance of a stress situation is a powerful tool of adaptation.

While impacts of atmospheric hazards on the human condition have been either recorded or implicitly recognized throughout written history (Lamb 1977), neither adaptations by physiological exposure nor personal decisions were considered to be of particular significance until academic geographers became the chief protagonists of climatic determinism. In USA, Ellen Churchill Semple (1911) and Ellsworth Huntington, and in Australia and then later Canada, Thomas Griffith Taylor (1959) while espousing less radical notions of climatic impacts than those of Hippocrates or Woodruff, believed that both human evolution and cultural development were interlinked and depended upon the characteristics of particular climates, some being more favorable than others. Tropical neurasthenia, ranging from alcoholism to lack of social control, affected all peoples.

In terms of academic, social and political influence, this softer climate determinism reached its apogee with the prolific writings and sweeping visions of Huntington (1907, 1916, 1924a, b, 1927, 1945). His works were both widely read and for a time were at the forefront of the modified paradigm. To his credit, Huntington attempted to advance Hippocratic ideas on health, performance and social development away from subjective qualitative statements of the preceding years. He tried to establish optimum temperature, humidity, weather and climate conditions in terms of quantitative data obtained for physical and mental work (e.g. factory production, library usage, college examination), and assess societies with objective indicators used even today in United Nations documents on development (e.g. birth rate, death rates, infant mortality, literacy, income). Huntington's analysis suggested that optimum performance temperature was near 18°C and then in comparing isotherm patterns to mapped distributions of the available social indicators, he arrived at a map of "climate energy". It revealed that with the epicenter on eastern USA, the Atlantic littoral was the most beneficial for human settlement. Thus, at the height of European colonialism, the philosophy of climate determinism had triumphed in many quarters, and especially in questions of settlement within Aristotle's long ago defined "torrid" climate zones.

Had Huntington stopped at this stage of his research, his version of climate determinism may have remained largely unchallenged. The problem was that Huntington was also unhappy with the previously vague concepts of affected levels of cultural achievement. Huntington decided to quantify levels of "civilization" by sending a detailed questionnaire to selected outstanding individuals across the world, asking for subjective assessments of national achievements on a large variety of criteria that he believed revealed levels of progress. It was this questionnaire, heavily biased towards technological advances and industrialization, his sampling procedures that were also weighted in favor of the developed European world, and lack of responses that made his map of aggregated civilization scores anathema to many social scientists, including fellow geographers. That this map well coincided with climatic energy unfortunately was interpreted as bogus evidence for climatic causality of group superiority. Indeed, within the then burgeoning quantification in social sciences and especially economic and human Geography, climatic determinism was considered to be totally discredited.

This determinist – free will schism, within the domain of the social sciences, has been extensively re-examined in climate – culture interactions by what can be described as multi-factorial historical studies (Toynbee 1945; Markham 1947; Carpenter 1968; Chappell 1970; Bell 1971; Ladurie 1972; Bryson and Murray 1977; Lamb 1977, 1982, Post 1977; Parry 1978; Pfister 1978, 1981; de Vries 1980; Fischer 1980; Lee 1981; Shaw 1981). The critical role of environmental forcing also has been reinterpreted within the new and exciting integrative fields of sociobiology (Wilson 1975) and evolutionary psychology (Cosmides and Tooby 1997).

In *Collapse*, Jared Diamond (2005) presents an original compromise to the ancient debate by demonstrating that society has choice, depending upon preference for appropriate action or inactivity in crisis situations, following which chance takes over. In this treatise, Diamond proposes five interdependent vulnerability factors that determine survival or demise of groups of peoples. Two are environmental quality and related decision making and management skill; two are human functions of information flows and hostility by neighbors, and the other adaptability to climate adversity and specifically its change. (Diamonds hypothesis framework is subsequently shown as part of the integrated model of adaptation (see Fig. 11.6 of this chapter).

To the above authors, whether implicitly or explicitly, critical to human social development and indeed survival, has been a question of their inability to adequately respond to extreme fluctuations in weather, changes in climate and earth resources. To Hans von Storch and colleagues (Stehr et al. 1996; Stehr and von Storch 2000; von Storch and Stehr 2006), there has been sufficient evidence to vindicate Huntington's multidimensional determinism.

11.2 The Homeothermic Imperative

The pitfalls of nineteenth century determinist versus twentieth century free will semantics were sidestepped by biologists who traced the evolutionary distribution of the human organism on the earth's surface in response to the demands of thermoregulatory processes (Burton and Edholm 1955; Dubos 1980; Newman 1956; Sargent 1963; Scholander 1955, 1956). While large scale determinism and questions of survival of civilizations are of interest in themselves, examination of the smaller scale level of the individual enables identification of the actual mechanisms involved, both biological and technological.

First order (direct environmental effect) studies of the homeothermic imperative came from several fields including medical biometeorology (e.g. classical works of Mills 1946; Petersen 1947; Tromp 1963), applied psychology (e.g. Mackworth 1950; Pepler and Warner 1968), and especially thermophysiology (Yaglou 1926; Gagge 1936; Gagge et al. (138); Winslow and Herrington 1949; Burton and Edholm 1955; Hensel 1959; Hardy 1961) and heating, cooling and ventilating practice and theory (e.g. Houghten and Yagloglou 1923 a, b; Bedford 1936; Fanger 1967 and many others).The model for these studies was largely one that envisaged a linear cascade:

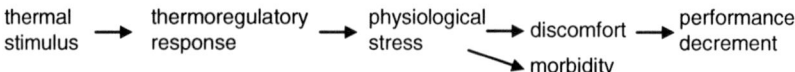

Although there were seemingly anomalous observations of seasonality and regional differences (e.g. Yaglou and Drinker 1928; Hickish 1955), until the late 1960s of the last century there was very little dissent from Gagge's (1936) renowned formulation that human responses could be well estimated from the thermal energy budget equation $+/- S = M +/- R +/- C - E$ where S is storage or stress, M is metabolism,

R is radiation, C is convection and E is evaporation. Each of the environmental terms could be measured instrumentally and metabolic rates could be approximated from empirically derived tables for work categories. Minimum stress occurs when metabolism is balanced by heat losses which maintains homeostasis at a constant core temperature, or homeothermy, at near 37°C. This particular core temperature is reached when environmental temperature is near 25°C. Following a series of controlled laboratory experiments with healthy young males, it was strongly argued by Fanger (1970) that this was a universal optimum for people engaged in tasks at near 100 Wm^2 metabolic rate, when at 25.5°C his "predicted mean vote" PMV (Fanger 1967) and S = 0. Any deviation from this could be simply explained through either metabolic variability or clothing insulation (ASHRAE laboratory based standards prescribed 24°C for summer and 22°C for winter clothing). Second order or higher level cultural adaptations were dismissed as being of little relevance.

As with Huntington's search for a universal optimum temperature, Fanger's insistence upon a simple deterministic explanation fails when different peoples and cultures are examined. Thermal comfort and real life preference surveys, using the 7 point ASHRAE or Bedford scales of subjective thermal sensations, have shown that there is no single "neutral" temperature. Rather thermal neutrality migrates with exposure towards ambient temperature. Empirically, such far reaching adaptation has been observed in repeated and comparable studies into differences of warmth perception in regression of group mean neutralities on monthly outdoor and prevailing indoor temperatures (Auliciems 1969, 72, 83; Humphreys 1975, 1976; Howell and Kennedy 1979; Berglund 1979; Auliciems and de Dear 1997; Nichol 1974; de Dear et al. 1997; Schiller et al. 1988; Humphreys and Nicol 2000a, b; Heidari and Sharples 2002; de Dear and Brager 2001; Soebarto et al. 2004; van der Linden et al. 2006).

The neutrality shift cannot be explained by clothing or other personal parameters, but rather by the "thermopreferendum", or more simply preference (or choice) resulting from thermal expectations elicited by current and past thermal experiences, and cultural and technical practices (Auliciems 1981). This "adaptive comfort model" is shown schematically in Fig. 11.1. Within air conditioned buildings with constant equable levels of warmth, however, seasonality and regional adaptability tends to fail (de Dear et al. 1997). This provides serious argument in favor of reducing air conditioner usage, which happily in turn would reduce energy consumption. At this time, the adaptive model is being espoused in amendments in ASHRAE Standards 55, ASHRAE database RP884, European directive EN 15,251 and various national standards publications.

Minimum neutrality values have been observed below 15°C for highly acclimatized British schoolchildren (Auliciems 1969) and also the elderly at 17°C (Fox 1973), and maximum preferred temperatures at 34°C and higher in hot countries (Nicol 1974). In brief, the shift of preferred indoor comfort temperatures at a monthly resolution (see Fig. 11.2) is at 0.31°C/1°C in the direction of prevailing outdoor (and indoor) temperature according to a generalized regression

$T_\psi = 0.31 T_M + 17.6$ where T_ψ is the preferred group temperature, and T_M is mean monthly temperature outdoors (Auliciems 1981). Despite earlier criticism of this

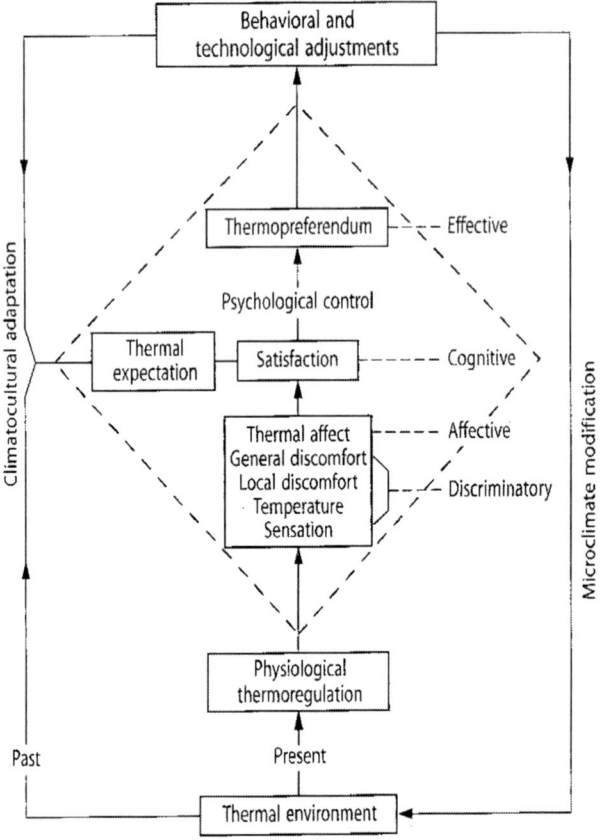

Fig. 11.1 The Adaptive Comfort Model (from Auliciems 1981, 1983)

seemingly simple relationship (de Dear et al. 1997), de Dear and Brager (2001) have revalidated the equation and the Auliciems (1981) adaptive model as the best predictor of indoor comfort for naturally ventilated buildings.

Table 11.1 attempts to generalize main characteristics or principles of human adaptations to atmospheric variability. In summary, since biologically humans are homeotherms, their life processes and infrastructures are ultimately centered to the maintenance of a constant core temperature. To do so, the core is protected (cocooned, enveloped) by a combination of thermoregulatory processes as listed in bold fonts, together with empirically observable impacts in Table 11.2. All are linked to thermoregulatory responses, and except for the adverse impacts and the involuntary physiological responses (item i), are controlled to some degree by cognitive and affective evaluations of thermal signals, with respect to their intensity, and to their subjectively evaluated desirability as determined by past experiences (Fig. 11.1). The processes in Table 11.2 are arranged in an approximate ascending time sequence and total impacts on a society, but probably any

11 Human Adaptation within a Paradigm of Climatic Determinism and Change

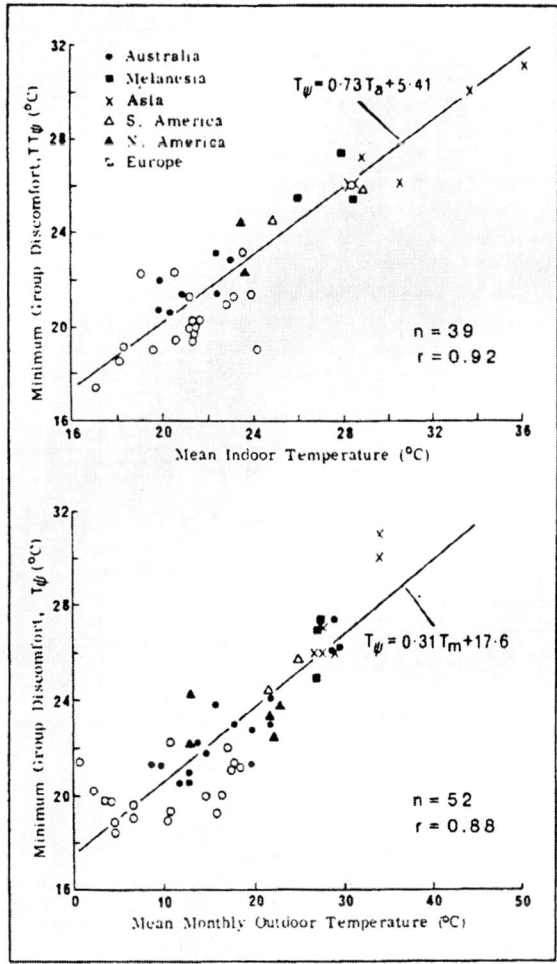

Fig. 11.2 Regression of Thermal Neutralities On Indoor and Outdoor Temperatures (from Auliciems 1981, 1983)

adaptation, or impact, can trigger those either above or below. The duration of any cascade would depend upon the strength of the signal and thresholds for impacts, the readiness and capacity of the particular mechanism, the fitness and vulnerability of individuals or groups being affected. Since all of the eight hazard categories in Table 11.2 are essentially the same but magnified physical avenues for metabolic heat dissipation, it is not surprising to see them impacting upon all human systems and their manifestations linked to thermoregulatory processes.

Where individuals are free to make decisions and respond, depending upon the signal strength, thermoregulatory processes may be triggered individually or as a system. The initial and reinforcing signals may be real or perceived,

Table 11.1 Principles of Human Adaptation to Atmospheric Variability

Atmospheric variability, at all temporal and spatial scales, provides *stimulus* that requires *response*, either spontaneous or voluntary.

A precondition to the well being of the homeothermic human, is a successful maintenance of a *constant core temperature* near 37°C. This requires continuous physiological and behavioural adjustments to balance energy exchanges between the body and the environment.

Most atmospheric stimuli may be thought as occurring on a continuum where "absence" or extremes of thermal or hydrometeor phenomena deviate from a biologically neutral intermediate state, or a subjectively perceived *comfort zone*. Within limits, deviations above and below induce increasing levels of response.

Repeated exposure to a particular stimulus promotes *habituation* that enables a reduction of response to that stimulus, but at the cost of narrowing the tolerance band or coping range.

In contrast, repeated and cyclical exposure to stimulating climatic variation and variability leads to *acclimatization* that promotes increased broadening of the tolerance band, both physiological and psychological[1].

Biological fitness results from a sustained ability to adapt to a changing resource base and an individual's or group's initial psycho-physiological condition (physiological fitness, mental health, including capacity for acclimatization and habituation)[2].

The relative position of any *"optimum" is not constant*, but shifts in the direction of the exposure[3,4].

There is a *limit to adaptation capacity* over time. Overexposure to atmospheric stimulus becomes hazardous: the resulting physiological and psychological impairment leads to reduced capacity in bodily and decision-making responses[5].

Involuntary exposure to atmospheric stimulus is particularly stressful[6].

Failure to adapt or exceeding adaptation energy results in individual and group *dysfunction* (ability to concentrate, vigilance, motor coordination and dexterity, moods, prison order, street riots, sexual aggression, domestic violence etc), morbidity and mortality[7,8].

Some human responses may over time prove to be *unsustainable maladaptation*.

Survivability is greater amongst the adaptable than the adapted[9,10].

Particularly relevant discussions can be found in: [1]Sargent (1963), [2]Medwar (1957), [3]Helson (1964), [4]Wohlwill (1974), [5]Selye (1957), [6]Starr (1969), examples of meteoroaversisms have been extensively reported in the International Journal of Biometeorology as well as specialist journals and outstanding medical reviews including [7]Petersen (1947) "Patient and Weather", and [8]Tromp (1963) "Medical Biometeorology"); [9]Dobzhansky (1962), [10]Sargent and Tromp (1964).

instantaneous or seasonal, deliberate or spontaneous. They also may reset new thresholds, such as perceptible cold or neutrality temperature. For example, with the advent of autumn, hearing an inclement weather forecast, a person may deliberately decide to wear a pullover, and perhaps in anticipation of the coming cold, light the fireplace. This voluntary action may lessen subsequent discomfort, or desire to carry out certain tasks, but at the same time initiate an involuntary alteration to seasonal acclimatization. The same complicated combination of processes may have been initiated simply by an unnoticed spontaneous increase in vasoconstriction without the benefit of the weather forecast, and the act of putting on of the pullover may have been quite involuntary. In other words, thermoregulation in an individual is a complex integrated biological and techno-cultural system that may be variously triggered at first, second or higher levels of impacts.

11 Human Adaptation within a Paradigm of Climatic Determinism and Change

Table 11.2 Atmospheric Hazards Impacts and Adaptations

Impacts in bold involve thermoregulatory processes, asterisks denote likely lowest impact order
Observed impacts of weather and climate hazards: 1 hot temperatures, 2 cold temperatures, 3 solar radiation, 4 wind, 5 rain, 6 snow, 7 moisture, 8 "weather", "climate". D / F Deterministic (involuntary) and / or Free-will (voluntary) responses

	hazard	D/F	impact or adaptation
i	1- 2	D	involuntary physiological adjustments[+]
ii	1- 5	D - F	postural adjustment[*]
iii	1- 6	D - F	seeking shelter[*]
iv	1- 7	D - F	subjective expressions of levels of discomfort[*]
v	1, 4- 5	D	increased nervousness, irritability, aggressiveness[*]
vi	1- 2, 7	D	reduced mental performance[*]
vii	2, 6, 8	F	avoidance of stimulus – hot baths, sauna, heated swimming pools[**]
viii	1, 3, 8	F	avoidance of stimulus – cold showers, swimming[**]
ix	1- 7	F - D	interposition of insulating clothing[**]
x	1- 8	F – D	construction of buildings[**]
xi	1- 2	D - F	acclimation (acclimatization / incidental habituation)[*]
xii	2- 8	F	space heating[**]
xiii	2	F - D	increasing metabolic/activity rate[*]
xiv	1, 3	F - D	decreasing metabolic/activity rate[*]
xv	1, 3, 7	D - F	consumption of cold food, drink[*]
xvi	2, 8	F - D	consumption of hot food, drink[*]
xvii	1, 7	F - D	air-conditioning, cooling[**]
xviii	1- 2, 8	F	deliberate fitness programs[***]
xix	1- 3	F - D	use drugs, alcohol[**]
xx	1, 7	F - D	antisocial behavior[**]
xxi	1, 4, 8	F - D	increased violence[*]
xxii	1- 6	F	metabolic alterations by rescheduling work/activities[***]
xxiii	1- 2, 6	D	elevated morbidity[*]
xxiv	1- 2	D	increased mortality[*]
xxv	1- 2, 8	F - D	temporary migration, holidaying[***]
xxvi	1- 2, 8	F - D	permanent avoidance of particular climate, emigration[***]
xxviii	1- 2	F - D	changed diet[***]
xxix	1- 8	F	climatic design, economic resource and infrastructure planning[***]
xxx	1- 8	F - D	cultural practices and patterns, philosophies[***]

[+] biological first order involuntary mechanisms- vasodilatation and vasoconstriction, perspiration, active sweating, thermogenesis – shivering
[*] biological "first order" largely involuntary responses,
[**] integrated first and second order responses and behavior
[***] third and higher order largely voluntary responses
excluded from list: extreme thermoregulatory behaviors (e.g. ice swimming) and obvious maladaptation (e.g. leaving open refrigerator door to cool house)

11.3 Human Biological Adaptation: The Thermal Coping Range

The most immediate of the cocooning layers are the bio-technological constructs as illustrated in Fig. 11.3. For the present purposes, those items that are boxed constitute minimal cost measures that are sustainable or contained within comfortable "coping ranges" (Burton et al. 1978). The aggregates in Fig. 11.3 show an evaluation of the temperature equivalents of comfort coping ranges for acclimated and healthy individuals below and above neutral temperatures: cool range 14–23°C, warm range 12–17°C.

The comfortable coping range as defined above is merely the tip of thermoregulatory response. Excluded from the comfort ranges, are the major emergency but exhaustible processes of sweating, active metabolic increases in shivering, and deliberately increased work rates (above the light sedentary category of <100 Wm^2 as assumed in comfort studies). Excluded also are active energy inputs into heating or cooling, deliberate relocation to less stressful locations, interposing

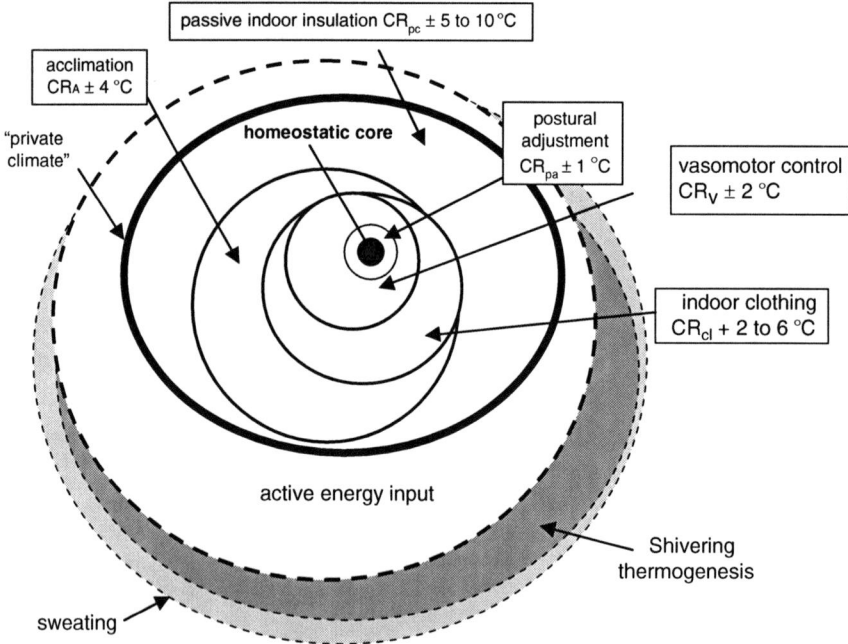

Σcool range $CR_{cl} \pm CR_V \pm CRA \pm CRpc \pm CRpa = 14\text{–}23$ °C
Σwarm range $CR\ V \pm CRA \pm CRpc \pm CRpa = 12\text{–}17$ °C

Fig. 11.3 Biotechnological Thermoregulatory Adaptation Comfortable Coping Rages (CR) (excluding sweating, shivering and active energy input)

temporary reflective material barriers, wearing specialized clothing. Such emergency measures have enabled humans to occupy and survive in all climate zones, and may do so again, but at this stage in development, are inconvenient to the modern urban dweller.

Here, for the moment we might reflect upon a highly deterministic theme that could be loosely labeled "urban metabolism". Thermoregulatory responses, and solutions to their demands, may be deeply, but not always obviously embedded in the fabric of urban infrastructure and cultural practices. Amongst these we might note that, depending upon the climate and sophistication in technology, 15–30% of national power consumption goes directly into air heating and cooling. Thus, at least a goodly proportion of power generation, transport and transmission of energy can be regarded as dedicated to the maintenance of homeothermy. The same and more can be claimed for house function, electricity grids, building materials, heating and cooling machinery, appliance shops and such unnoticed everyday phenomena of all-weather proof transport, and the construction of increasingly large buildings (cocoons) that ensure thermal comfort conditions irrespective of inclemency of weather beyond. Beside the hardware and obvious physical infrastructures are the vast thermal behavioural networks that may be represented by mundane everyday activities as in Table 11.2: hot and cold food provision, increased or decreased shopping frequencies, weather forecasts, determining temperature standards, organizing of weather related trips, timing of holidays, lingering in pleasantly shady or sheltered nooks… The use of the term "urban metabolism" in itself is indicative of the all pervasive role of homeostasis and thermoregulation in the lives of all human beings.

11.4 Deterministic Disaster: Exceeding the Coping Range

Elevated mortality rates during naturally occurring summer heat waves have received considerable attention since Ellis et al. (1975). A common observation in subsequent studies is that absence of seasonal and short term acclimatization is a major determinant of increased mortality (Kilbourne 1989; Kalkstein 1991, 1997; Kalkstein and Davis 1989; Kalkstein and Smoyer 1993; Smoyer 1998), and that especially age and poverty are significant contributory factors (Kilbourne 1989; Smoyer 1996).

The definition of heat wave and any other atmospheric hazard is therefore a matter of both the intensity of the parameter, which can be quantified by complex heat budget indices such as Klima-Michel (Jendritzky and Nübler 1981), STEBIDEX (de Freitas 1985) and PET (Höppe 1999) models, and population risk characteristics. As found by Karen Smoyer (1998), however, when related to mortality data, the best index utilizes adaptation parameters: her best predictors of heat deaths in St Louis were duration of heat waves and time sequence within the hot season.

During the decades around the turn of the twenty-first century, the global community has been struck by severe weather, including exceptional summer heat waves and pronounced death rate increase at the time. The European summer of especially 2003 was abnormally warm. In Paris maximum temperatures rose to above 35°C for 10 days until the 13th of August and for 4 days were above 38°C. Given that in urban areas heat islands develop, in the larger cities the levels of heat stress would have been more elevated in particular locations. For example, satellite imagery showed a systematic 4°C difference between parks, industrial and residential areas of Paris (Doussett 2007).

In much of Western Europe, emergency services, both fire departments and hospitals, showed abrupt increases in the number of interventions. Excess deaths for Europe in the first 2 weeks of August 2003 were around 35,000 (New Scientist 2003). Fouillet et al. (2006) examined the 15,000 causes of death in France using the International Classification of Diseases (ICD10) and an additional category for deaths reported to be directly related to the heat wave by the physician, i.e. dehydration, hyperthermia and heatstroke. The biggest cause of excess death was directly heat related: heatstroke, hyperthermia and dehydration + 3,306 deaths (a 20-fold overall increase in the number of deaths, and an even greater increase in certain age-group categories), circulatory system diseases + 3,004 deaths, ill-defined morbid conditions + 1,741 deaths, respiratory system diseases + 1,365 deaths and nervous system diseases + 1,001 deaths. By far the greatest increases in mortality were indeed in the older age groups. Fouillet et al. (2006) extrapolated " no segment of the population may be considered protected from the risks associated with heat waves".

Bouchama et al. (2007) analyzed details in reported heat stress cases both in Europe and USA as related to factors contributing to excess deaths as listed in web Medline, WHO and EU Centre for Environment and Health, and national disease control center databases for the period from January 1966 to March 2006. Analysis of 1,065 cases through "case-control" or "cohort studies" of odds ratios (ORs) established statistically significant associations, both beneficial and detrimental.

The beneficial were the mitigation by technological devices. The availability of working air conditioning ($P<0.01$) was the strongest protective factor, followed by access to an air conditioned place for some hours ($P<0.001$). There also seemed to be a trend that showed taking extra showers or baths and use of a fan during a heat wave reduced the risk of dying. Of benefit was also participation in social activities ($P<0.01$).

The detrimental associations were in the area of health. This included poor general health: being confined to bed ($P<0.001$), unable to adequately care for self ($P<0.001$) or to leave home daily ($P<0.001$), or having a preexisting cardiovascular ($P<0.01$), pulmonary ($P<0.001$), or psychiatric ($P<0.01$) condition. In addition in the USA, but not in France, living alone greatly increased the risk of dying ($P<0.001$) during a heat wave.

The excess 2003 French death increases are graphically presented in Fig. 11.4, which also shows average daily maximum (T_{max}) and minimum (T_{min}) temperatures, estimated thermal neutrality (T_ψ) and accumulated heat stress values (AHDD) for the period. The latter were calculated to allow for psycho-physiological acclimatization as estimated by T_ψ and for Smoyer's (1998) duration*time factors by

11 Human Adaptation within a Paradigm of Climatic Determinism and Change

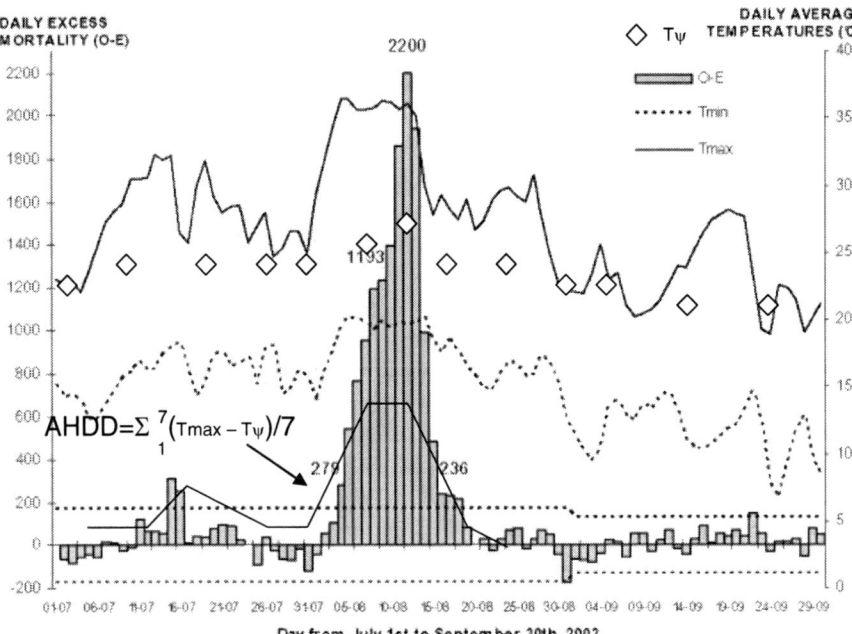

Fig. 11.4 Excess Deaths in France During the Heat Wave of 2003. (Based on Fouillet et al 2006) Average daily maximum (T_{max}) and minimum (T_{min}) temperatures recorded in Paris, and estimated accumulated daily heat stress in degrees calculated by summation of maximum temperatures in excess of thermal neutrality for the preceding two weeks, which becomes the base value for this accumulated heat stress "degree day" $AHDD = \Sigma_1^7 (T_{max} - T_\psi)/7$

summation of daily mean maximum temperatures in excess of variable thermal neutrality calculated for the preceding 2 weeks, which become the base values for this accumulated heat stress "degree day" method:

$$AHDD = \Sigma_1^7 (T_{max} - T_\psi)/7 \qquad (11.1)$$

Three weeks of July, August and September showed mean daily AHDD>4. Those week ending on 17/7 AHDD=8, 7/8 AHDD=14 and 14/8 AHDD=13, while the monthly accumulated totals for July were 134, August 254 and September 30. The correlation between excess deaths and AHDD values as shown in Fig. 11.4 is high, but perhaps surprisingly, these daily averages and $T_{max} - T_\psi$ seem to fall well within the estimated warm coping range of 12–17°C (Fig. 11.3).

This particular coping range, however, was estimated for healthy and relatively young people, while the excess European death rates showed victims to be the elderly, the ill and the less well off with decreased mobility. It is likely that the victims were poorly acclimatized, and their coping ranges would have been reduced by perhaps 4°C. Moreover, amongst them, many would have been homeless, or in poorly ventilated buildings that might not have been adequately cooled down during night time. Without such cooling down, and without the benefits of day time insulation, the CR would have been reduced by a further 5–10°C. In other words, it is probable that the coping rages of the victims were no more than the sum of $CR_V + CR_{pa}$ that is no more than 2–3°C.

On the one hand, the immediate cause of death was at the first order biological adaptation level due to reduced coping ranges, while on the other, the Bouchama et al. (2007) analysis points to life saving measures within second order technology, as controlled within third social systems level. The Fouillet et al. (2006) extrapolation that no segment of the population may be considered protected from the risks associated with heat waves may have been somewhat hasty: the victims were not those cocooned by the higher order adaptation mechanisms.

Previous American experience had found that mitigation of heat wave stress carried implications for policies of social intervention. In Chicago, elevated temperatures had led to some 700 excess death rates in 1995. In a press interview Eric Klinenberg (2002) is reported: "Yes, the weather was extreme. But the deep sources of the tragedy were the everyday disasters that the city tolerates, takes for granted, or has officially forgotten "In 1999, when Chicago experienced another severe heat wave, the city issued strongly worded warnings and press releases to the media, opened cooling centers and provided free bus transportation to them, phoned elderly residents, and sent police officers and city workers door-to-door to check up on seniors who lived alone. That aggressive response drastically reduced the death toll of the 1999 heat wave: 110 residents died, a fraction of the 1995 level but still catastrophic". The Chicago applications are in tune with the findings of Bouchama et al. (2007) who agreed that "the notion that withdrawing this distinct population at risk from heat, even for a short time, is the cornerstone of any public health response during a severe heat wave."

Clearly, as in the case of Chicago, simple and temporary intervention at higher levels could be successful, but further modeling might suggest more permanent solutions within issues of social equality. The tragedy of the European 2003 episode is exacerbated by the fact that effective models for proactive third order adaptations to heat waves were already available through appropriate warning forecasts as pioneered by Brezowsky (1960) and their translation by Larry Kalkstein (1997), as in the Philadelfia system, which was designed for intervention by the Department of Public Health through improving communications, and especially telephone hot-lines, between the public and agencies such as public utilities, aged care centers, and actions by proactive, reactive and buddy teams of professionals and volunteers.

As a postscript to the 2003 heat wave disaster, Fouillet et al. (2008) have analyzed the subsequent 2006 episode in France. No data are given on social conditions, but in terms of the earlier episode, death increases are a half of those expected. Fouillet et al. (2008) conclude that this can be "interpreted as a decrease in the population's vulnerability to heat, together with, since 2003, increased awareness of the risk related to extreme temperatures, preventive measures and the set-up of the warning system". Leaving aside the question whether that particular French population has actually become less vulnerable and more aware, or had been culled in size by the 2003 episode, heat wave warning systems have been now either activated or proposed for a large number of urban areas, including all 17 major cities in France (Kovats and Ebi 2006). Perhaps the most encouraging development is that warnings can be based simply on average daily temperatures (Nicholls et al. 2008), which can be reliably forecast for most locations 3–5 and more days in advance.

Finally, in the light of such observations, should the lens of modern thermoregulatory adaptation be enlarged back again and applied to tropical settlement, what were once perceived as negative characteristics within colonists and indigenous peoples, may now be interpreted as appropriate adaptive responses within given conditions. Tropical locations, as for example in lowland India, the Philippines, and Queensland Australia, where winter temperatures are often warmer than those in Britain in summer, were not benevolent to Europeans determined to maintain the lifestyles, schedules, fashions and customs evolved within much cooler climates (Auliciems and Deaves 1988). Woodruff and Huntington were not mistaken in observing that the effects of the elevation of work metabolism, whilst wearing heavy Victorian clothing either outdoors or in poorly designed uninsulated and unshaded buildings of that period, could indeed do no other than promote severe discomfort, improper behavior or ill health. As early on recognized by Raphael Cilento (1925); Thomas Griffith Taylor (1959) and Sargent (1963), successful settlement requires profound adaptive changes in custom and technology.

11.5 Anthropogenic Climate Change: A Consensus Paradigm

The 1960s and 1970s of the last century saw remarkable advances in space exploration and computer technology that vastly facilitated acquisition of data and capacity for numerical analysis. In the new electronic age, multidisciplinary developments flourished through a leap in the capacity for information transfer between environmentally concerned individuals, national and international integrative research centers, philanthropic and conservation groups, conservation clubs, NGO's, and a variety of lobby groups. New concepts and environmental consciousness were boosted by highly organized and prestigious WMO, UNEP and United Nations sponsored programs and conferences, and coordinated leadership, as provided for example by the International Council of Scientific Unions (ICSU) and its SCOPE Scientific Committee on Problems of the Environment. Particularly important also became publication series devoted to natural hazards and climate change issues that have culminated in the Assessment Reports of the Intergovernmental Panel on Climate Change (IPCC). These tumultuous new developments had also bypassed many of the abstractions about determinism and free-will.

That climate change *per se* could be dominant in the survival of groups of people and civilizations themselves is a relatively new realization. First recognition of climate change at the decadal scale, according to Stehr et al. (1996), can be ascribed to Eduard Brückner (1890), who is also credited with being one of the first to recognize the potential human role in climate change. A possible "Greenhouse" mechanism of carbon dioxide for this hypothesis was proposed by Arrhenius (1896). Subsequently ground level thermometric observations have provided quantitative proof of short term climate changes, or temperature trends at least (Jones et al. 1986a, b), The UNEP/WMO/ICSU conference at Villach/Vienna (Austria) in 1985, chaired by the author of the carbon cycle model Bert Bolin (Bolin et al. 1986), provided modern day legitimacy to these hypotheses.

This pivotal Villach meeting seems to have been more political than scientific: there is more concern with generation of hypothetical scenarios of climate change (in terms of increased temperatures, shifts in storm strengths and tracks, heights of clouds, elevation of the tropopause and rising sea levels) than establishing irrefutable scientific validity of recent observations. Green light was given to further modeling of radiative gas generation and effects, other meetings and to new programs, most notably to the establishment of IPCC in 1988.

According to its public announcements, IPCC is a scientific intergovernmental body set up by the World Meteorological Organization (WMO) and by the United Nations Environment Programme (UNEP). Its mandate is to provide an objective source of information about climate change. Its role is to assess the latest scientific, technical and socio-economic literature of the risk of human-induced climate change, to report its observed and projected impacts, and to put forward options for adaptation and mitigation. IPCC conducts plenary Sessions where main decisions are made, its program and reports are accepted, adopted and approved. The responsibility of IPCC is to coordinate climate change research across the globe, and hundreds of scientists contribute as authors, contributors and reviewers.

In this vast enterprise, there are three main working groups: *Working Group 1 The Physical Science Basis*, *Working Group II Impacts Vulnerability Adaptation, and Working Group 3 Mitigation of Climate Change*. Each of these provides Assessment Reports which according to IPCC are variously made available to decision makers, scientists and the public, which "immediately become standard works of reference, widely used by policymakers, experts and students." The first Assessment Report of 1990 was decisive in developing the 1997 Kyoto Protocol – the United Nations Framework Convention on Climate Change (UNFCCC), which became the overall policy framework for addressing the climate change issue.

If the aftermath of the Villach meeting had seen a consolidation of the notion of anthropogenic climate warming, there also has been a consolidation of dissent both with the rationale of Greenhouse economics (Lomborg 2004) and the simpler interpretations of solar variability, oceanic circulation and climate process controls (e.g. Soon and Baliunas 2003; Gagosian 2003; Svensmark and Calder 2007; Robinson et al. 2007). The scientific backlash to aggressive IPCC pronouncements has also found expression in web pages *Accuweather, Climate Science, Copenhagen consensus, CSCCC* and in disturbing reports of a lack of coherence within IPCC itself (e.g. von Storch and Stehr 2005; Christy 2007; Harris and McLean 2007; Haag 2007).

However, without reviving the argument about scientific responsibility, integrity and methodology here, it can be legitimately claimed that political and social awareness of global climate change has reached unprecedented heights. Financial and institutional support to global change related research funding is on par. The reputation of IPCC itself has probably peaked, but with some caution that avoids overzealous claims of scientific objectivity and suppression of legitimate scientific discourse (see Crook 2007), the organization is likely to retain political support and prestige for the next decade at least. The IPCC view remains that there is accelerated global climate change which can be modeled, and the observed average 0.74°C/century temperature rises by virtue of "consensus" to have been proven as anthropogenic. The IPCC (2007) Fourth Assessment Special Report on Emissions Scenarios (SRES) advises that global temperature, depending upon emissions scenario, best estimate rises will by the end of the century be within a range of 0.6–4.0°C.

The consensus approach and IPCC itself have been massively backed by WMO, UN, national weather services, and sundry committed groups such as *Union of Concerned Scientists* (ucsusa 2007). Its publicity and undeniable contribution to the coordination of scientific climate change studies earned the 2007 Nobel Peace Prize. The Committee's award citation to IPCC was "for their efforts to build up and disseminate greater knowledge about man-made climate change, and to lay the foundations for the measures that are needed to counteract such change." Terjung's (1970) and Hare's (1985) urging for an anthropocentric focus within climate studies seems to be fulfilled, and in words, and perhaps in deeds, the Marxist Feuerbach thesis appears to have been vindicated. Some scientists have switched their primary allegiance from the ideal of objective hypothesis testing, to commitment for betterment of the environment. Not surprisingly, to an environmentally concerned public, the public media, governmental agencies, and scientific institutions dependent upon grants from mission oriented funds, "Greenhouse" warming has become an acknowledged and almost tangible reality. Within a human generation, highly speculative models and hypotheses have been elevated by consensus to theory status. In Kuhn's (1970) terms, a paradigm shift has taken place.

Within this paradigm shift, humans are seen both as causal agents of large scale environmental change, and as the potential victims of this change. On the one hand we can see a triumph of human activity over nature, on the other, possible climate induced human demise. In either case, it seems that we may have more choice, but only limited freedom to curtail radiative gas emissions, intervene in some sensitive regulator within the solar cascade, or adapt to climatic changes. By comparison to the earlier and simpler constant climate paradigm, the main difference may not be one of proportions of free-will and determinism, but one of increase in the swing amplitudes of several free-will/determinism pendulums.

11.6 Integration and Adaptation Definition

To return to the European heat wave narrative, continental European summers and the Mediterranean winters are expected to become considerably warmer: heat waves are "very likely" at the <90% level of confidence to increase in frequency and severity (IPCC 2007). Medical researcher Tony McMichael (2007) is reported "I welcome the growing emphasis on the modeling of adaptive strategies to lessen the health risks. Recent evidence from extreme heat waves, cyclones and droughts has shown how widely the health impacts vary between old and young, rich and poor, and those with strong versus weak social institutions and supports. Human societies and communities are very varied in resources, culture and behavior. Stronger linkages between natural and social sciences are now essential if we are to develop realistic integrated models that connect projected climatic change, social change and human vulnerability."

The issue of a globally integrative and numerically predictive model linking physical and human systems is likely to remain unresolved (Zillman 2004). At a more modest scale, however, within the physical domain the spatial resolution of scenario estimates by atmosphere-ocean coupled global circulation models (AOGCMs) have achieved good several hundred kilometer resolution, and while computer modelers are understandably cautious about scenario downscaling, coupling at the regional scale to topographic and terrain data is already producing significant improvement. There can be little doubt that this capacity will greatly increase with further advances in data acquisition and processing. This potential for spatial and temporal resolution is already ahead of our understanding of global atmospheric processes themselves, and certainly those of human systems at the regional scale. Discussion of integration of the several systems, however, seems to require some change in definition, if not adjustment to philosophical construct.

Firstly, adaptation is defined broadly by IPCC (2001) as "adjustment in natural or human systems in response to actual or expected climatic stimuli or their effects which moderates harm or exploits beneficial opportunities". Despite a vast amount of time spent in IPCC discussions on definitions, the present still fails to make adequate provision for those situations where a seemingly successful adjustment becomes a detrimental maladaptation only over time. This may include thermoregulatory functions such as active energy usage and air conditioning, which ultimately may not prove to "moderate harm" or be "beneficial" (Auliciems 1989; Auliciems and de Dear 1998; Auliciems and Szokolay 1998).

Of the six IPCC offered category definitions of adaptation as "anticipatory", "planned", "private", "public", and "reactive" are inapplicable to thermoregulation as a first order process. "Autonomous" adaptation is referred to as "spontaneous", "invariably reactive" and "adaptation that does not constitute a conscious response to climatic stimuli but is triggered by ecological changes in natural systems and by market or welfare changes in human systems". This definition fails to include deterministic biological processes as in Tables 11.1 and 11.2 and in Fig. 11.3, and is at odds to the thermoregulatory system's voluntary control mechanism, the thermopreferendum mobility (Fig. 11.1), and even the short term vasomotor processes.

Clearly the universality of homeostasis, and the ubiquitous nature of thermoregulation, does not sit well with the soft determinism semantics or concepts of either mainstream IPCC (2001) or its progenitor the Chicago school of Natural Hazards (as in Burton et al.1978). Marginalization of the significance of active biological impact – adaptation processes, can be seen in the seminal contribution by Bob Kates (1985) to SCOPE 27. Here links between orders of interaction are shown, in a much used model from Ingram et al. (1981), to flow from hazards to "direct" first order impacts (crops/plants, micro-organisms and animals), which according to Wigley et al. (1985), "may have economic or social (second-order) significance, affecting food and raw material supplies derived from agriculture or animal husbandry", which "may have an effect on biophysical processes important to human health" which is shown as a second order entity (together with food supplies, transport/ communications and wind/water power), to culminate in this model with wider economy and society. That is, as underlined by the apology from Kates (1985) for even suggesting deterministic overtones, there is no first order hazard impacts upon humans.

This surprising anachronism probably stems from the determinist-free will schism within Geography and from Natural Hazards work which had focused on differentiating cultural attitudes and perceptions. According to Burton et al. (1978) "the process of biological adaptation is generally slow; it cannot play a significant role in the short term responses to natural hazards" (p. 36), but "biological adaptations may also involve numerous mechanisms for temporary physiological responses in the face of hazards" (p. 39). Seemingly over time human biological processes have atrophied to insignificance, although in a subsequent section on coping, both cultural and biological processes seem to win some reprieve as parts of "loss absorption", but outside the main purposeful sets of adaptive processes. In any case, it appears that a society, and presumably its individual members, actually may be unaware of this hazard absorption. Overall the suggestion seems to be that humans may be considered biologically as less sensitive or adaptable than animals, and not be subject to first order deterministic impacts, and that thermoregulatory adaptation is little more than "routine" (Carter 1996) or "incidental" (Burton et al. 1978).

In part, the problem with the above underestimation of the role of dynamic biological processes may also relate to a critical period of computerization within weather forecasting. Natural Hazards concepts and semantics were formulated at a time when weather forecasts were for tomorrow, and the onset of most hazards was by definition sudden, i.e. less predictable, and therefore impacts were more severe. Nowadays, with advances in satellite technologies, numerical forecast skills and electronic communication networks, warnings of the likely onset of hazards is considerably enhanced. And anticipation and preparedness for an event reinforces the more recent concept of psycho-physiological adaptations within integrated first-second order impacts.

At the same time, that human adaptation and homeothermy is at the heart of biometeorology is implicitly accepted in health studies. In essence, the appeal of McMichael for the establishment of "stronger linkages between natural and social science" is also a call for rationalization in semantics and a bridging between artificial distinctions of first, second and third order adaptations. An integrative thermoregulatory system seems to provide a suitable template to fill this gap. Quite appropriately, the proposed system spans the extremes of each of the semantic dichotomies characterizing and differentiating adaptations to climate change (e.g. Smit et al. 1999: Autonomous – Planned, Passive – Active, Proactive – Reactive, Instantaneous–Cumulative).

Being mindful of the concern of Glantz (2007) that critical terms including adaptation can be misleading, for the present purposes adaptation is regarded no more than "adjustment in natural or human systems in response to actual or expected climatic stimuli". Adopting this simplified definition also avoids the use of the term autonomous altogether, in favor of treating the set of human homeothermic adaptations as an *Integrated Adaptive System*.

The significance of a need for redefinition can be underlined by overall changes in the patterns of death rates from extreme climate events. US death percentages (CSCCC 2007) lean increasingly away from major disaster events towards thermal or first order extremes. In percentages terms for the period 1992–2002, thermoregulatory deaths accounted for 82% of the total 1275/annum: extreme cold 53%, heat 28%, flood 9%, lightning 5%, tornado/hurricane 5%. Since reductions in death rates is largely a matter of economic development and research and investment, a more sophisticated understanding of first order thermoregulatory processes may have the potential for relatively large net returns.

Irrespective of realities in emission rates, future temperature trends and meteorological risks, at the smaller scale, human vulnerability and adaptability to specific impacts already can be realistically modeled on basis of established biometeorological relationships in most third order human systems. Undoubtedly, there also exist untested public and private records on every aspect of human life during the time of specific events, as well as routinely monitored data on environmental conditions, workforce and infrastructure anomalies or at least deviations from their norms. Initially accessing such records and organizing them in suitable database format no doubt would be a major task, but the information that could be rapidly obtained by suitable interrogation procedures, might vastly enhance human system inputs into models to match those of the physical world, and especially so in emergencies.

To illustrate, heat wave experience indicates that excess mortalities are caused by first order impacts and first order biological failures, while prevention is achieved by second order technologies as implemented at voluntary third order levels. There can be little doubt that the observed impacts of the 2003 European heat wave episode did precipitate impact and adaptation cascades well beyond the elevated mortality and shortfalls within hospital systems. Implications would have been considered throughout other third order human systems, including those of appropriate adjustments within social services, education, information, transport, emergency capacities, future urban planning and so on. The tragedy of 2003 had been of a sufficient magnitude to also alert French politicians, but their immediate suggested solution, however had been Woodruffean. Online encyclopedia Wikipedia reports that the administration of President Chirac and Prime Minister Raffarin laid the blame on the 35-h workweek, which affected the amount of time doctors could work, and in any case family practitioners took their vacations in August. Here we might be tempted to ask for a deeper investigation of the politician and administrator perceptions and attitudes towards optimizing solutions other than the hint of simply transferring some of the risk from one population to another.

Most of the higher order systems, in responding to events such as elevated heat stress, may have shared the initial first order trigger mechanism, but probably not coping ranges of particular groups of people or even the operation of second order technological infrastructures. Integrating information from these and other techno-cultural, spatial and socioeconomic systems and matching them to mortality data and medical service responses may have produced even more valuable insights into causes and solutions. Such projects as Monitoring and Measurement in the Next Generation Technologies (MOMENT 2008) may well resolve the vexing issue of accessing suitable and at times sensitive third order data, with fulfillment of their stated aims of "integrating different platforms for network monitoring and measurement to develop a common and open pan-European infrastructure. The system will include both passive and active monitoring and measurement techniques via a common web services interface and ontology that allows semantic queries."

11.7 Towards an Integrated Adaptive Model

The envisaged models below are based on Figs. 11.1 and 11.3, the summaries in Tables 11.1 and 11.2, and discussions of the "homeothermic imperative". As discussed in Auliciems (1981, 1983), and Auliciems and de Dear (1997), cognitive-affective-effective control is a main interface between the biological and technological response, and one that should be central to the field of natural hazards risk-management research as defined in Burton et al. (1978), and Whyte (1985), and to coping with natural disasters (Alexander 1993). Indeed, given the earlier listed impacts and adaptations in Table 11.2, and the urban metabolism analogy, it would be surprising if at least temperatures or their thermal equivalents did not feature prominently in most third order systems within the human domain, in addition to the specific attributes of the hazard under consideration. Summaries of the three orders are as follows.

First order adaptation is the initial reduction of hazard impacts (upon specified and definable core parameter/s). Human first order defence mechanisms consist of integrated biological and behavioral adaptive processes as augmented by second order technological mechanisms. These processes may be a mixture of predetermined or probabilistic elements which may have been modified by experiences and decisions as based on perceptions and choices that had resulted from earlier biological and/or techno-cultural adaptations.

The second order is one within which impacts and adaptations are behavioral and technocultural, with responses ranging from routine maintenance of infrastructures, to emergency activation of resources and technologies. Its control may be corporate within the third impact level or the integrated first-second order system itself. The larger adaptation costs are usually transferred to third order socioeconomic systems, and as is the case with atmospheric pollution, to the environment and the future.

The third order of adaptations includes all processes above the integrated first-second order impacts-adaptations, and depending upon the nature and scope of a particular investigation can be conceptualized as a single level or to multiple nth level structures. This level contains socioeconomic systems that variously cope with functions of security, health, education, environmental, amenity and resource availability etc. Depending upon the process or event being investigated, the third order is least deterministic, and presents the widest choice in methodology. Irrespective of the problem, scope or particular methods used in analysis, there needs be a core construct within the decision making module/s.

The form of this construct will depend upon specific requirements or methods used, and may consist of objective criteria, aims, goals or mission statements.

In many cases, the research would be concerned with adaptations and mitigation to some dimension of climate change, especially the assessment of vulnerability of particular populations and environments. In many areas literature search would reveal that core constructs can be expressed as interactions in terms of correlation and regression coefficients, probability and confidence levels, percentages occurring to non occurring, observed to expected ratios. These would permit the definition of functions and interfaces by criteria such as "critical thresholds", "coping ranges", "optima" or "adaptation deficits".

A possible conceptualization of the three order adaptation processes is shown in the Euler – Venn type of representation (Ruskey and Weston 2005) in Fig. 11.5,

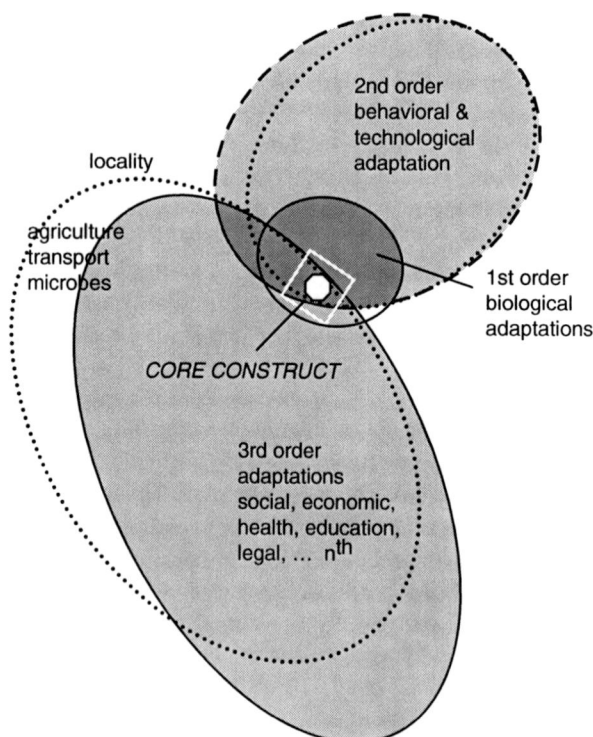

Fig. 11.5 A Conceptual Model of Integrated Human Adaptation to Climate Hazards and Change

which is essentially an extension of coping range model in Fig. 3. Ideally this version of the model should be visualized as three and more dimensional Borromean rings anchored in place by a lynch pin – the homeostatic core, and its ever present control of Fig. 1. As in Fig. 3, the core is central and cocooned from the outside world by all level, deterministic and less deterministic adaptations. The inner core and its intelligent control mechanism, is at the centre of all the webs. If valid, such conceptualization must come close to an integration of human adaptive systems.

Whatever the cascading manifestation, the most immediate concern is the impact, or potential threat, to "life and limb", that is the essential homeostatic core. Its defences at the higher orders are coordinated through the corporate decision making control module shown in Fig. 11.6, which in effect is a reinterpretation of the adaptive model as depicted in Fig. 11.1, with the core construct replacing the thermopreferendum entity. Its framework should apply to any hazard or level of impact and adaptation, including its use for linkage to very different systems (in this example referring to Diamond's neighbour and environment concepts).

An alterative representation to Fig. 11.5 is that in Fig. 11.7. Here, the model is conceived as three interrelated adaptation entities through which climate hazard impacts cascade. The three entities are first order adaptations, in which is embedded

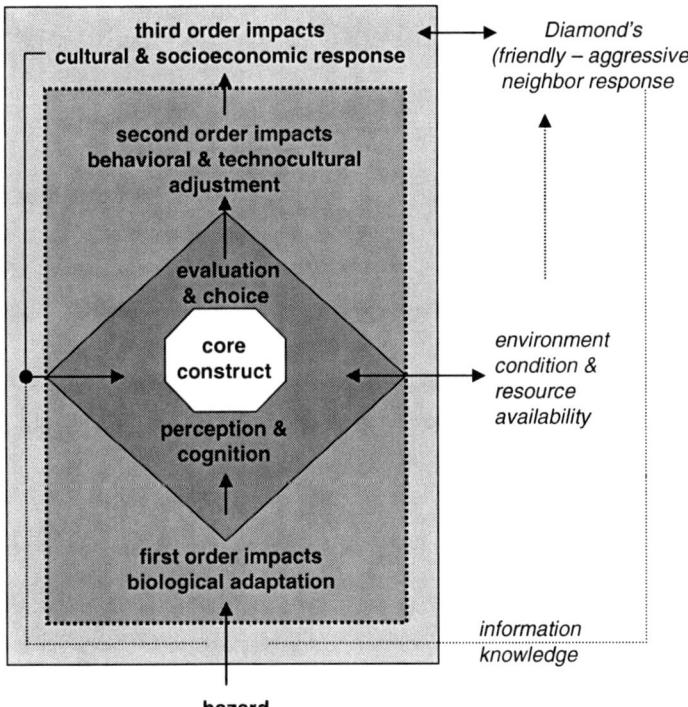

Fig. 11.6 Generalized Format of the Control Mechanism for a Model of Integrated Human Adaptation to Climate Hazard

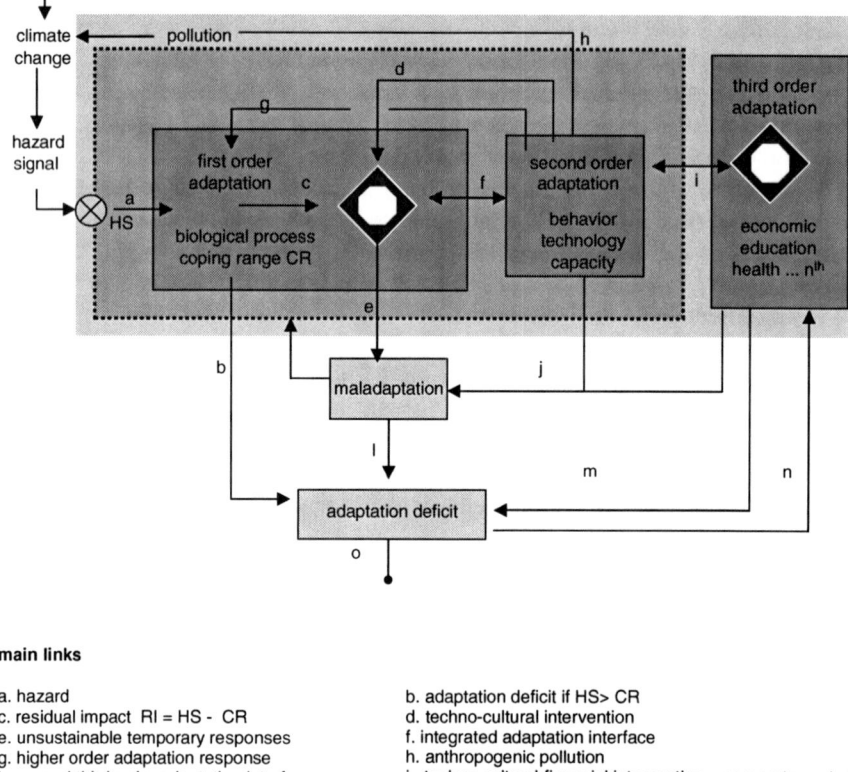

Fig. 11.7 Integrated Human Adaptation to Climate Hazard

the homeostatic core and its attendant control, its cocooning second order techno-cultural adaptations, and third order adaptations that contain the corporate decision making module, as generalized in Fig. 11.6. Depending upon the impact cascade, all components in Fig. 11.7, irrespective of the order, can be at risk. Decision making can be purposeful or *ad hoc*, coordinated or not, and both first and third order decisions can draw upon the resources of the second order. The incidental shelter, which represents a naturally occurring, but essential defensive mechanism which is employed either spontaneously or by considered selection.

Although mostly responding as an integrated whole, adjustments to the three impact levels shown in Fig. 11.7 could be loosely described as synoptic, seasonal scale and deterministic in the first, deliberate "management" of the physical resource and human technological and strongly probabilistic response systems in the second, and decadal and secular term ranging anywhere from highly deterministic to probabilistic, including climate related lifestyles and social structures within the third.

A simple illustration of the model in Fig. 11.7 may be for assessing the efficacy of proposed heat wave management by provision of public shelters with air cooling facilities. A late summer heat wave hazard (link 'a' – e.g. specify hazard, intensity, duration, thresholds), results in first order impacts (e.g. estimate environmental stress values, identify population at risk, specify vulnerabilities and biological thresholds, determine acclimatization levels and rates of change, estimate coping ranges, expected morbidity and death rates). Information on impacts, absorbed and residual hazard, need for altered behavioural and technological support – emergency services is forwarded (via 'c') to integrated first and second order control and (via 'i') to higher order societal decision making, which responds according to available facilities, alternatives, age needs and medical capabilities and predetermined or ad hoc procedures via 'I', 'f' and 'g'. In the meantime, information on adaptation deficits (morbidity and mortality) has been passed on via 'b' and further advanced for action at the higher levels of response via 'n'. Third order processes (action or deferred response) are initiated via 'm'. This also promotes a greater demand (feedback 'i') for more technological support (space cooling) and thus for fossil burning (second order impact), which in turn, however, may decrease the availability of funds for alternative purposes (and technologies), and over time tend to lower the overall well-being of the whole society (third order impacts via link 'j' to the maladaptation entity, and 'm' to adaptation deficit). In the short term, at least, the second order technological and behavioral adjustments enable increased space cooling and thereby reductions in heat stress impacts (feedback 'g'), but there has been an increased generation of radiative gasses ('h'), which in the long term may contribute to global warming, and also ultimately to warmer summers that may actually tend to increase the heat stimulus (link 'a').

The "adaptation deficit" entity in Fig. 11.7 was suggested by Ian Burton's (2004) concept of a simple tool for estimating investments required to reduce adverse climate impacts. This is a useful concept also to estimate the shortfall in adaptation capacity in individuals and groups. In the present model, the concept would require four estimates:

1. The existing stress demand or residual impact RI (hazard HS via 'a' minus coping range CR of biological adaptations i.e. signal via 'c')
2. Adjustment achieved by techno-cultural adaptations, consisting of (2_i) sustainable adaptations via 'f' and (2_{ii}) adaptations which are only temporarily effective and over time become counterproductive, maladaptive or unsustainable adjustments, or stress demand minus sustainable adaptations via 'e'
3. The potential for future sustainable adaptation for specific hazards and climate trends and/or change, or estimated augmentation in RI ('g' and 'd')

4. The remaining amount of the unadaptable residual transferred to entity adaptation deficit via 'b', or new RI minus sustainable adaptation

Pragmatically (4) and (2_{ii}) above would become the potentially unsupportable and predetermined maladaptation "lost cause" cases. Type (4) also allows recognition of the law of diminishing returns–there never was or will be a perfectly adapted society. For funding purposes, such as providing for mitigation of future heat wave impacts, however, (2_{ii}) may provide a useful criteria for action or otherwise for economically developed localities and (3) for less developed. Optimal investment decisions would probably provide for (2_i)+(3). The (2_{ii}) category would include poor design (urban morphologies etc.), energy inefficiencies and waste (lack of insulation, misuse of air cooling technologies etc.), lack of provision for enabling physiological adaptations.

The notion of maladaptation is simple enough. An entity within any of the three adaptation levels may with time prove to cause more problems than those it resolves. Its assessment as such becomes a matter of deliberate decision making and in turn the maladaptive nature of the entity may be rectified or not, depending upon control decisions for initiating rectification within higher order processes, depending upon its adaptation deficit classification. Leaving aside issues relating to the needs of special populations such as those in hospitals or the aged, air conditioners are an obvious example. Air cooling has provided comfortable conditions, saved lives and enabled a continuation of cooler-zone behavioural patterns within tropical environments. However, given that this technology has proliferated also within temperate climatic zones, more often than not, as panacea for otherwise poor building design or as a convenient enticement to shoppers. Here, in terms of degradation of the environment, and in loss of acclimatization, the costs are far in excess to benefits of non-essential luxury both at the local and global scales.

11.8 Conclusion: Philosophies, Responsibilities and Adaptability

Clearly human-atmosphere interrelationships are not simple, and both the well being of people and condition of the indoor and outdoor atmospheric environment, are parts of a multifaceted system that need to be viewed (a) holistically, (b) in terms of probabilities and (c) specific time horizons for (d) particular locations. Over-deterministic expressions of human responses, fail to recognize the power of human perceptions, and their adaptability as the basis for sustainable choices and rational control decisions. At the same time, it is essential that we recognize the imperatives of our biological human legacy.

At any stage of development, human ability for maximum energy generation or production of food, even if in a seemingly sustainable way, does not necessarily indicate an optimum condition for survivability. The possible resulting build-up of populations, or even of expectation of resource availability, may with change create instability in the "fitness" of individuals or groups to survive. That is, what may be regarded as successful adaptation at a particular time and location may prove to be unsustainable maladaptation in the longer term. Paradoxically, the currently prevailing global warming paradigm is an attestation of faith in human free will. If climate change is anthropogenic, then humans can claim superiority over the forces of Gaia. If by deliberate reductions of radiative gases we can halt our own runaway climate change juggernaut, our superiority will have been triumphant once more. The rub is that we cannot be certain that our consensus belief in a dominant anthropocentric paradigm, and our abilities to adapt, may be no more than a placebo that eventually proves to be an expensive maladaptation.

The suggested integrative model illustrates the complexity of the adaptation system at different levels of human organization. Nowadays, there can be no definition of climatic determinism in simple terms of causality. As before, there remain deterministic nodes, especially within the first level, and indeterministic ones, especially at the third level of decision-making and management. Here, most relationships can be regarded as probabilistic, and ones that are particularly sensitive to cultural preference, resource availability and socio-economic capacity. At the largest temporal and spatial scales, beyond the medium range time horizon, the prediction of the human condition and adaptation measures become increasingly less reliable.

Deliberations on determinism and predictability versus free-will become even more complicated within the new consensus paradigm that also carries implications of active involvement by the scientist in adaptation as dependent upon political, social and economic decision making. The notion of activist "concerned scientists" is at least a superficially attractive response, but extending scientific advice to include lobbying for intervention with major climate controls (see Kerr 2007) seems to carry responsibilities beyond the scientific mandate. We may require yet again another re-examination of what constitutes validity and reliability within the present issue.

Stepping back from such largely semantic issues, no responsible scientist will pretend to know beyond some testable probability that potentially dominant and naturally occurring hazard singularities within solar emissions or tectonic events will not occur. In any case, the present-day era of Holocene warmth appears to have reached the usually expected twelfth to thirteenth century long span: there is no method available, beyond trend extrapolation, to forecast future solar developments within the historical time frame. The Milankovitch (1941) solar radiation cycles will sooner or later lead to global cooling.

There is little certainty in the prediction of natural climate change, its impacts and long term human responses *per se*. No matter what the personal conviction of the causes of global warming, advocacy of single purpose adaptation to a warmer world cannot be supported. Scientific responsibility is not a matter of loyalty to a consensus paradigm, but to objectivity that points towards uncertainty and the possibility of climate change trends either towards net warming or cooling. Recommendations for survival strategies should give preference to flexible measures that encourage adaptability to change, rather than adaptation specifically to a warmer world. The success or otherwise of present-day philosophies will only be tested with time, but survivability will be enhanced in the adaptable, not the adapted.

References

Accuweather http://www.accuweather.com/global-warming/index.asp.
Alexander D (1993) Natural Disasters. UCL Press, London.
Arrhenius S (1896) On the Influence of Carbonic Acid in the Air upon Temperature of the Ground. Philosophical Magazine and Journal of Science, 41 237–276 Fifth Series, London, Edinburgh and Dublin.
ASHRAE American Society of Heating, Refrigeration and Air – Conditioning Engineers (ASHRAE),
Auliciems A (1969) Some group differences in thermal comfort. Heat. Vent. Engr 42, 562–564.
Auliciems A (1972) The Atmospheric Environment. A study of comfort and performance. University of Toronto Department of Geography Research Publications, University of Toronto Press 166 pp.
Auliciems A (1981) Towards a psycho-physiological model of thermal perception. International Journal of Biometeorology 25, 109–122.
Auliciems A (1983) Psycho-physiological criteria for global zones of building design (pp. 69–86), in Overdieck O, Muller J, Lieth H eds. Proceedings of the 9th International Society of Biometeorology Conference (pp. 69–86), Part 2, Biometeorology 8, Swets and Zetlinger, Lisse.
Auliciems A (1989) Human Dimensions of Air-Conditioning. Chp. 5 (pp. 73–87), in Ruck NC ed. Building Design, Human Performance. van Nostrand, Reinhold, NY.
Auliciems A and Deaves S (1988) Clothing, heat and settlement. Ch 8 in Heathcote RL ed. The Australian Experience. Essays in Australian Land Settlement and Resource Management, Longman Cheshire, Melbourne.
Auliciems A and de Dear RJ (1997) Thermal Adaptation and variable indoor climate control (pp. 61–86), in Auliciems A ed. Bioclimatology 5. Advances in Human Bioclimatology, Springer, Heidelberg.
Auliciems A and Szokolay SV (1998) Thermal Comfort. PLEA Notes 3, Brisbane.
Bedford T (1936) The Warmth Factor in comfort at work. Rep. ind. Hlth Res Bd no 76. London.
Bell B (1971). The dark ages in ancient history, 1: The first dark age in Egypt. American Journal of Archaeology 75, 1–26.
Berglund L (1979) Thermal acceptability. ASHRAE Transactions 85, 825–834.
Bolin B, Doos BR, Jager J and Warrick RA eds. (1986) The Greenhouse Effect, Climatic Change, and Ecosystems, SCOPE 29, Wiley, New York.
Bouchama A, Dehbi M, Mohamed G, Matthies F, Shoukri M and Menne B (2007) Prognostic Factors in Heat Wave–Related Deaths. A Meta-analysis. Archive of Internal Medicine 167, 2170–2176.
Brezowsky H (1960) Über die pathogene Belastung durch Wettervorgänge. Med. Klin. 50, 2235.
Brückner E (1890) Klimaschwankungen seit 1700 Nebst Bemerkungen über die Klimaschwankungen der Diluvialzeit. Wien und Olmütz: Hölzel.
Bryson RA and Murray TJ (1977) Climates of Hunger. Mankind and the World's Changing Climate. Wisconsin University Press, London.
Burton I (2004) Climate Change and the Adaptation Deficit. Adaptation and Impacts Research Group. Meteorological Service of Canada, Environment Canada, Occasional Paper 1.
Burton AC and Edholm OG (1955) Man in Cold Environment, Edwards Arnold, London.
Burton I, Kates RW and White GF (1978) The Environment as Hazard. Oxford University Press, New York.
Carpenter E (1968) Discontinuity in Greek Civilization. Norton, New York.
Carter TR (1996) Assessing climate change adaptations in The IPCC Guidelines p 42 in Adapting to Climate Change: An International Perspective, Smith JB, Bhatti N, Menzhulin GV, Benioff R, Campos M, Jallow B, Rijsberman F, Budyko MI, and Dixon RK eds. Springer, New York.
Chappell JE (1970). Climatic change reconsidered: Another look at 'The Pulse of Asia'. Geographical Review, 60, 347–373.
Christy JR (2007) My Nobel Moment. Wall Street Journal November 1.

Cilento R (1925) The White Man in the Tropics. With Especial Reference to Australia and its. Dependencies. Government Printer: Melbourne, Climate Science, http://climatesci.colorado.edu.

Copenhagen Consensus (2008) http://www.copenhagenconsensus.com/Default.

Cosmides L and Tooby J (1997) Evolutionary Psychology: A Primer www.psych.ucsb.edu/research/cep/primer.html

Crook C (2007) The steamrollers of climate science. Financial Times FT.com August 1.

CSCCC (2007) Civil Society Report on Climate Change. http://www.csccc.info/

Darwin C (1859) On The Origin of Species by Means of Natural Selection, or The Preservation of Favoured Races in the Struggle for Life. First Edition www.literature.org/authors/darwin-charles/the-origin-of-species/

de Dear R and Brager G (2001) The adaptive model of thermal comfort and energy conservation in the built environment. International Journal of Biometeorology, 41, 100–8.

de Dear RJ, Brager G and Cooper D (1997) Developing an adaptive model of thermal comfort and preference. Final Report ASHRAE RP–884, March.

de Freitas CR (1985) Assessment of human bioclimate based on thermal response, International Journal of Biometeorology 29, 97–120.

Diamond J (2005) Collapse. How Societies Choose to Fail or Succeed. Penguin Books, New York.

Dousset B (2007) Satellite thermography of urban temperature variability. Heat islands and extreme events. ESA meeting Athens June 8.

Dubos R (1980) Man Adapting. 2nd ed. Yale University Press. New Haven, CT.

Dobzhansky Th. (1962) Mankind Evolving. Yale University Press, New Haven, CT.

Ellis FP, Nelson R and Pincus L (1975) Mortality during heat waves in New York City, July 1972 and August and September 1973, Evironmental Research 10, 1–13.

Fanger PO (1967) Calculation of thermal comfort: Introduction of a basic comfort equation. ASHRAE Transactions 73, III.4.I—III.4.20.

Fanger PO (1970) Thermal Comfort. Danish Technical Press, Copenhagen.

Fischer DH (1980). Climate and history: Priorities for research. Journal of Interdisciplinary History 10, 821–830.

Fox RH (1973) Body Temperatures in the elderly. British Medical Journal 1, 200–6.

Fouillet A, Rey G, Laurent F, Pavillon G, Bellec S, Guihenneuc-Jouyaux C, Clavel J, Jougla E and Hémon (2006) Excess mortality related to the August 2003 heat wave in France. Journal of International Archives of Occupational and Environmental Health, 80, 1, 16–24.

Fouillet A, Rey G, Wagner V, Laaidi K, Empereur-Bissonnet P, Le Tertre A, Frayssinet P, Bessemoulin P, Laurent F, De Crouy-Chanel P, Jougl E and Hémon D (2008) Has the impact of heat waves on mortality changed in France since the European heat wave of summer 2003? A study of the 2006 heat wave. Int J Epidemiol. 2008, 37, 2, 309–17.

Gagge AP (1936) The linearity criterion as applied to partitional calorimetry. American Journal of Physiology 116, 656.

Gagge AP, Winslow CEA and Herrington LP (1938) The influence of clothing on the human body to varying environmental temperatures. American Journal of Physiology 124, 30.

Gagosian RB (2003) Abrupt Climate Change: Should We Be Worried? Woods Hole Oceanographic Institution, Paper Prepared for a panel on abrupt climate change at the World Economic Forum Davos.

Glantz MH (2007) 1 February. Lost in Translation: Society's "Adaptation" to Climate Change www.fragilecologies.com

Haag AL (2007) What's next for the IPCC? Nature Reports Climate Change http://www.npg.nature.com/climate/2008/0801/full/

Hardy JD (1961) Physiology of temperature regulation. Physiological Reviews 41, 521–606.

Hare FK (1985) Climatic Variability and Change. Chapter 2 In Kates RW, Ausubel JH and Berberian M eds. (1985), "Climate Impact Assessment: Studies in the Interaction of Climate and Society", SCOPE Volume 27, Wiley, Chichester, UK.

Harris T and McLean J (2007 14 December) The UN Climate Change Numbers Hoax. http://canadafreepress.com/index.php/article/968

Heidari S and Sharples S (2002) A comparative analysis of short-term and long-term thermal comfort surveys in Iran. Energy and Buildings 34, 607–614.

Helson H (1964) Adaptation – Level Theory. Harper and Row, New York.

Hensel H (1959) Heat and cold. Annual Rev Physiol, 211, 91.

Hickish DE (1955) Thermal sensation of workers in light industry in summer: a field study in southern England. Journal of Hygiene (London) 153, 112–123.

Hippocrates (circa 400 BC) On Airs Waters and Places. Translated by Francis Adams http://classics.mit.edu/Hippocrates/airwatpl.html

Höppe P (1999) The physiological equivalent temperature – a universal index for the biometeorological assessment of the thermal environment. International Journal of Biometeorology 43, 71,–75.

Houghten FC and Yagloglou CP (1923a) Determining Lines of Equal Comfort. Transactions of the American Society of Heating and Ventilating Engineers 29, 163.

Houghten FC and Yagloglou CP (1923b) Determination of the comfort zone. Transactions of the American Society of Heating and Ventilating Engineers 29, 361.

Howell WC and Kennedy PA (1979) Field study of the Fanger comfort model. Human Factors 21(2), 229–39.

Humphreys MA (1975) Field studies of thermal comfort compared and applied. UK Dept of Environment, Building Research Establishment current paper 76–75, Watford BRE.

Humphreys MA (1976) Comfortable indoor temperatures related to the outdoor air temperature. UK Dept of Environment, Building Research Establishment note PD117/76, Garston, Watford.

Humphreys MA and Nicol, JF (2000a) Outdoor temperature and indoor thermal comfort: raising the precision of the relationship for the 1998 ASHRAE database of field studies ASHRAE Transactions 206(2), 485–492.

Humphreys MA and Nicol JF (2000b) The effects of measurement and formulation error on thermal comfort indices in the ASHRAE database of field studies. ASHRAE Transactions 206(2), 493–502.

Huntington E (1907) The Pulse of Asia. Houghton and Mifflin, Boston, MA.

Huntington E (1916) Climatic variations and economic cycles. Geographical Review 1, 192.

Huntington E (1924a) The character of races as influenced by physical environment, natural selection and historical development. Scribner's Sons, New York.

Huntington E (1924b) Civilization and Climate. Third Edition. Yale University Press, New Haven, CT.

Huntington E (1927) The Human Habitat. Van Nostrand, New York.

Huntington E (1945) Mainsprings of Civilization. Wiley, New York.

ICSU SCOPE Volumes 1–59 (1971–2008) downloadable http://www.icsu-scope.org/downloadpubs/scope..../

Ingram MJ, Farmer G, and Wigley TML (1981) Past climates and their impact on Man: A review. In Wigley TML, Ingram M J and Farmer G eds. Climate and History (pp 3–50), Cambridge University Press, Cambridge.

IPCC (2001) Contribution of Working Group II to the Third Assessment Report 18. Adaptation to Climate Change in the Context of Sustainable Development and Equity http://www.ipcc.ch/ipccreports/tar/wg2/index.htm

IPCC (2005) Expert Meeting on Emission Scenarios Washington, DC, 12–14 January http://www.grida.no/climate/ipcc_tar

IPCC (2007) Fourth Assessment Synthesis Report, Special Report on Emissions Scenarios (SRES).

Jendritzky G and Nübler W (1981) A model analysing the urban thermal environment in physiologically significant terms. Arch. Meteor. Geophys. Bioclimatol. Serial B 29, 313–326.

Jones PD, Raper SCB, Kelly PM, and Wigley TML, Bradley RS and Diaz HF (1986a) Northern Hemisphere Surface Air Temperature Variations: 1851–1984, Journal of Applied Meteorology 25, 161.

Jones PD, Raper SCB and Wigley TML (1986b) Southern Hemisphere Surface Air Temperature Variations: 1851–1984, Journal of Applied Meteorology 25, 1213.

Kalkstein LS (1991) A new approach to evaluate the impact of climate on human mortality Environmental Health Perspectives 96, 145–150.

Kalkstein LS (1997) Climate and human mortality: relationships and mitigating measures (pp. 161–177), in Auliciems A ed., Bioclimatology 5. Advances in Human Bioclimat18. Adaptation to Climate Change in the Context of Sustainable Development and Equityology, Springer, Heidelberg.

Kalkstein LS and Davis RE (1989) Weather and human mortality: an evaluation of demographic and interregional responses in the United States. Annals of the Association of American Geographers 79, 44.

Kalkstein LS and Smoyer KE (1993) The impact of climate change on human health: some international implications. Experientia 49, 969–979.

Kates RW (1985) The Interaction of Climate and Society. Chapter 1 in Kates RW, Ausubel JH and Berberian M eds. (1985), "Climate Impact Assessment: Studies in the Interaction of Climate and Society", SCOPE Volume 27. Wiley, Chichester, UK.

Kates RW, Ausubel JH and Berberian M eds. (1985), "Climate Impact Assessment: Studies in the Interaction of Climate and Society", SCOPE Volume 27. Wiley, Chichester, UK.

Kerr RA (2007) How Urgent Is Climate Change? Science 23, 1230–1231.

Kilbourne EM (1989) Heat Waves. in Gregg MB ed. The public health consequences of disasters. Centre for disease Control, Atlanta, 51–61.

Klinenberg E (2002) Heat Wave. An social autopsy of disaster in Chicago. University of Chicago Press. Also http://www.press.uchicago.edu/Misc/Chicago/443213in.html

Koppe C, Jendritzky G, Kovats PS and Menne B (2004) Heat-waves: Impacts and Responses. Health and Global Environmental Change Series, No. 2.World Health Organization, Copenhagen, 123 pp.

Kovats RS ad Ebi K (2006) Heatwaves and public health in Europe. European Journal of Public Health 16(6), 592–599.

Kuhn T (1962/1970) The Structure of Scientific Revolutions (2nd edition), University of Chicago Press, Chicago, IL.

Ladurie Le Roy E (1972). Times of Feast, Times of Famine: A History of Climate Since the Year 1000. George Allen and Unwin, London.

Lamb HH (1977) Climate: Present, Past, and Future. London. vol 2 Climatic History and the Future, Methuen.

Lamb HH (1982) Climate, History and the Modern World. Methuen, New York.

Lee R. (1981). Short-term variation: vital rates, prices and weather. In Wrigley, EA and Schofield RS (eds.) The Population History of England (pp. 1541–1871), a Reconstruction (pp. 356–401), Edward Arnold, London.

van der Linden A, Boerstra AC, Raue AK, Kurvers SR and de Dear RJ (2006) Adaptive temperature limits: A new guideline in The Netherlands: A new approach for the assessment of building performance with respect to thermal indoor climate, Energy and Buildings 38, 8–17.

Lomborg B (ed.) (2004) Global Crises, Global Solutions. Cambridge University Press, Cambridge.

Mackworth NH (1950) researches on the measurement of human performance. Medical Research Council Special Report no 268, HMSO, London.

Markham SF (1947) Climate and the Energy of Nations. Oxford University Press, London.

Marx K (1845) part XI to Appendix to Ludwig Feuerbach and the End of Classical German Philosophy, 1886 http://www.marxists.org/archive/marx/works/1845/theses/theses.htm

McMichael A (2007) Reported in Latest IPCC Report Highlights Geosphere-Biosphere Programme April 6. Need For Integrated Climate/Human Behavior.

Medwar PB (1957) The Uniqueness of the Individual. Methuen, London.

Milankovitch M (1941) Kanon der Erdbestrahlung und seine Anwendung auf das Eiseitenproblem, Königlich Serbiche Akadamie, Belgrad.

Mills CA (1946) Climate Makes the Man. Gollan, London MOMENT (2008) Http://en.wikipedia.org/wiki/

Newman MT (1956) Adaptation of Man to Cold Climates. Evolution, 10,101.

New Scientist (2003) European heatwave cause 35,000 deaths, 10th October.

Nicol FP (1974) An analysis of some observations of thermal comfort in Roorkee India and Baghdad Iraq. Annals of Human Biology 1, 211–16.
Nicholls N, Skinner C, Loughnan M and Tapper N (2008) A simple heat alert system for Melbourne, Australia, International Journal of Biometeorology, Accepted for publication 15 October 2007.
Parry M L (1978) Climatic Change, Agriculture and Settlement. William Dawson & Sons, Folkestone, UK.
Pepler RD and Warner RE (1968) Temperature and learning: an experimental study. ASHRAE Transactions 74, 211–19.
Petersen WF (1947) Man, Weather and Sun. Thomas. Springfield Illinois.
Pfister C (1978) Climate and economy in eighteenth century Switzerland. Journal of Interdisciplinary History 9, 223–243.
Pfister C (1981) An analysis of the Little Ice Age climate in Switzerland and its consequences for agricultural production. In Wigley TML, Ingram MJ and Farmer G (eds.) Climate and History (pp. 214–248). Cambridge University Press, Cambridge.
Pielke RA Jr (2004) What Is Climate Change? www.sciencepolicy.colorado.edu/admin/publication_files/resource
Pielke RA Sr, Davey CA, Niyogi D, Fall S, Steinweg-Woods J, Hubbard K, Lin X, Cai M, Lim Y-K, Li K, Nielsen-Gammon J, Gallo K, Hale R, Mahmood R, Foster S, McNider RT and Blanken P (2007) Unresolved issues with the assessment of multidecadal global and surface temperature trends. Journal of Geophysical Research Vol. 112(d24): D24S08, 26.
Post JD (1977) The Last Great Subsistence Crisis in the Western World. John Hopkins University Press, Baltimore, MD.
Riebsame WE (1985) Research in Climate Society Interaction Ch 3 in Kates, RW., Ausubel, JH and Berberian, M. (eds.) "Climate Impact Assessment: Studies in the Interaction of Climate and Society", SCOPE Volume 27. Wiley, Chichester, UK.
Robinson AB, Robinson NE and Soon W (2007) Environmental Effects of Increased Atmospheric Carbon Dioxide. Journal of American Physicians and Surgeons 12, 3.
Ruskey F and Weston M (2005) A Survey of Venn Diagrams.www.combinatorics.org/Surveys/ds5/VennSymmVariants.html
Sargent F II (1963) Tropical neurasthenia: giant or windmill ? Arid Zone Research. Environmental psychology and physiology. Proceedings Lucknow Symposium. UNESCO, Paris.
Sargent F II and Tromp SW (eds.) (1964) A Survey of Biometeorology. WMO Technical Note No 65, Geneva.
Schiller GE, Arens EA, Bauman PE, Benton C, Fountain M and Doherty T (1988) A Field Study of Thermal Environments and Comfort in Office Buildings, ASHRAE Transactions 94(2), 280–308.
ScienceDaily (2007) 6 April Latest IPCC Report Highlights Need For Integrated Climate/Human Behavior Models http://www.sciencedaily.com/releases/2007/
Selye H (1957) The Stress of Life. Longman, London.
Shaw BD (1981) Climate, environment, and history: The case of Roman North Africa. In Wigley TML, Ingram MJ, and Farmer G (eds.) Climate and History (pp. 379–403). Cambridge University Press, Cambridge.
Scholander PR (1955) Evolution of climatic adaptation in homeotherms. Evolution 9, 15–26.
Scholander PR (1956) Climatic rules Evolution 10, 339–340.
Schumacher F (2002) "Under the Weather": Climatic Anxiety, Environmental Determinism, and the Creation OF the American Empire – A Preliminary exploration. Erfurter of contributions to North American history No. 4 http://www.uni-erfurt.de/nordamerika/erfurterbeit/umwelt geschichte.html
Smit B, Burton I, Klein RJT, and Street R (1999) The science of adaptation: a framework for assessment. Mitigation and Adaptation Strategies for Global Change 4, 199–213.
Smoyer KE (1996) Environmental risk factors in heat wave mortality: implications for mortality projections under climate change scenarios, Proceedings of the 14th International Congress on Biometeorology part 2 vol 3, Llubljana, pp. 150–157.

Smoyer KE (1998) A comparative analysis of heat waves and associated mortality in St Louis, Missouri 1980 and 1995. International Journal of Biometeorology 42, 44–50.

Soon W and Baliunas S (2003) Proxy climatic and environmental changes of the past 1000 years. Climate Research 23, 89–110.

Soebarto V, Williamson T, Radford A and Bennetts H (2004) Perceived and prescribed environmental performance of award winning houses. Published Proc of 38th ANZAScA Conference R. Fay (ed.). The University of Tasmania Launceston: Australia and New Zealand Architectural Science Association.

Starr C (1969) Social benefit versus technological risk. Science 165, 1232–1238.

Stehr N, von Storch H and Flügel M (1996) The 19th century discussion of climate variability and climate change: analogies for present day debate? World Research Review 7, 589–604.

Stehr, N. and von Storch H (2000) Von der Macht des Klimas. Ist der Klimadeterminismus nur noch Ideengeschichte oder relevanter Faktor gegenwärtiger Klimapolitik? Gaia 9, 187–195.

von Storch H and Stehr N (2005) Klima inszenierter Angst. SPIEGEL 4/2005 160–161.

von Storch H and Stehr N (2006) Anthropogenic climate change – a reason for concern since the 18th century and earlier. Geographical Analysis 88A(2), 107–113.

Svensmark H and Calder N (2007) The Chilling Stars: A New Theory of Climate Change. Totem Books, 256 p.

Taylor GT (1959) Australia: A Study of Warm Environments and Their Effect on British Settlement. 7th Ed. Methuen, London.

Terjung WH (1970) The energy balance climatology of a city-man system. Annals A.A.G. 60, 466–492.

Toynbee AJ (1945) A Study of History. Vol. 2, 3rd ed, Oxford University Press, London.

Tromp SW (1963) Medical Biometeorology. Elsevier, London.

ucusa (2007) http:/www.ucsusa.org/global_warming/

United Nations (2004) UNFCCC The First Ten Years. Climate Change Secretariat, Bonn Germany.

de Vries, J (1980) Measuring the impact of climate on history: The search for appropriate methodologies. Journal of Interdisciplinary History 10, 599–630.

Wigley TML, Huckstep NJ, Ogilvie AEJ, Farmer G,. Mortimer R and Ingram MJ (1985) Historical Climate Impact Assessments. Kates RW, Ausubel JH and Berberian M (eds.) "Climate Impact Assessment: Studies in the Interaction of Climate and Society", SCOPE Volume 27. Wiley, New York.

Whyte AVT (1985) Perception. in Kates RW, Ausubel JH and Berberian M (eds.) "Climate Impact Assessment: Studies in the Interaction of Climate and Society", SCOPE Volume 27. Wiley, New York.

Wikipedia (2007) http://en.wikipedia.org/wiki/2003_European_heat_wave

Wilson, EO (1975) Sociobiology: The New Synthesis. Harvard University Press, Cambridge, MA.

Winslow CEA and Herrington LP (1949) Temperature and Human Life. Princeton University Press, Princeton, NJ.

Wohlwill JF (1974) Human adaptation to levels of environmental stimulation. Human Ecology 2, 127.

Woodruff CE (1905) The Effects of tropical light on white men. Rebman Co., New York.

Yaglou CP (1926) Thermal index of atmospheric conditions and its application to sedentary and industrial life. Journal of Industrial Hygiene 8, 5.

Yaglou CP and Drinker P (1928) The summer comfort zone: climate and clothing. Journal of Industrial Hygiene 10, 350.

Zillman JW (2004) Statement made in launching William Kininmonthns book on Climate Change: A Natural Hazard at' 'Morgans', 401 Collins St, Melbourne on 22 November.

Chapter 12
The Status and Prospects for Biometeorology

Kristie L. Ebi, Glenn McGregor, and Ian Burton

12.1 Introduction

Biometeorology has a long tradition in the biophysical sciences, and has for sometime focused on what are essentially reductionist questions in its various subfields, with less thought given by its practitioners to the relevance of its science to society at large. Biometeorologists have begun to reflect upon the utility of their science, as society has demanded greater accountability on the societal value of their outputs. In particular, questions such as how can an understanding of the relationships between the atmosphere and biophysical systems provide insights into how environmental and social well-being can be sustained or improved have become important in the face of significant environmental changes. This is because it is more than apparent that the environment is subject to increasing pressures from society, and many human activities are sensitive to variation and change in environmental conditions. Accordingly, there has been critical reflection in the field of biometeorology, an outcome of which has been an unashamed push to be more applied in its orientation while not neglecting a strong engagement with its theoretical base.

Over the last few decades the subfields of biometeorology have accumulated a vast body of knowledge about the interactions between the atmosphere and biophysical systems; much of which has been published in the International Society of Biometeorology's *International Journal of Biometeorology*. That knowledge has formed a firm foundation for gaining insights into how human-induced climate change might impact biophysical systems and thus human activi-

---

I. Burton
Professor Emeritus, University of Toronto, Toronto, Canada

K.L. Ebi
ESS, LLC, Alexandria, VA

G. McGregor
Director, School of Geography, Geology and Environmental Science, University of Auckland, Auckland, New Zealand

ties dependent on the ecosystem services provided by those systems. Although not explicit in its knowledge base, a theme that runs through biometeorology is that of adaptation, the process by which living organisms adjust to variable and changing environmental conditions. Climate change, including shifts in the mean climate as well as variability and extremes, will pose series challenges for human and biophysical systems. In many ways, the degree to which humans can sustain their relationship with climate, as one aspect of the environment, will depend on the ability of societies to adapt to the changing conditions associated with climate change.

This volume, by surveying the vast knowledge base in biometeorology, has brought together for the first time the understanding of the subfields of biometeorology in relation to adaptation to climate change. Emergent are current directions and future avenues of research required for enhancing our understanding of how biophysical and human systems might adapt to climatic variability and change, the development of adaptation theory, and the advancement of climate change related managerial and policy responses in social, economic, and cultural systems.

As clearly shown in the preceding chapters, climate underlies the structure and function of human, animal, plant, water, and recreational systems. Climate patterns are primary determinants of the geographic range of plant and animal species, pathogens associated with plant, animal, and human diseases, the availability of water, and whether the location has the conditions necessary for specific recreational activities. Biometeorology has provided significant contributions to better understanding the relationships between climate variables and humans, animals, and ecosystems. Such understandings are critically needed to help people, and the social and economic systems in which they live, adapt to the projected impacts of climate variability and change. Most research in biometeorology has been carried out prior to the recognition of anthropogenic climate change. A major task for the field is to revise, or in some instances reinvent, its practical knowledge to take actual and projected climate change into account.

This volume set out to:

1. Communicate some of the basic ideas and concepts of the sub-fields of biometeorology as they relate to adaptation to climate change
2. Explore ideas, concepts, and practice that may be developed in common and
3. Begin to converge on a new vision for biometeorology that will help to communicate its understanding and expertise, as well as enhance its utility

Lessons offered in the previous chapters can be categorized into:

- Driving forces of impacts, including climate, often interact in complex and surprising ways.
- Global environmental changes are increasing the complexity of challenges to which human, animal, and plant systems have to adjust.
  - Current levels of adaptation are uneven across vulnerable regions and sectors, with many poorly prepared to deal with projected changes in climate and climate variability.

- Understanding of the interactions of climatic factors with human, animal, and plant systems can be used to increase adaptive capacity.
  - Early warning systems are increasingly important.
- Effective adaptation to climate variability and change will come from adjustments in social and economic systems.
- Significant opportunities exist for biometeorology to contribute to policy development so that societies can live effectively with climate variability and change.
  - Models are needed to study the behavior of complex systems.

These lessons and the findings presented in this volume not only consolidate but also advance our knowledge concerning climate change and adaptation. Moreover, the material presented demonstrates that biometeorology is entering a new era as it explores ways for society to adapt to an uncertain future climate and deal with the climate crisis.

12.2 Driving Forces of Impacts Often Interact in Complex Ways

Human, animal, and plant systems have co-evolved over the past millennia within the context of particular climatic conditions. Life, from microbes to the largest flora and fauna, depend on and interact with each other in ways that are still not well understood even under conditions of an assumed constant climate. All have evolved to operate within a relatively narrow range of climate conditions. There is limited understanding of how systems will respond to changes in climate conditions, as well as changes in extremes.

One of the lessons learned from biometeorology is that system vulnerability moderates or exacerbates the impacts arising from climate anomalies and extreme weather events. High ambient temperatures, or heavy precipitation events, are problems when the system of interest (human or animal health, water resources, agriculture, tourism, etc.) is unable to cope or respond effectively. As noted by Auliciems in this volume, because of the uncertainty about the rate and extent of climate change, increasing the adaptive capacity of systems will likely moderate the impacts observed.

Human populations are, in general, acclimatized to the weather patterns in their local region (Kalkstein and Sheridan; Jendritsky and deDear). Tolerance to thermal extremes depends on the interaction of personal characteristics (age, fitness, gender, chronic diseases, etc.), behavioral choices (level of activity during heatwaves, etc.), and infrastructure (how much hotter buildings become than the surrounding environment). Morbidity and mortality during heatwaves also depends on whether a community has an effective and timely heatwave early warning and response system. Changing one of these driving factors can affect the impacts observed during a heatwave. This illustrates the need to view adaptation itself as a complex system, where changing one action can alter the timeliness and effectiveness of a warning or adaptive strategy.

As discussed by Sofiev et al., allergic diseases result from the complex interactions of genes, allergens, and co-factors; these factors vary across and within regions. For example, allergic diseases are more common in urban than rural areas in Africa, possibly because parasites protect against atopic diseases. The reverse pattern is observed in most other regions, suggesting non-allergic co-factors are important in the development of sensitization and symptoms. Future patterns of allergic diseases are likely to differ from current patterns as climate and other environmental changes alter vegetation, timing and magnitude of flowering, and atmospheric transport.

Under climate change the phenology, productivity, and spatial extent of crop systems is expected to change. As emphasized by Orlandini et al., the situation for any one crop is likely to be complex because of the multiple drivers of productivity and potentially competing effects. For example the yields of potatoes, as well as other root and tuber crops, are expected to increase in many regions due of CO_2 enrichment. However, warming may reduce the growing season in some species and increase water requirements in regions where water availability (and soil moisture) is projected to decrease. This clearly points to a spatially incoherent response of cropping systems to climate change and non-linear effects. Accordingly, effective crop management strategies will need to be place-specific, highlighting the fact that adaptation policies can not be spatially invariant and need to recognize system complexity.

Tourism-recreation is highly influenced by climate, from the local scale where the climate defines the length and quality of outdoor recreation seasons, to the global scale where climate drives some of the largest tourism flows. Climate also affects environmental resources, such as sea temperatures and bathing water quality, coral reefs, snow quantity and quality, wildlife and other attractions that are critical to (eco-) tourism. Climate, climate variability, and climate change affect tourists, tourism businesses, and destination communities. As emphasized by Scott et al., adaptation within the tourism-recreation sector includes a wide variety of measures undertaken by diverse stakeholders. These measures are often taken in isolation, without coordination and collaboration across affected stakeholders. Actions taken in other sectors will affect the tourism-recreation sector, such as coastal management plans, building design standards, emergency management, wildlife management, water quality standards, and environmental impact assessments. This clearly points to the fact that in addition to biophysical complexity, social, economic, political and cultural complexity may play a major role in determining the effectiveness of adaptation strategies. Accordingly, biometeorologists need to engage with the challenge of how biometeorological knowledge can be used most effectively in the complex decision environments within which adaptation policies are developed.

12.3 Global Environmental Changes Are Increasing the Complexity of Challenges and Responses

Emergent from the chapters is that system properties in many ways determine vulnerability of a system and therefore the impact of a given event. In the case of human systems, if vulnerability is viewed as a product of the interaction between exposure (frequency, magnitude, or probability of event occurrence, sometimes referred to as biophysical vulnerability) and the sensitivity of a system (social or inherent vulnerability, which is independent of biophysical vulnerability), then one way of managing the risk associated with an event, such as climate change, is to modify the system sensitivity. This is because different social, economic, and environmental conditions lead to different degrees of impacts. In short, event outcomes are context specific. Therefore, for biometeorogical knowledge to be useful as input into adaptation policy, the context within which the knowledge has been generated and is to be applied needs to be understood. In other words, an underlying assumption is that knowledge is generated from within stationary systems where there is a stable relationship between system components. Effective adaptation policy therefore needs to be based on science that takes into account the possibility of non-stationarity of system relationships that may be determined by changing boundary conditions such as global greenhouse gas emissions.

Another common issue is distinguishing between possible adaptation options and the capacity of communities and states to develop and apply decision-making frameworks that lead to successful adaptation as outlined by de Freitas. A third issue is the value of models for studying the complex interactions between sector(s) and climatic changes, and for offering insights to understand the consequences of possible responses.

Climate change is projected to profoundly affect the availability and quality of water; this will impact other sectors (de Freitas). Any change in water availability can have indirect social and economic consequences, such as affecting agricultural productivity, availability of renewable hydroelectric power, and municipal water supplies. Trans-boundary water security issues may also lead to political conflict. However, most sectors (i.e. human and animal health, agriculture, infrastructure) traditionally act on the assumption that water resources will remain relatively constant. The growing awareness of the decline in reliability of water resources is not always understood as related to climate change. De Freitas makes the point that the capacity of a country or region to adapt to climate change is determined by the availability of skilled personnel, legal frameworks within which water is managed, money and resources, appropriate technology and technical ability, hydrological and climate data, and analytical tools for determining environmental suitability. Therefore, deploying effective adaptation measures, from education of water users on conservation to watershed management, requires understanding of the local capacity and constraints to implement specific activities.

Domestic animals can be affected by changing climatic patterns through impacts on feed-grain availability and prices; pasture and forage crop production and quality; health, growth, and reproduction; and distributions of pests and diseases (Gaughan et al.). Increases in the frequency and intensity of heatwaves are projected to affect milk and meat production in a range of animals, as well as affect immunity that could alter the success of vaccine interventions. Because of the need to increase production to meet rising demand, there has been a shift in some developing countries from indigenous and lower-yield animals to imported animals with higher yields; however, these imported animals may not be as tolerant to the increased heat load in tropical and sub-tropical countries.

Agriculture represents a good example of the different temporal and spatial scales over which adaptation can take place (Orlandini et al.). Short-term adjustments are efforts to optimize production without major system changes; examples include changes in planting dates and cultivars, changes in external inputs, and practices to conserve soil fertility and moisture. Long-term adjustments could include changes in land allocation; breeding crops with increased tolerance to changing conditions; crop substitution; and changing farming systems.

12.4 Understanding of the Interactions Between Climatic Factors and Human, Animal, and Plant Systems Can Increase Adaptive Capacity

One of the themes running across the chapters is the promise of using understanding of the interactions between climate factors and human, animal, and plant systems to design systems to provide advance warning of potentially adverse weather conditions using seasonal and short-range forecasts. The initial focus has been on understanding weather conditions that can lead to a significant change in response, particularly for human health. For example, Kalkstein and Sheridan, and Jendritzky and de Dear review some of the research aimed to better understand the thresholds for determining when high ambient temperatures lead to adverse health outcomes. Different approaches have been used that aim to describe when humans go from being uncomfortable in hot weather to when they are at risk for heat stress or a heat-related illness. This information has been used to design heatwave early warning systems worldwide. The weather conditions that put humans at risk vary geographically and are driven by factors contextual to a population (i.e. housing stock, cultural clothing and behavioral preferences, etc.). Therefore, significant local knowledge is needed to help identify thresholds for the identification of a heatwave and the issuance of warnings and advisories (i.e. calling a heatwave).

Over the past decade, as approaches to identifying thresholds for action have become more standard, there is increasing interest in understanding how to effectively intervene once a heatwave warning is declared. Although there is evidence that heatwave early warning systems save lives, there is limited understanding of which components are critical to a system's success. As demonstrated in innumerable public health campaigns, changing people's behavior is difficult. A challenge for biometeorologists is to establish the extent to which behavioral change is needed to ensure maximum effectiveness of a policy that has been informed by biometeorological research. For example, research is needed on effective approaches for motivating those most at risk during a heat wave, including adults over the age of 65, diabetics, and people taking certain drugs, to change their behavior. Surveys suggest that approximately half of those at increased risk do not alter their behavior during a heatwave, even when they know what actions should be taken (Kalkstein and Sheridan). Further, Stewart in this volume, in describing the psychological constraints to effective response to warnings, quotes Mileti (1999): "people typically are unaware of the hazards they face, underestimate those of which they are aware, overestimate their ability to cope when disaster strikes, often blame others for their losses, underutilize pre-impact hazard strategies, and rely heavily on emergency relief when the need arises." Clearly establishing the barriers to advice and policy uptake is an area in which further collaboration between biometeorologists, psychologists, public health professionals, and social scientists would prove valuable; no one discipline has the training and expertise to identify, implement, and evaluate possible approaches.

In another example, Ebi reviewed recent advances in malaria early warning systems using seasonal forecasts and remotely sensed variables. A significant advantage of these systems is that they can provide several months lead time of a pending epidemic, allowing local authorities to ensure there are sufficient drugs and insecticide treated bednets, and to initiate indoor residual spraying programs either before or at the start of a potential epidemic. However, malaria early warning systems have not been in place long enough for evaluation of their efficacy over time. Constraints to their wider implementation include the limited skill of seasonal forecast models in some regions, particularly where there is significant variability over relatively small spatial scales in the microclimates that influence malaria transmission. Anomalous years may be difficult to forecast, with the result that the early warning system may miss conditions conducive to a malaria epidemic. Even if malaria early warning systems are highly predictive of a pending epidemic, there are significant constraints in many malarious regions in the ability of local health care systems to identify that an epidemic has started and to quickly implement appropriate actions. Increasing local capacity will be necessary to ensure full utilization of early warnings.

12.5 Effective Adaptation Will Come from Adjustments in Behavior, and Social and Economic Systems

As noted in several chapters, increased understanding is needed of how climatic risks are perceived, in the short and long term, to better target adaptation efforts. Efforts are now turning to understanding how to most effectively motivate at-risk individuals to take necessary actions to reduce the possibility of heat stress. Early warning systems have used traditional media, particularly the radio and television, and are now beginning to use the internet and cellular telephones to communicate more rapidly and personally.

Adaptation in the tourism-recreation sector will include changes in behavior (such as adjusting activities, timing of visits, or destination) and business management (such as developing tourism attractions that are not climate sensitive (i.e. health and wellness spas, study tours, indoor entertainment, shopping) and developing conventions or exhibitions for business travelers) (Scott et al.). Larger companies may more successfully adapt through spreading the risk of adverse weather conditions across multiple locations so that those experiencing poor conditions can be supported financially or perhaps even relocated.

Societal vulnerability is to a significant extent a reflection of individual vulnerability (Stewart). Therefore, a greater understanding is needed of individual perception of risk and effective measures for motivating appropriate change. Stewart uses the Protection Motivation Therapy (PMT) to examine psychological constraints to an individual's adaptation to climate variability. Climate events are individually and socially construed, which affects the information that people extract and their uses of this information. People evaluate both whether a particular weather pattern is dangerous, as well as their vulnerability to the event. As a consequence, the types of adaptation actions undertaken will vary across people, groups, and organizations. Although people gather information from a range of intra- and interpersonal sources, there is evidence that they are more likely to gather and trust information from friends and family members than from government representatives. Increasing the ability of an individual to accurately perceive the risks of a weather pattern and their vulnerability to that pattern will decrease the probability of maladaptation. However, typically, people act on the basis of their biases, suggesting barriers to increasing the effective use of climate information. A further barrier is that people tend to underestimate the likelihood of rare events and exaggerate the likelihood of more common events. A particular challenge then is to communicate low-probability, high consequence climatic events. In relation to this, there are opportunities for biometeorologists to work with risk communication specialists so that information generated by research can be turned into effective messages that risk managers can use to communicate with a wide range of stakeholders, ranging from, for example, public health officials to the public.

12.6 Opportunities for Biometeorology to Contribute to Adaptation

Significant opportunities exist to increase resilience to the risks of climate change. Auliciems discusses three levels of adaptation. First order adaptation is the initial reduction of hazard impacts. Second order adaptations are behavioral, technical, and cultural responses from routine maintenance of infrastructure to disaster risk reduction activities. Third order adaptations include social, economic, education, legal, and other factors. These levels of adaptation interact to determine the coping range (and adaptation deficit) of the system. Traditionally biometeorology has concerned itself more with the science associated with first order adaptation. The challenge exists to elevate our interest as a scientific community to the level of second and third order-related research questions. As these are at increasing distances from the core activity of biometeorology and outside the comfort zone of most within the biometeorological community, collaborative efforts with "other" scientists will be required if biometeorologists are to provide insights concerning the factors that influence coping range and adaptation deficit.

Biometeorology can make effective contributions to the management of energy resources and thus adaptation to changing energy supplies. For example Jendritzky and de Dear, in their chapter on the thermal environment, suggest that a better understanding of acceptable indoor thermal conditions, and how they relate to outdoor temperatures, have the potential to save large amounts of energy and to avoid carbon dioxide emissions. Behavioral and infrastructure adaptations are needed to increase the ability of humans to live in warmer climates with limited thermal discomfort. Occupants of naturally ventilated buildings have been shown to adjust to a wider range of temperatures than occupants of centrally air-conditioned buildings, suggesting that increasing the widespread use of air conditioning may not be adaptive for continuing increases in average summer temperatures, although the use of air conditioning and cooling centers are important adaptations for people particularly vulnerable to heatwaves.

Biometeorology has a key role in facilitating the rational management of animals to meet the challenges of changes in the thermal environment (Gaughan et al.). Increasing the resilience of domestic animals to climate change through the breeding and identification of animals who have acquired genes for thermo-tolerance are two possible adaptation responses. Other options include physical modification of the environment (such as using water for cooling) and nutritional modification to reduce heat stress.

Considerable adaptation (and evolution) has taken place for plants, animals, and humans to live in current climatic zones. Weather patterns are changing faster than they have in the past 10,000 or more years (IPCC 2007), challenging the ability of individuals and communities to adjust fast enough. Additional research is needed to gain deeper insight into the process of acclimatization, to inform the development of models to quantify acclimatization and so to better understand the degree to which faster, better, and more effective acclimatization could reduce projected impacts due to climate change.

Heatwave, vectorborne disease, and pollen and allergen early warning systems are examples of adaptive responses to climate variability and change. There is no doubt that biometeorologists will continue to be involved in the type of scientific activity that underpins the development of these systems. Opportunities exist to work more closely with numerical weather prediction and seasonal climate forecast model developers to identify the type of forecast products that are most appropriate for end-users of biometeorological forecasts and to diagnose forecast model problems. To capitalize on these opportunities, it will be necessary to understand the dynamics and complexity of end-user systems so that weather and climate information can be optimized for a given early warning system situation and for biometeorologists to be familiar with forecast diagnostic tools and the science of skill evaluation. Evaluation of early warning system effectiveness and the opportunity cost of system presence or absence are areas for future collaboration between biometeorologists and economists, which will assist with costing adaptation strategies. Couching biometeorological responses or warnings in a risk management framework by using a probabilistic approach is also an applaudable challenge for the field and will assist with decisions related to adaptation.

Opportunities for adaptation in agriculture include increasing the skill of weather forecasts to improve farm management; increasing understanding of how farmers perceive climate-related risks and how they respond; research and modeling of the integrated impacts of climate change on crop systems, not just individual crops, and on mixed farming systems that take into account non-climatic conditions; and developing technological and policy strategies for improving the sustainability of farming systems under uncertain climate projections.

Increasingly climate change is likely to result in significant changes in the tourism-recreation sector, from increasing ecotourism to visit changing landscapes and species projected to disappear, to decreasing visitors to parks if glaciers or other environmental changes occur (Scott et al.). Better understanding of how climate change could affect tourism would facilitate identification of possible adaptations.

Comparatively, the production of scientific knowledge is far easier than its consumption by end users, whether individuals or institutions. As outlined by Stewart, increasing the effectiveness of adaptation to climate variability and change will require greater understanding of ways to motivate appropriate changes in the behavior of individuals. The same can be said for institutions. To understand the controlling factors that determine the uptake of their science, biometeorologists may increasingly have to turn to other fields such as psychology for help. Biometeorologists could effectively work with psychologists in assessing individual's response to a range of climate change or adaptation scenarios. Using a range of psychological variables, the role of emotional processes in affecting decisions that involve weather and climate risk and uncertainty (and how to effectively intervene), and thus adaptive behavior, could be established. Stakeholders need to be involved in the identification and implementation of adaptations, to ensure their values and concerns are taken into consideration.

In conclusion, this survey of the recent literature indicates that the status and prospects of biometeorology in the field of climatic variability, change and adaptation are excellent. It has also provided us with the opportunity to reflect on the relevance of the more traditional view of biometeorology as a discipline concerned with examining the relationship between atmospheric processes and living organisms; in many ways a deterministic view of the discipline. The material contained in this volume begins to suggest a new vision for biometeorology as the science of understanding the interactions and feedbacks between atmospheric conditions (as codified in the sciences of meteorology and climatology) and biophysical and human systems. We believe that biometeorologists working within this paradigm will be well placed to facilitate improved management and policy choices in a world seriously threatened by the "rapid and extensive climate change". Lastly, the content of this volume affirms that a central ethos of biometeorology is that adaptation to climatic variability and change can be achieved by realigning human use systems with changing environmental conditions; society should be discouraged from engineering its way to adaptation.

References

Intergovernmental Panel on Climate Change. (2007) Climate Change 2007: Synthesis Report. Summary for Policymakers. http://www.ipcc.ch/pdf/assessment-report/ar4/syr/ar4_syr_spm.pdf

Mileti DS. (1999) *Disasters by design: A reassessment of natural hazards in the United States.* Joseph Henry Press, Washington, DC, 351 pp

Index

A

Acclimatization, 10–12, 37, 38, 139, 145, 242, 245, 246, 259, 260, 277
Adaptation, 1–278
Adaptation (behavioural, technological, management), 174
Adaptation deficit, 256, 259, 260, 276
Adaptive comfort, 27, 28, 239, 240
Adaptive comfort model, 239
Adaptive model, 10, 27, 239, 240, 255, 257
Aerobiology, 76, 94
Agriculture, 110, 111, 114, 115, 120, 124, 132, 133, 135, 172, 190, 198, 200, 253, 271, 273, 274, 277
Agroclimatic indices, 108
Air conditioning, 10, 28, 36, 45, 152, 183, 190, 246, 252, 277
Air quality, 76
Animal, 2–4, 76, 80, 97, 120, 131–136, 138–140, 143–158, 253, 270, 271, 273, 274, 277
Animal adaptation, 139–151
Architecture, 11
Atolls, 206, 207

B

Behavioural adaptive measures, 199 200
Beliefs, 221
Biofuel crop, 116, 117
Biological adaptation, 2, 3, 139, 244–245, 248, 253, 259
Biometeorology, 1–278
Biotechnology, 120, 124

C

Cereal, 85, 111, 113, 115, 117, 119
Climate and health, 9, 71

Climate change, 2–5, 11, 18, 20, 26, 28, 29, 51, 53, 66, 67, 70, 71, 95, 98, 109, 111, 112, 114–124, 132–136, 138–140, 145, 149, 151, 152, 156, 157, 172, 173, 175–178, 180, 182–186, 188–191, 196, 198, 201–208, 213, 215, 216, 219, 236, 249–251, 253, 256, 261, 270–273, 276–278
Climate change analogue, 176, 177
Climate determinism, 236, 237
Climate variability, 2, 3, 50, 110, 114–115, 123, 157, 172, 175, 176, 179–182, 213, 219, 227, 270–272, 276–278
Climatology, 2, 11, 15, 17, 21, 36, 278
Coastal tourism, 178, 186, 189
Cognition, 213, 216, 220
Comfort preference, 10
Conceptual models, 16, 208, 256
Coping range, 244, 245, 247, 248, 255–257, 259, 276
Core temperature, 12, 19, 134, 135, 239, 240, 242
Crop, 96, 108, 111–121, 123, 133, 198–200, 203, 253, 272–274, 278

D

Demand side options, 198
Dr. Tromp, S.W., 1, 238, 242

E

Early warning systems, 4, 49–71, 111, 271, 274, 275, 277
Ecology, 4, 52
Economic adaptive measures, 200
Emotion, 213, 217, 224–228
Energy balance, 12
Enhanced CO_2, 112–113

Equivalent temperature, 16
Excessive heat events, 34, 41, 42, 45

F
Farming system, 121, 123, 274, 278

G
Global cooling, 261
Global warming paradigm, 261

H
Health impact, 10, 11, 18, 19, 70, 71, 99, 252
Heat balance, 15, 24, 25, 27, 28, 149
Heat budget model, 15–17
Heat exchange theory, 9
Heat health warning system (HHWS), 11, 15, 22–24, 29, 44
Heat wave, 11, 18–21, 23, 24, 33, 37, 40, 44, 110, 111, 114, 134–136, 138, 145, 149, 245–249, 251, 252, 254, 259, 260, 274
Homeostasis, 148, 239, 245, 252
Homeothermic adaptation, 254
Horticultural crop, 116
Human, 2–5, 10, 12–19, 23, 24, 26–28, 33, 34, 36, 50, 53, 56, 66, 77, 79, 87–89, 95, 99, 110, 135, 136, 139, 148, 196, 199, 213, 219, 220, 227, 235–238, 240–242, 244, 245, 250–261, 269–274, 277, 278
Human health, 18, 24, 28, 36, 99, 253, 274

I
Impact levels and controls, 255
Impact potential, 196, 203, 205, 208
Industrial crop, 116
Interdisciplinary, human, animal and plant, 2
International Congress of Biometeorology, 1
International Journal of Biometeorology, 1, 270
International Society of Biometeorology (ISB), 2, 17, 172

L
Last change tourism, 183
Legal adaptive measures, 198, 200–201
Leisure, 80, 173, 175
Long-term adaptation, 11, 24–28, 180
Long-term adjustment, 274

M
Maladaptation, 190, 252, 259–261, 276
Malaria, 4, 49–71, 275
Managing climate risk, 134, 158
Meteorology, 1, 94, 269, 278
Migration, 2, 10, 172, 199
Model, 5, 10, 11, 15, 16, 18, 19, 24, 25, 27, 28, 35, 36, 52, 54–64, 66, 70, 71, 77, 85, 86, 89–93, 95, 98, 108–110, 114, 115, 119, 123, 138, 146, 147, 151, 152, 181, 182, 197, 202, 204, 206, 208, 213, 215–217, 219, 220, 224–228, 238–240, 245, 248, 250–257, 259, 261, 271, 273, 275, 277
Motivation, 121, 215–217, 224, 225, 276
Multi-node model, 18

N
New Zealand, 176, 187, 205

P
Pacific, 70, 179, 206, 207, 236
Permanent crop, 117–118
Physiological response, 10, 20, 25, 143, 240, 253
Plant, 2–4, 26, 28, 76, 77, 79, 82–84, 86–90, 95–98, 107, 108, 110–117, 180, 181, 185, 186, 200, 203, 205, 253, 270, 271, 274, 277
Pollen allergy, 76
Pollen modeling, 76
Pollination, 76, 83–89, 114
Precautionary planning, 18, 29
Precipitation, 50, 51, 55, 56, 84, 98, 110, 114, 116, 118, 119, 133, 180, 197, 199, 200, 203, 204, 206, 207, 218, 271
Protection motivation, 215–217, 224, 276
Psychological adaptation, 213, 214
Public education, 177, 184, 186

R
Recreation, 5, 171–176, 178, 179, 181, 184–191, 272, 275, 278
Resorts, 175, 176, 178–183, 189, 190
Response surface, 203–206
Risk-taking, 224, 225
Root crop, 112, 115

S
Seed crop, 113, 115, 117
Sensitivity assessment, 202

Short-term adaptation, 22–24, 36, 145
Short-term adjustment, 118–120, 274
Ski industry, 173, 179, 182, 184, 185, 188–190
Social adaptation, 3
Strain, 9, 10, 12, 15, 23, 139, 150
Stress, 1, 5, 9, 10, 12, 18, 19, 23, 34, 36, 37, 39, 86, 98, 114, 120, 131, 132, 134–140, 143–150, 152, 153, 155, 156, 187, 236, 238, 239, 246–248, 254, 259, 275, 277
Supply-side options, 198
Sustainability, 11, 123, 132, 206, 278
Sustainable livestock production, 132
System evaluation, 33, 92

T
Technological adaptive measures, 198–199
Temperature, 10–12, 14–16, 18–22, 24, 25, 27, 28, 34–39, 50–52, 55, 56, 65, 71, 76, 78, 83–87, 92, 95–98, 108, 110, 113–120, 133–138, 140, 142–145, 147–156, 176, 177, 180, 187, 203, 206, 219, 237, 239–251, 254, 255, 271, 272, 274, 277
Thermal assessments, 11, 14–16
Thermal comfort, 10, 11, 27, 239, 245
Thermal environment, 4, 9–29, 145, 146, 149, 157, 277
Thermal exposure, 10, 11
Thermal preference, 239
Thermo physiology, 238
Tourism, 5, 18, 171–191, 271, 272, 275, 278
Tourist, 172–178, 183, 186, 189, 190, 272
Tuber crop, 115, 272

U
UHI. *See* Urban heat island
United Nations Framework Convention on Climate Change (UNFCCC), 250
United Nations World Tourism Organization (UNWTO), 172, 173, 185
Universal thermal climate index (UTCI), 4, 11, 16–19
Urban, 1, 10, 11, 18, 20, 22, 24, 25, 29, 33, 37, 39, 41, 43–46, 76, 81, 87, 114, 174, 185, 190, 199, 212, 245, 246, 249, 254, 255, 260, 272
Urban climatology, 2, 11
Urban heat island (UHI), 20, 24
Urban planning, 11, 24, 25, 254
Urban response to heat, 33
UTCI. *See* Universal thermal climate index

V
Vulnerability, 1–3, 33, 37–39, 47, 53, 55, 56, 110, 182, 190, 202, 203, 205, 208, 212, 213, 216, 224, 225, 238, 241, 248, 250, 252, 254, 256, 269, 271–273, 276

W
Water, 5, 11, 14, 52, 92, 97, 98, 108, 110, 111, 113–116, 118–123
Water balance, 108, 206
Water resources, 5, 190, 195–198, 200–203, 205–208, 271, 273
Weather and health, 33, 47, 70
Weather forecast, 11, 18, 23, 65, 124, 242, 245, 253, 277

Printed in the United States
144913LV00003B/47/P

DATE DUE